FLOOD RISK MANAGEMENT

Edited by **Theodore Hromadka**
and **Prasada Rao**

Flood Risk Management
http://dx.doi.org/10.5772/66850
Edited by Theodore Hromadka and Prasada Rao

Contributors

Kai Wang, Mingkai Qian, Mohammad Mojtahedi, Sidney Newton, Faham Tahmasebinia, Ali Assani, Mohd Talha Anees, Khiruddin Abdullah, Mohd Nawawi Mohd Nordin, Nik Norulaini Ab Rahman, Mohd Omar Ab Kadir, Muhammad Izzuddin Syakir Ishak, Mohammad A. Hijji, Saad Amin, Wayne Harrop, Rahat Iqbal, Muhammad Saleem Ashraf, Sebastian Goers, Friedrich Schneider, Horst Steinmüller, Andreas Zauner, Elżbieta Jarosińska, Katarzyna Pierzga, Benjamin Kofi Kofi Nyarko, Adams Osman, Felipe I. Arreguin-Cortés, Claudia Elizabeth Cervantes-Jaimes, Barry Hankin, Milena Marília Nogueira De Andrade, Iris Celeste Nascimento Bandeira, Paulo Eduardo Silva Bezerra, Ádanna Andrade, Rodrigo Oliveira, Dianne Fonseca, Oscar Dena, Griselda Janeth Obeso Cortez, Diane Irene Doser, Luz Estella Torres Molina, Avelino Isaias Mondlhane

Published by InTech
Janeza Trdine 9, 51000 Rijeka, Croatia

Publishing Process Manager Iva Simcic
Technical Editor SPi Global
Cover InTech Design team

First published September, 2017
Printed in Croatia
Legal deposit, Croatia: National and University Library in Zagreb

Additional hard copies can be obtained from orders@intechopen.com

Flood Risk Management, Edited by Theodore Hromadka and Prasada Rao
p. cm.
Print ISBN 978-953-51-3465-7
Online ISBN 978-953-51-3466-4

For EU product safety concerns:
IN TECH d.o.o., Prolaz Marije Krucifikse Kozulić 3, 51000 Rijeka, Croatia,
info@intechopen.com or visit our website at intechopen.com.

Meet the editors

Theodore Hromadka, PhD, PH, PE, has extensive scientific, engineering, expert witness, and litigation support experience. His frequently referenced scientific contributions to the hydrologic, earth, and atmospheric sciences have been published in the peer-reviewed scientific literature including 26 books and over 400 scientific papers and book chapters. His professional engineering experience includes supervision and development of over 1500 engineering studies. He is currently a faculty at West Point Academy and California State University, Fullerton.

Prasada Rao, PhD, is a professor in Civil and Environmental Engineering Department at California State University, Fullerton. His current research areas relate to climate change, surface and subsurface flow modeling, and computational mathematics. He is also the associate director for International Institute for Computational Engineering Mathematics. The topics of climate change, weather prediction, atmospheric sciences, and other related fields are gaining increased attention due to the possible impacts of changes in climate and weather upon the planet. Concurrently, the increasing ability to computationally model the governing partial differential equations that describe these various topics of climate has gained a great deal of attention as well. In the current book, several aspects of these topics are examined to provide another stepping stone in recent advances in the fields of study and also focal points of endeavor in the evolving technology.

Contents

Preface

The topic of flood risk management is an important subject to practitioners and researchers specializing in a variety of technical disciplines. The audience of people that deal with topics in flood risk management include urban planners, municipal engineers, civil engineers, flood plain managers, as well as the individuals living within the floodplain of a nearby watercourse. Key topics of interest include rainfall, storm runoff, storage of runoff, transport of storm runoff, flood infrastructure, and several other focal points including computational engineering mathematics, computational hydrology and hydraulics, statistical methods for use in assessment of future rainfall quantities as well as runoff quantities, ice flows and their impacts, and general engineering principles involving related subjects such as sediment transports, coastal engineering, and the more recent endeavors addressing topics in global climate change and modeling. Many of the topics covered in the typical University STEM programs are particularly relevant.

In this book, contributions from several experts are assembled into a single volume that addresses many of the components of flood risk management. Application and testing of numerical and statistical models that can simulate the complex reality along with effective flood management strategies that are being implemented in various nations are presented in depth. These chapters will advance the knowledge in flood risk management, which has now become a conversational topic across all nations.

This collection of topics will provide an update to the reader as to the state of the art in this important technical field.

Dr. Theodore Hromadka
United States Military Academy, West Point, NY, USA
California State University, Fullerton, CA, USA

Dr. Prasada Rao
California State University, Fullerton, CA, USA

Chapter 1

Strategies for Testing the Impact of Natural Flood Risk Management Measures

Barry Hankin, Peter Metcalfe, David Johnson,
Nick A. Chappell, Trevor Page, Iain Craigen,
Rob Lamb and Keith Beven

Additional information is available at the end of the chapter

http://dx.doi.org/10.5772/intechopen.68677

Abstract

Natural Flood Management (NFM) is an approach that seeks to work with natural processes to enhance the flood regulating capacity of a catchment, whilst delivering a wide range of ecosystem services, from pollution assimilation to habitat creation and carbon storage. This chapter describes a tiered approach to NFM, commencing with strategic modelling to identify a range of NFM opportunities (tree-planting, distributed runoff attenuation features, and soil structure improvements), and their potential benefits, before engagement with catchment partners, and prioritisation of areas for more detailed hydrological modelling and uncertainty analysis. NFM measures pose some fundamental challenges in modelling their contribution to flood risk management because they are often highly distributed, can influence multiple catchment processes, and evidence for their effectiveness at the large scale is uncertain. This demands we model the 'upstream' in more detail so that we can assess the effectiveness of many small-scale changes at the large-scale. We demonstrate an approach to address these challenges employing the fast, high resolution, fully-distributed inundation model JFLOW, and visualisation of potential benefits in map form. These are used to engage catchment managers who can prioritise areas for potential deployment of NFM measures, where more detailed modelling may be targeted. We then demonstrate a framework applying the semi-distributed Dynamic TOPMODEL in which uncertainty plays an integral role in the decision-making process.

Keywords: natural flood risk management, uncertainty

1. Introduction

Natural Flood Management (NFM), commonly referred to in the UK as Working with Natural Processes (WWNP), has been defined as taking action to manage flood risk by protecting, restoring and emulating the natural regulating function of catchments, rivers, floodplains and coasts [1]. NFM can integrate improvements to the local landscape and ecology, thereby contributing to meeting environmental goals (such as European Water Framework Directive objectives). Compared to hard-engineered Flood Risk Management (FRM), NFM is becoming attractive to policy makers and catchment managers due to lower upfront costs, the potential to create multiple ecosystem benefits (e.g. carbon storage, diffuse pollution and sediment risk regulation), and for its flexible scale of deployment, which may also help to stimulate or encourage community involvement. The mechanisms to achieve such aims include run-off storage, increasing soil infiltration, slowing surface water movement and reducing flow connectivity.

Across a catchment, there can be many different opportunities for NFM such as upland mire restoration, revised/modified land management and land use, woodland creation, sediment management, built water storage, river restoration and development of Runoff Attenuation Features (RAFs) to intercept overland flow. The key characteristics shared by these measures are that they are typically small-scale and highly distributed, and potentially alter a wide range of catchment processes. As a consequence of their small scale and local impact, they are mostly likely to be effective in reducing downstream flooding when implemented widely in the headwaters of catchments to reduce streamflow reaching downstream floodplains.

Catchment models routinely applied by regulatory agencies and water authorities for flood risk assessments, forecasting, water resources planning or water quality management, tend to represent upstream areas as discrete sub-catchments with uniform inputs. This can be very effective for simulating flows in the river further downstream near urban settlements, but averages over the effects of small-scale interventions and thus loses information into their impacts on the various hydrological processes of interest. For highly distributed, smaller scale measures and interventions, it will be vital to consider new approaches that not only scale their impact up more accurately but also consider the uncertainty in the representation of how catchment processes might change. There are still large evidence gaps and a need to test the effectiveness of these distributed measures against observational data. This will require detailed monitoring of catchment processes. A recent survey [2], however, showed that as few as 6% of NFM schemes in the UK have intensive hydrological monitoring.

The approaches demonstrated here aim to address the scaling-up problem in a way that reflects the uncertainty in the representation of small-scale hydrological processes and flood mitigation methods. In doing so we meet the challenge that was put to environmental modellers by Beven in 2009 [3], to give decision makers:

> ***'...a realistic evaluation of uncertainty since this might actually change the decision that is made'***

By means of an example, we examine here a tiered approach taken to planning NFM strategies within the headwaters of the Eden, Kent and Derwent catchments in Cumbria, UK (**Figure 1**), that aimed to prioritise where different measures are likely to be most effective.

http://dx.doi.org/10.5772/intechopen.68677

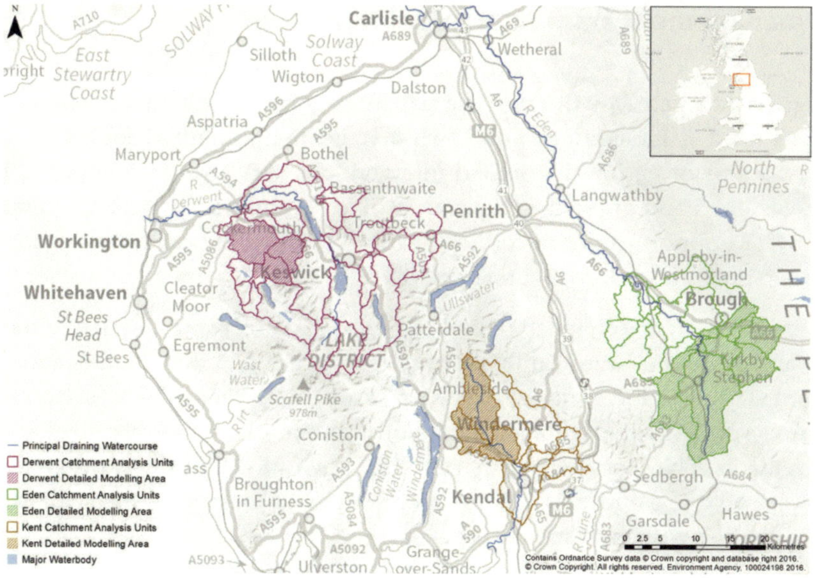

Figure 1. Overview of the study area.

This was followed by an analysis of uncertainty in the environmental model parameters and of the fuzziness in the evidence behind the changes applied to these models to represent effects of NFM measures.

We show how the effectiveness of very distributed NFM measures such as tree planting and enhanced storage in the headwater catchments can be appraised in a framework similar to established FRM practice. Advances in high-resolution modelling (here we explicitly model runoff on a 2 × 2m resolution grid within a 100-million-cell model [4]), have unlocked the potential to model 'upstream' at high resolution, and enable us to *test the aggregated impacts of very small scale measures at the larger scale*. This delivers a deeper understanding of the effectiveness of potential NFM plans in reducing peak runoff, but requires new strategies to consider 'synchronisation' issues [5], whereby flooding can be made worse by slowing the time of arrival of a flood peak in one tributary such that it interferes constructively with that of the receiving watercourse. It also opens up the ability to undertake continuous modelling through sequences of events, such that the antecedent wetness is taken into account.

Since NFM measures could be deployed in very many spatial configurations, it follows we should test for synchronisation effects against a wide range of plausible extreme loading conditions. The authors demonstrated such an approach in a recent UK government Flood Modelling Competition, where their winning entry [6], combined high resolution modelling of NFM with advancements in spatial joint probability analysis of extremes [7, 8], whereby multiple extreme rainfall scenarios were simulated, with realistic spatial patterns based on the long-term records at rainfall gauges around the catchment. The 'average' effectiveness of NFM across the upper part of the 2,300km^2 Eden catchment was then tested against 30 simulated extreme rainfall events selected to span a range of spatial patterns.

2. A risk management framework for NFM

Figure 2 highlights a pathway through the risk management cycle that attempts to integrate core elements needed for modern flood risk management, adapted for NFM with its distinguishing features of highly distributed interventions and uncertain impacts. This framework was developed and applied in the course of the Cumbrian project for identification of the types of NFM opportunities introduced earlier, with whole catchment modelling being applied to map potential risk reduction benefits.

This first step was undertaken for the Cumbrian project using rapid overland flow modelling approach using a 2 m resolution 2d JFLOW model to identify where modification of features in the landscape to slow and store surface water flows might make the most difference. These included the use of RAFs at locations of high flow accumulation, tree-planting, and soil structure improvements. The speed of set-up, very high resolution and rapid run times of JFLOW meant we were able to rapidly assess the effectiveness of these very distributed NFM measures, before sharing with catchment partners.

Our work in Cumbria has involved a wide range of catchment partners including landowners, flood management agencies, voluntary groups and farming groups, who used the whole-catchment mapping to refine potential opportunities for NFM deployments and also prioritise areas for more detailed modelling. The sub-catchments containing the most promising opportunities

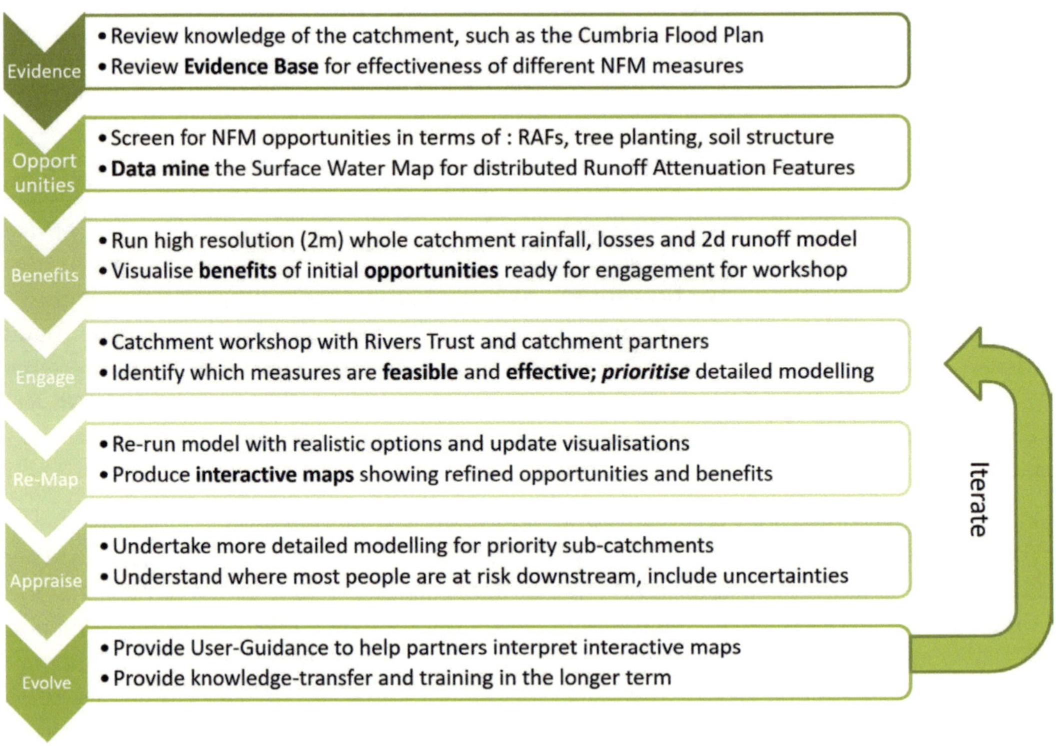

Figure 2. The risk management cycle for NFM.

based on the JFLOW modelling were prioritised for more detailed uncertainty investigation using Dynamic TOPMODEL [9, 10] to compute the projected benefits of NFM.

Dynamic TOPMODEL provides a tool to help understand the effects of NFM on the total hydrograph, including contributions from overland flow and subsurface flow. The detailed models were calibrated against observed data collected during the period November–December 2015, in order to capture the flows arising from a sequence of prolonged and intense rainfall events associated with Storms Abigail and Barney prior to the most severe storm, Desmond (4–6 December 2015), which led to extreme rainfall and flooding in Cumbria [11, 12].

An uncertainty framework was used to represent not only the uncertainties in the parameters used in the model but also the gaps in the scientific evidence on how different NFM measures influence catchment processes (**Figure 3**). A large number of simulations were undertaken for each catchment, and a set of model parameterizations showing 'acceptable' performance on

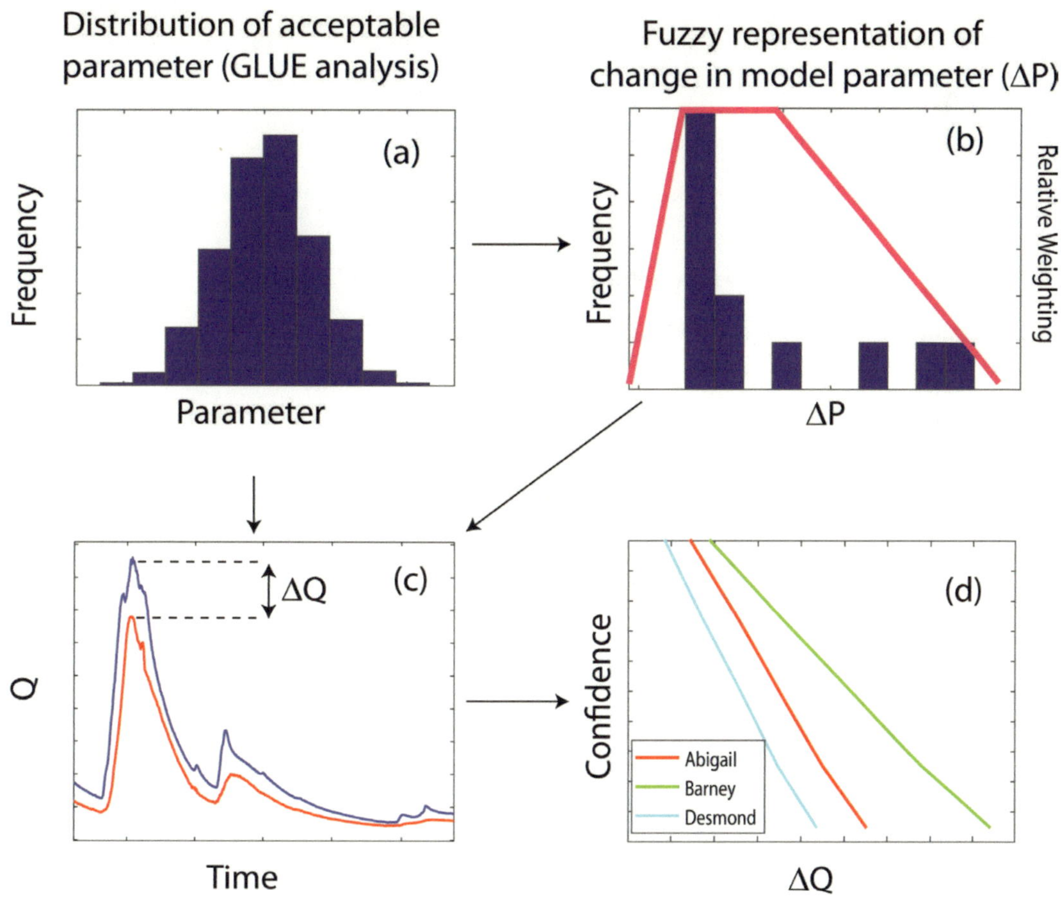

Figure 3. Stratified sampling of parameter uncertainty and fuzziness in evidence parameter changes to reflect NFM interventions. (a) Weight of evidence for effective parameter values (b) Weight of evidence for change in effective parameter values to reflect NFM intervention (c) Resulting change to peak flow in hydrograph response as a result of NFM (d) Confidence in magnitude of peak flow change based on combined weightings.

the basis of the evidence, were identified based on a range of measures including the ability of the models to reproduce the observed peak flow for Storm Desmond.

The next sections cover the broad steps in this flow chart.

3. Evidence for effectiveness of NFM

This section reviews evidence of the effects of NFM on catchment processes, and how this evidence was mapped onto the effective parameters that are used in the different modelling strategies adopted in the Cumbrian opportunity mapping project. This section gives a flavour of how we gathered evidence of *changes* to physical processes in the UK and tried to map these changes onto already uncertainly modelled processes.

For the JFLOW overland flow modelling, the physical influences considered were:

- Increasing roughness through land cover changes: grasses and mosses to shrubs and trees
- Increasing localised depression storage through construction of RAFs
- Increasing infiltration through improvements in soil structure

For Dynamic TOPMODEL, we consider the above processes and also those influencing sub-surface flow, for example:

- Reducing surface overland flow velocities through woodland planting.
- Increasing surface storage associated with RAFs using modified surface routing and a maximum storage parameter.
- Increasing soil transmissivity through woodland planting.
- Increasing wet canopy evaporation through woodland planting.

The evidence for the above changes linked to tree-planting was considered with respect to deciduous woodland, which brings the most habitat and biodiversity benefits in this environment, and is in line with current activity within NFM schemes across the UK.

3.1. Surface roughness

Considering the roughness of the ground covered by improved pasture, heathland or deciduous woodland, many different elements combine to produce an effective roughness for a whole hillslope. Within a deciduous woodland, roughness contributions come from: (1) roughness of the litter layer, (2) roots at the ground surface running across the slope, (3) obstructions to flow caused by tree stems/understory, (4) paths and tracks, (5) fence lines or walls, and (6) sub-grid topographic irregularities in slope. All components need to be characterised to provide an accurate measurement of the effective roughness of a whole hillslope. Direct measurements of the roughness components across a range of surface vegetation conditions at floodplain sites in Florida (USA) [13] produced Manning's n [14] values that ranged from 0.030 to 0.061 for forest areas against a range of 0.013–0.050 for other surfaces including barren land and grasslands.

Chow states that floodplains covered by pasture should be ascribed a roughness value of 0.035, while those covered by light brush and weeds 0.050, dense brush 0.070 and dense forest 0.1–0.2 [14]. Thus in direct comparison with the direct measurements of roughness undertaken in Ref. [13], the differences between woody vegetation and pasture-cum-barren land are considerably less. Given: (1) the limited number of studies directly measuring hillslope roughness [13], (2) the discrepancies between the field-measured and tabulated (or estimated) values, combined with (3) the large variability in roughness values measured even within the same vegetation types, a large uncertainty should be placed on the range of possible roughness values used in models.

In this study, a comparison is made between with and without woodland, that amounts to a maximum increase of 50% in Manning's roughness, over the broad-scale 'upland' roughness value of 0.1 used to provide national flood maps in the UK. *Thus a maximum of 0.15 was used compatible with the engineering tables of Chow, but which requires more research to resolve the underlying physical processes contributing to frictional losses.*

3.2. Peatland management

Similar issues arise when estimating the effects of peatland management on the effective roughness of hillslopes as with the comparisons between the effects of forest versus pasturelands. For example, peatland restoration may involve replacing patches of bare peat with Sphagnum spp. moss, changing micro-scale roughness, but also adding small obstructions within the artificial drains. Holden et al. [15] used 256 bounded overland flow plots (0.5 m x 6 m) in the Upper Wharfe catchment (UK) and found that the roughness was greater when *Sphagnum spp.* moss rather than bare ground was present, resulting in a reduction of overland flow speeds by a factor of 3.3. *This equates to an increased roughness of ~0.3, which was used in the JFLOW modelling and the reduced wave speed in Dynamic TOPMLODEL,* albeit starting from the relatively high generic roughness factor used in the national mapping, of 0.1 and increasing it to 0.13. Here we have used the roughness increase only, as the effect of drain blockage is not as clear and can be strongly dependent on the orientation of the drains, i.e. whether they run across slope or down slope [16].

3.3. Runoff attenuation features (RAFs)

Small ponds to capture and temporarily store overland flow on its way to stream channels have been described as 'overland flow interception RAFs' or 'overland flow disconnection ponds', where RAFs are 'runoff attenuation features'. Temporary storage of the overland flow on slopes could delay this component of the flow so that it reaches streams after the peak of the hydrograph has passed. However, if overland flow on a particular slope is generated on the rising stage of a stream hydrograph [17], delaying it could have the unwanted effect of adding the overland flow contribution to the channel at the time of the peak in the streamflow. Clearly understanding precisely when the overland flow is being added to stream channels, and doing so within a spatial frame of reference, is critical for understanding how it should be managed. Only a few direct measurements of overland flow using plot studies are available in the UK (e.g. Ref. [18]) to help quantify the timing of this process. The Belford Catchment Solutions Project in Northumberland has demonstrated how such features are able to retain and hence attenuate the initial phase of overland flow generation. Overland flow interception RAFs have been constructed at many locations in the UK and individually range from 20 to 1000 m^3 in capacity (see Refs. [19–21]). *RAFs added into the simulations for this study are 100–5000 m^3 in area* and so are broadly similar in capacity.

3.4. Wet-canopy evaporation

Deciduous woodlands in the early winter (i.e. November-December) in Western Europe, when the overstory is leafless, exhibit only very small rates of transpiration [22]. Potentially, these rates may be marginally higher than those for improved grasslands, if the woodland is open and accompanied by a leafed understory vegetation of shrubs and/or longer grasses [23–26]. This effect is, however, likely to be insignificant when compared with the contrasts in wet canopy evaporation (also called 'interception loss' or Ewc).

As Reynolds and Henderson [27] noted '*...although sometimes there is a measurable reduction of interception losses in winter due to leaf fall, the effect is commonly surprisingly small...*' Combined Ewc and transpiration losses from grasslands in the early winter are likely to be small, for example, 5.6% of gross rainfall (100 [16.3/289.5 mm] for months of December 1975–1985 [28]). However, in some contrast, the wet canopy evaporation rate for deciduous woodland when the overstory is leafless in winter is likely to be within the range 10–20% of the gross rainfall for the conditions prevailing in the UK or for similar situations in continental Europe (**Table 1**), but perhaps less (as a percentage) for high magnitude events. *The first column of*

% P (by rank)	Dominant species	Reference	UK/Europe
40–50	Hawthorn (hedge)	Herbst et al. [82]	UK
36	Oak/birch	Noifalise [83]	Continental Europe
29	Hornbeam	Leyton et al. [84]	UK
22.5	Oak	Vincke et al. [22]	Continental Europe
19.8	Oak/birch	Herbst et al. [30]	UK
15.1	Beech/hornbeam	Aussenac [85]	Continental Europe
14	Beech	Reynolds and Henderson [27]	UK
12.1	Mixed	White and Carlisle [86]	UK (Cumbria)
12	Oak coppice	Thompson [87]	UK
11	Oak	Dolman [88]	Continental Europe
10.5	Hornbeam/oak	Schnock [89]	Continental Europe
10	Oak/beech	Staelens et al. [90]	Continental Europe
9.9	Oak	Carlisle et al. [91]	UK
7	Beech	Gerrits [92]	Continental Europe

Table 1. A wet canopy evaporation range (over 1–3 months) that also encompasses most of the extremes in observed behaviour of leafless vegetation canopies would be 7–50% of gross rainfall through winter storms for deciduous trees, although likely to be at the lower for an extreme event so a range of 4–20% was used in the modelling.

http://dx.doi.org/10.5772/intechopen.68677

Table 1 was used to define the fuzzy set of wet canopy evaporation rates (top right panel of **Figure 3***) used in the Dynamic TOPMODEL but constrained to the smaller range of 4–20% considering the extreme nature of Desmond.*

3.5. Antecedent moisture status

The higher rate of wet canopy evaporation during winter leafless periods, combined with the higher combined rates of wet canopy evaporation and transpiration from leafed deciduous trees in the proceeding summer and autumn in comparison to grassland [29–31], means that UK woodland soils are likely to be drier during the winter. A drier subsurface condition will reduce the proportion of rainfall delivering fast streamflow responses thereby reducing peak flood flows [32, 33]. Finch [34] observed drier soil moisture profiles (by 250 mm) beneath sweet chestnut and larch woodland and grasslands in Pang basin (Berkshire, UK) through December in 1997. Indeed, the profile did not reach its maximum saturation until April 1998. Similarly, Calder et al. [35] show soil moisture deficits through December 2000 that are drier by 30 mm in the soil (0–0.90 m) beneath oak (*Quercus robur* L.) of Clipstone Forest (Nottinghamshire, UK) than beneath adjacent grassland. A scenario of 80 mm of additional soil moisture deficit beneath deciduous woodland compared to grassland in the early winter is within the 30–250 mm range of the two UK studies noted but is clearly associated with a highly uncertain range.

NFM might feasibly give wetter antecedent conditions in a sequence of winter events if the increased infiltration effects of tree planting on soil moisture are larger than those of enhanced wet canopy evaporation ('infiltration trade-off hypothesis'). *Here we attempted to account for this through detailed modelling of several consecutive storms, and although the drier antecedent soil moisture was taken into account, the deficit was reduced considerably after the first storm in the series.*

3.6. Woodland on slowly permeable, gleyed UK soils

Overland flow on hillslopes may be caused by rainfall intensities (mm/hr) exceeding the saturated hydraulic conductivity (K_S; mm/hr) of a topsoil or other surface horizon (equivalent to the 'infiltration capacity' or 'coefficient of permeability of the topsoil'). This rapid pathway of rainfall towards stream channels is called 'infiltration-excess overland flow' [36]. If rainfall is reaching the ground at a rate less than the saturated hydraulic conductivity of the topsoil, but cannot infiltrate because the topsoil is already saturated as a result of drainage from upslope areas, then the rainfall onto these saturated areas will move as overland flow across the surface. This pathway is called 'saturation overland flow by direct precipitation' or SOF by direct precipitation [37]. If the downslope subsurface flows exceed the ability of the downstream soils to discharge them directly into a stream channel, then subsurface water may emerge from the topsoil onto the ground surface as 'return flow' [38]. This return flow may then travel overland towards a stream as so-called 'saturation overland flow by return flow' (SOF by return flow).

Soil types that typically have a lower saturated hydraulic conductivity have a greater likelihood of generating 'infiltration-excess overland flow', and were present in downslope areas, also a greater likelihood of generating surface flows by 'saturation overland flow by direct precipitation' and 'saturation overland flow by return flow'. The soil type called a Gleysol using the international soil classification system [39] or gley within the Soil Survey of England and Wales (SSEW) soil classification system [40] typically exhibits lower saturated hydraulic conductivity values throughout UK soil profiles [41]. **Table 2** shows an example K_S profile for a gley in the Lune Valley, Northwest England (UK). These measurements were undertaken in the field with a ring permeameter [42], a technique demonstrated to give accurate values, even for disturbance-sensitive gley soils [43].

Depth (m)	Mean K_S (cm/hr)	Range (n = 56)
-0.10	9.1	1.31–30.7
0.10–0.20	21.8	8.98–57.0
0.20–0.50	0.11	0.021–3.02
0.50–1.00	0.002	0.0007–0.21

Table 2. Horizon-specific saturated hydraulic conductivities (cm/hr) of a Humic Gleysol near Farleton, Lancashire (UK) after [39, 40].

Soils in England and Wales that are classified as gley cover a range of soil associations based on the SSEW [44]. As a result of their greater likelihood of generating overland flow, these gley soils are classified as having an SPRHOST[1] value in excess of 50% [45]. Enhancing the permeability of such soils could have the greatest impact on reducing overland flow across catchments and thereby have the greatest potential to reduce flood peaks in rivers [46]. As a result, tree planting to increase soil permeability includes areas with such gley soils.

Consequently, for NFM modelling a key need is to represent the permeability effects of planting deciduous trees on gley soils. Very few UK studies are available that quantify the difference in soil Ks of gley soils beneath deciduous trees relative to that beneath adjacent grasslands [47].

These few studies are summarised in **Table 3** and give a range of 1.5–3.5 factor increase in permeability for deciduous tree planting on gley soils. *The Ks factors in the first column of* **Table 3** *were then used to provide the fuzzy set of parameter changes in the Dynamic TOPMODEL scenarios.*

This observed increase in permeability for gley is smaller than the factor of five difference between predominantly deciduous woodland and improved pasture recently observed on well drained Eutric Cambisol (SSEW Brown Earth) soils in Scotland by Archer et al. [48, 49]. The observed effect on gley is also smaller than the effects observed on other soil types across the globe, the majority of which are between 2 and 20, with some outliers much greater than this, which is likely to be due to macropores along root channels. In any event, if the roughness of these high runoff areas can be increased through roughening up, then it may be possible to attenuate quick-flow from these soils until they become saturated by deficit filling.

[1]*Standard percentage runoff based on hydrology of soils types*

F/G[1]	Tree age	Soil type[2]	Location	Reference
1.81	2 years	713e-Brickfield-1	Tebay Gill Cumbria	Mawdsley, Chappell & Swallow [93]
2.43	10 years	721d-Wilcocks-2	Pontbren Mid-Wales	Marshall et al. [94][3]
3.40	107 years	713f-Brickfield-2	Lancaster Lancashire	Chandler and Chappell [47]

[1]Factor difference in KS, i.e. KS below deciduous trees/KS below grassland.

[2]SSEW Soil Association from Soil Survey of England and Wales [44].

[3]Results of related study of Carroll et al. [81] rejected on basic quality assurance criteria (i.e. absent sampling size per land-cover; absent information on frequency distribution, etc.).

Table 3. Ratio of Ks measured for deciduous trees to that grassland growing on gley soils in the UK.

Critically, it should be remembered that most of the stream hydrograph during floods comprises water that has primarily travelled to the stream via subsurface pathways (effects of trees on Ks might not be isotropic–but we have even less information on downslope Ks). Even within very flashy, but undisturbed tropical streams, only small proportions of flow within the basin have been directly measured as overland flow (e.g. <10% streamflow [50]). Therefore, while the overland flow pathways are important given their speed and sediment transport aspects, especially during flood events, simulated flow pathways are likely to be dominated by subsurface responses (which can also be fast–subsurface celerities in wet soils can even exceed overland flow velocities [51]), pathways either close to the surface in soils or deeper within the surficial or solid geology [52, 53].

4. Opportunity mapping of NFM

Opportunity maps can be developed from local knowledge, land cover maps, flood modelling outputs, or a combination of all three, as described here. In this chapter, opportunities for three core types of NFM (tree-planting, RAFs and soil structure improvements) were developed from different national strategic maps, and then through consultation at an engagement event. Ideally, engagement would be a continuous process of refinement where more local knowledge of the landscape and opportunities are built in through time, and evidence is co-produced [54]. The following sections explain how these opportunities were identified and refined.

4.1. Runoff attenuation features

Research on RAFs [55–57] such as storage ponds, bunds, in-stream storage through woody debris dams and disconnecting drain flow pathways has shown that these features have the potential to reduce flood peaks and increase the time to peak for overland flows and streamflows. Applying RAFs within the headwaters of a catchment, therefore, has the potential to attenuate sudden short duration storm events and reduce the subsequent flood risk to more urbanised areas of the catchment downstream.

Opportunities to deploy RAFs can be identified from areas of high flow accumulation in surface water flood maps, and comprise small areas such as natural depressions within the landscape, or small in-channel storage as shown in **Figure 4**. The JRAFF model identifies the

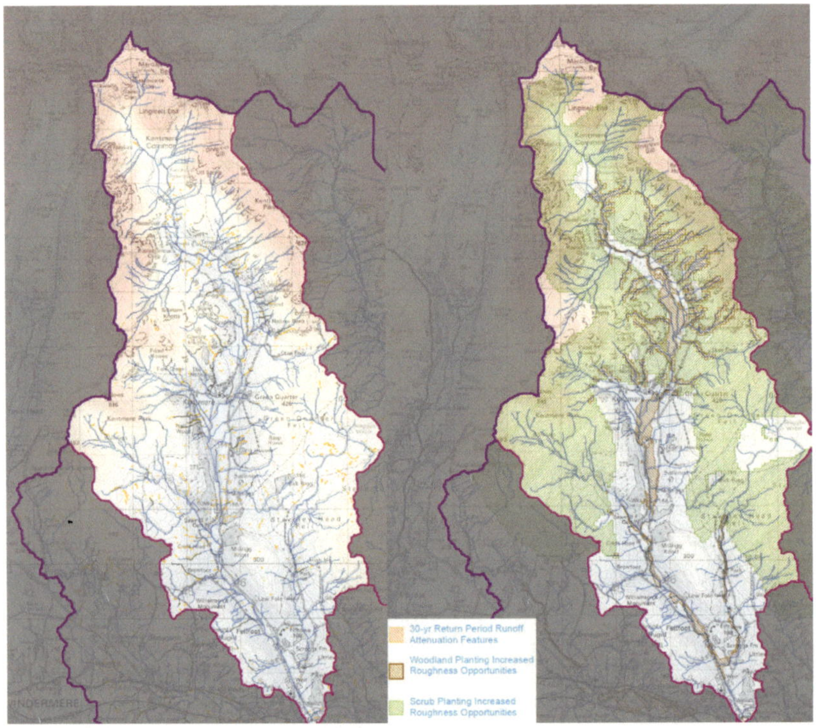

Figure 4. Runoff Attenuation Features (RAFs) and shrub and approximate woodland planting opportunities in the upper Kent (for illustration).

orange areas of isolated flow accumulation, such as ponds and small channels, which may be appropriate to excavate or bund, or disconnect from flow pathways through gully or ditch blocking. An additional storage of 1 m in depth at these locations is represented through burning the Digital Terrain Model (DTM) deeper by 1 m in the RAF model scenario.

The JRAFF tool places a set of constraints on the size and location of these accumulations:

- Area threshold between 100 and 5000 m^2
  - This was considered suitable for local land management alterations, and well below the threshold of capacity that would fall under the UK Reservoirs Act (10,000 m^3).
- CORINE land cover 2012 dataset and a 2 m buffer of OS OpenData buildings and roads deemed unsuitable to runoff attenuation features.
  - A 2 m buffer of roads results in a 4 m wide exclusion zone. This threshold has been derived based on typical road widths and ensures that opportunity features within any potential adjacent ditches are retained.

In the application to Cumbrian catchments, the JRAFF model was used to calculate the additional storage volume if such areas were to be deepened[2] by a further 1 m before summarising

[2]*Equally representing a bund around an existing flow accumulation area*

these volumes within priority sub-catchments defined earlier. Within the Kent catchment only, any RAFs identified within peat soils were excluded where hillslopes were greater than six degrees. This constraint is based on current peat restoration practices [58]. This process identifies a very large number of opportunities which can be incorporated into a model. These opportunities represent large scale, long term NFM delivery and provide the evidence required to take a strategic approach to optimising the benefits of providing additional distributes storage within the catchment.

4.2. Identification of tree-planting opportunities

Restoring the riparian zone and planting woodland within the floodplain has been simulated to provide the potential for significant flood attenuation (see Refs. [59, 60]). A combination of improvements in wet canopy evaporation and transpiration enhanced soil drying and soil infiltration together with increases in hydraulic roughness which arise from woodland creation can lead to reductions in flood peaks together with delaying and spreading of tributary hydrographs.

The Woodlands for Water (WfW) opportunity EA dataset was supplied by the Environment Agency for this project [61] and it was modified with the local knowledge and through inspecting soil series maps to give potential tree-planting opportunities (**Figure 4**). The dataset typically comprises a set of woodland plantings opportunity areas such as riparian zones and floodplain areas together with a number of constraints such as urban areas, existing woodland and inland water. Whilst the source dataset infers opportunities to plant and enhance woodland areas, this scenario rather reflects a more general improvement in planting density between scrubland and mature forest as it is understood that conversion to mature woodland would not be appropriate across all land covers.

The modified WfW opportunity maps represent large, long-term NFM delivery within each catchment. The incorporation of this opportunity into the model provides the evidence required to take a strategic approach to optimising the benefits of providing additional 'natural roughness' within the catchment.

4.3. Identification of opportunities for soil structure improvement

This scenario pertains to the fact that many soils have been compacted through more intensive farming practices over a long period of time, and if de-compacted, improved soil structure has potential to take in and store considerably more of the incident rainfall [62]. This can help reduce overland flow and reduce downstream flood risk, although could potentially have limited benefits in wet winters.

For the third type of opportunity, soil structure improvement, the JFLOW modelling targeted a particular land cover (improved grassland) which was identified as one of the most common land covers within each catchment based on the Land Cover Map 2007. For these areas, the catchment descriptor BFIHOST [45], which influences amount of runoff routed over the landscape in the modelling (see below), was increased by 10% resulting in an approximately equivalent increase in maximum soil moisture storage and reduction in initial soil moisture storage capacity for these land cover areas across the catchment. Users can then assess the improvement relative to that for this type of land cover by scaling up by relative area compared to improved grassland.

This more targeted approach to improving soil permeability avoids overestimating the impact of soil improvement by applying an unrealistic blanket improvement across the catchment. Soils can only be improved if they are damaged. Palmer and Smith, 2013 [63] estimated that approximately 38% of soils were structurally damaged, the percentage was higher under arable and lower under pasture, in a survey in the SW of England.

4.4. Strategic modelling and estimating benefits

The strategic modelling was undertaken using a fast 2d hydrodynamic modelling software JFLOW [4], which has been benchmarked against other 2d inundation models against a wide range of test cases [64]. The approach illustrated in **Figure 5**, builds on the blanket rainfall approach (e.g. Ref. [65]), which was developed further to include the ReFH losses model [66], and used to develop a national SW flood map RoFSW [67]).

The model integrates spatially varying rainfall, with the representation of both rural infiltrations using the ReFH rainfall to overland flow calculated losses (i.e. infiltration) model and urban sewer loss rates (a national average of 12 mm/hour was used). Rural ReFH losses are controlled by the maximum soil moisture storage capacity (Cmax) which is estimated using the catchment descriptors BFIHOST and PROPWET whilst urban losses are based on estimated sewer capacity losses and percentage overland flow.

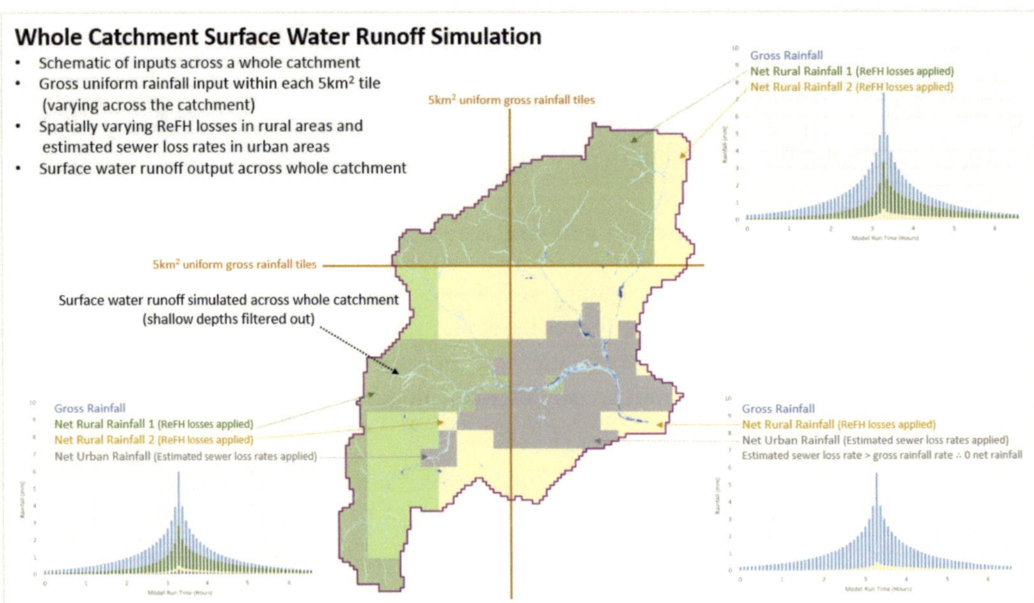

Figure 5. The rainfall and losses approach to whole catchment modelling.

The floodplain is represented using a 2 m Digital Terrain Model (DTM) based on a combination of filtered LiDAR, filled in with coarser scale photogrammetry or SAR data. Sinks and dams are removed in order to maximise hydrological flow pathway continuity. Spatially varying hydraulic roughness coefficients were adopted throughout, based on land cover and the same roughness coefficients adopted in national maps. The baseline scenarios were for the 10-year and 30-year return periods with a 6-hour storm duration.

Much consideration is given to the placement of virtual monitoring locations around the catchment that monitor the hydrographs for different sub-catchments and their modelled response to NFM interventions (**Figure 6**). For urban areas, culverts and 'cut-throughs' can be added to allow the passage of water downstream more realistically.

The percentage change in the peak runoff response for each sub-catchment was then visualised in a second set of 'benefit maps' (**Figure 7**), for use at the engagement workshops. These were colour themed in different ways, with the perimeters shaded to reflect the magnitude of the modelled peak runoff reduction and the fill based on the extent of the opportunity. Benefits are cumulative, in that every opportunity must be implemented within upstream sub-catchments in order to obtain the visualised benefits.

These maps represent the potential benefits of large-scale, long term NFM delivery across a catchment. They are most appropriately used in relative mode, allowing the user to identify particular sub-catchments which provide greater benefit than other sub-catchments and the different interventions that should be targeted to provide these benefits.

Other approaches to showing benefits or damages avoided have been developed, whereby the average property damages [68] per 1 km tile are shown with and without NFM, and this

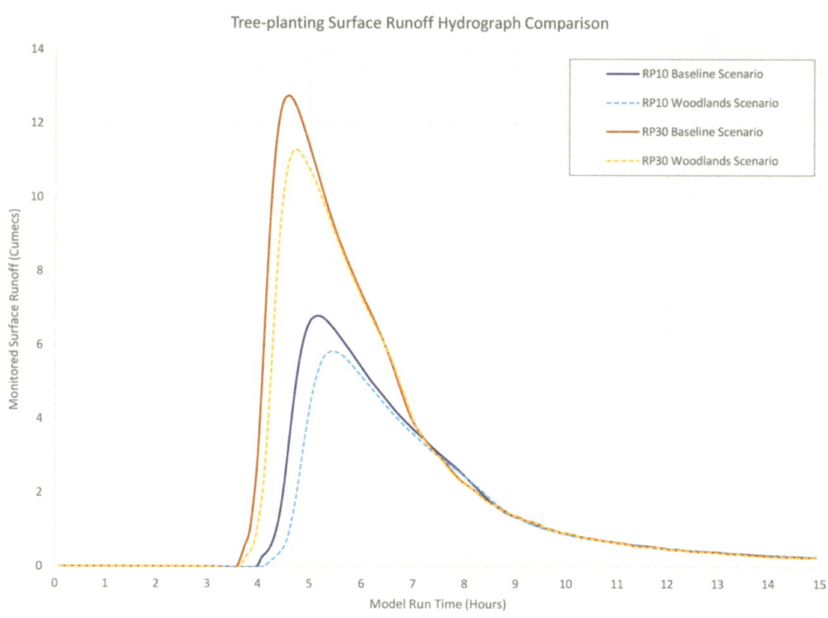

Figure 6. Example peak fast-flow reduction based on modelling NFM.

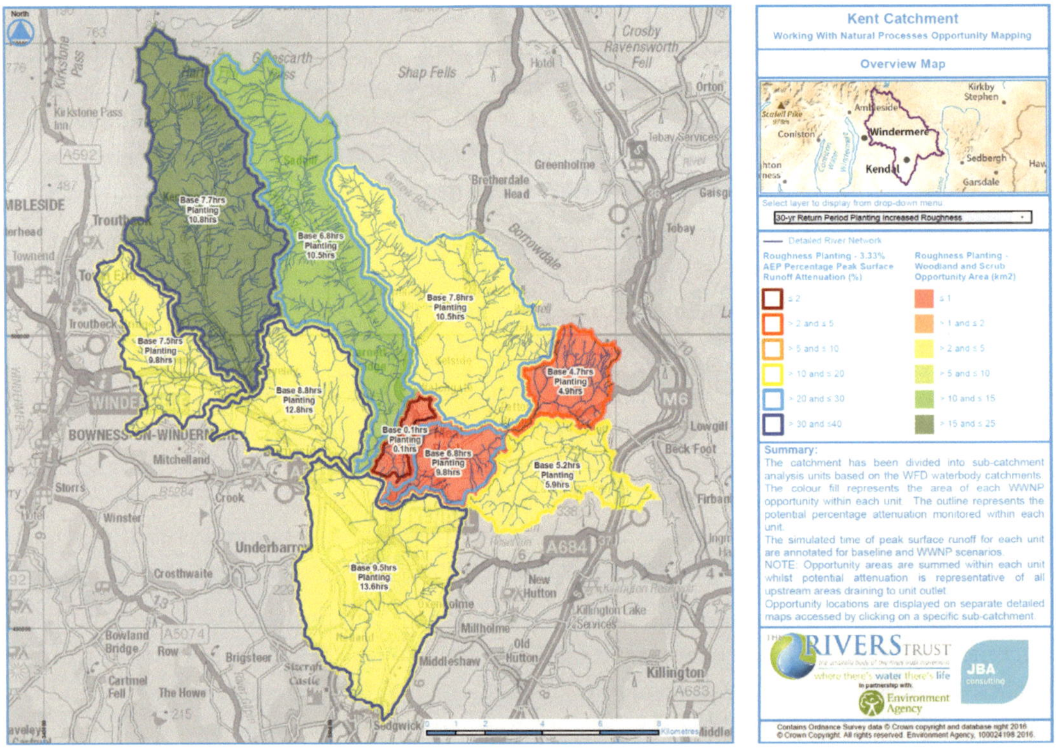

Figure 7. Visualisation of potential runoff reduction if all NFM opportunities taken up.

helps put appraisal of NFM on a footing with more established methods and identify potential dis-benefits due to, for example, backwater effects causing upstream flooding.

5. Engagement and refinement

Visualisations such as **Figures 4**, **7** and **8** have been found useful at engagement events, whereby catchment partners can mark the maps and alter the opportunities based on their local knowledge, then see the potential benefits based on the broad-scale modelling. Many of the identified opportunities will not be feasible due to land ownership and access issues. There may already be interventions planned which can also be added to the maps. Following the engagement, the models were re-run and the maps and interactive PDFs regenerated for the whole catchment.

One key benefit of the engagement approach is the development of a shared understanding of flood generation, routing and accumulation within a catchment. The opportunity for planners, water company engineers, land managers, FCRM, water quality and biodiversity specialists to elicit hydrological understanding from modellers including issues of synchronisation is of enormous value. The engagement process also enables the modelling team to appreciate and incorporate a more credible set of opportunities and parameters back into the model. This

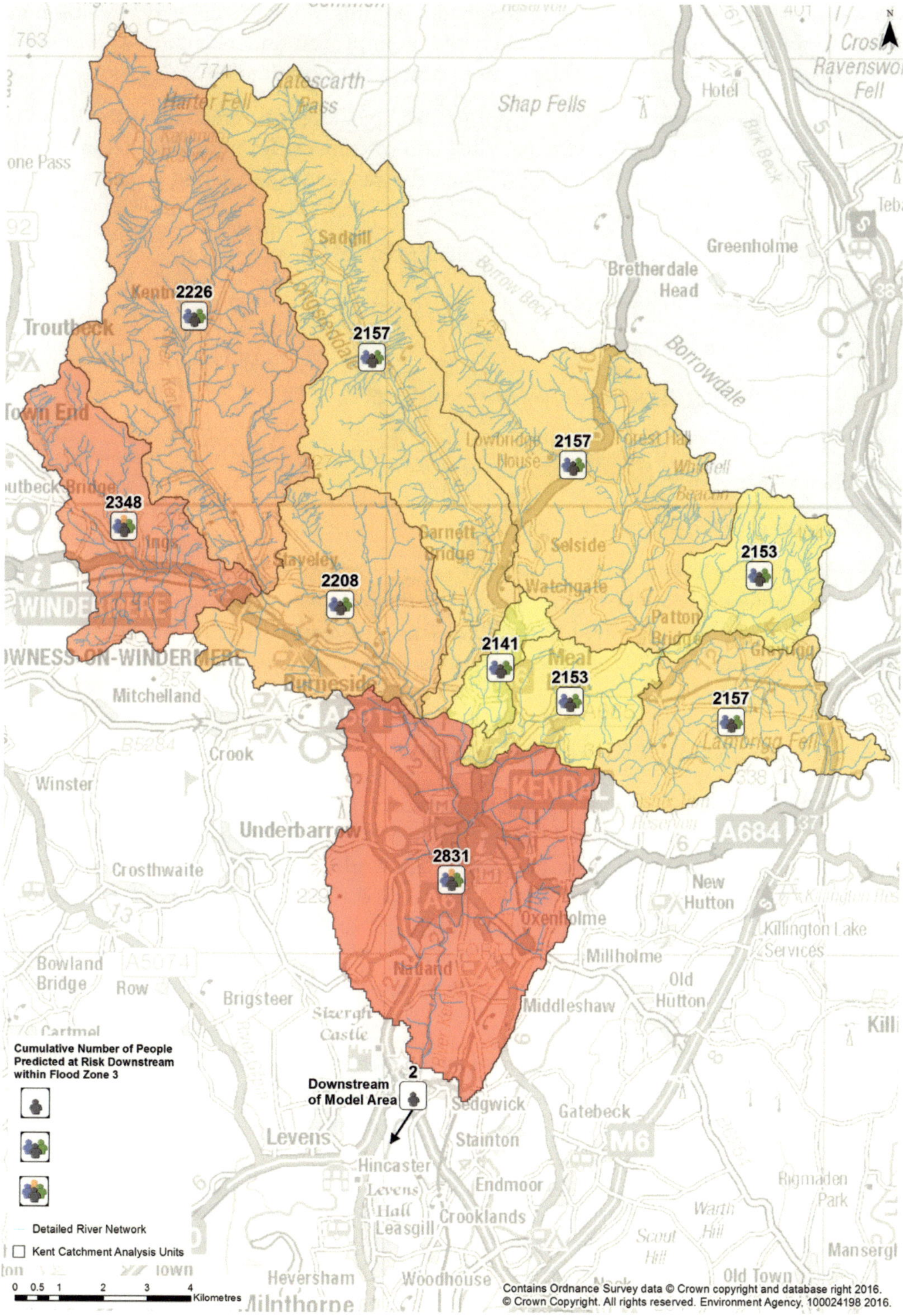

Figure 8. Downstream risk visualised for the Kent.

two-way process builds confidence in the model outputs, allowing delivery organisations to make appropriate use of the modelling, based on improved understanding, as part of the weight of evidence required to develop a more strategic approach to NFM delivery.

In addition to the refinement of the opportunities, and through remodelling the benefits, the catchment partners were asked to prioritise sub-catchments for more detailed modelling to build confidence and implement NFM. A matrix was developed, based on the following criteria:

- Land ownership and access—are opportunities feasible?
- Observations based on local knowledge
- Observations from strategic maps
- Scale of downstream risk
- Existence and location of monitoring
- Catchment size
- Preferences counted in the workshop

The upper and mid-Kent and Gowan sub-catchments were prioritised, and it can be seen in **Figure 8** that there are different centres of risk and strong reason to supplement any established FRM measures with NFM if possible. The catchments were also identified in the Cumbria Flood Plan for NFM-type interventions, largely for peat or bog restoration in the upper Kent.

5.1. Modelling of NFM with dynamic TOPMODEL

Given limited observational evidence effectiveness of NFM at scales >10 km^2, the prioritisation of areas to model in more detail and through using Monte-Carlo simulations was considered essential to understand more about uncertainty and build on the screening-level modelling with JFLOW (despite its high resolution). The detailed modelling was therefore calibrated against an observed series of extreme events, Nov-Dec 2015, within three severely-impacted catchments up to 223 km^2 in the area within Cumbria, UK. The results from one of these catchments, the upper Kent (90 km^2) are used in the following sections to illustrate the approach that was taken for all three catchments.

Lancaster University has recently developed an extended and flexible implementation of the Dynamic TOPMODEL model [9, 10], first implemented in FORTRAN by Beven and Freer [9]. It is an extension of the popular TOPMODEL [69] applied in many studies [70]. Dynamic TOPMODEL employs the efficient parameterisation scheme of TOPMODEL, but allows a more general approach to a grouping of points in a catchment for calculation purposes, based on overlays of characteristics rather than simply the map of the topographic index. All the model parameters needed to run the model are shown in **Table 4**, along with typical ranges of values applied in this project. **Figure 9** provides a perceptual overview of Dynamic TOPMODEL.

Parameter	Description	Units	Lower	Upper
V_{of}	Overland flow velocity	m/hr	1	150
m	Form of exponential decline in conductivity	m	0.0011	0.033
SRZ_{max}	Max root zone storage	m	0.1	0.3
SRZ_0	Initial root zone storage	%	20	100
V_{chan}	Channel routing velocity	m/hr	500	5000
$ln(T_0)$	Lateral saturated transmissivity	m²/hr	3	12
$Sdmax$	Max effective deficit of saturated zone	m	0.5	0.5
T_d	Unsaturated zone time delay	m/hr	1	10

Table 4. Parameter ranges/typical values within dynamic TOMODEL (after Ref. [10]).

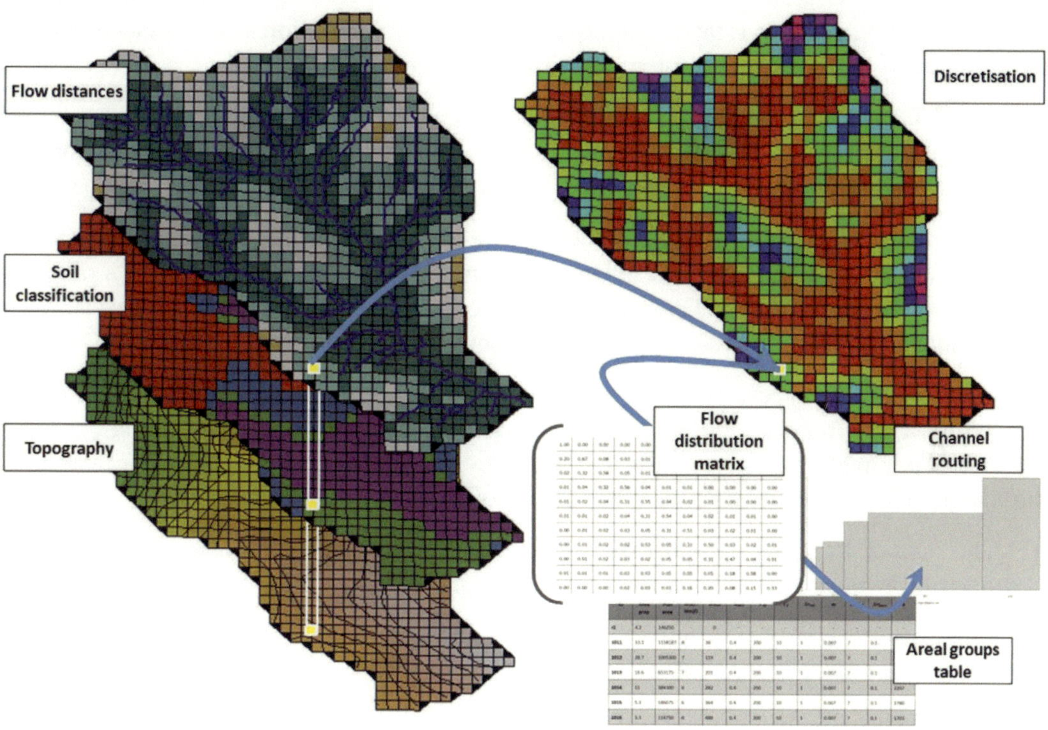

Figure 9. Schematic of dynamic TOPMODEL.

It is a semi-distributed hydrological model that utilises areas of similar hydrological behaviour called HRUs (Hydrological Response Units). These units may be in the evaluation of NFM divided to represent distributed interventions within the landscape. It has sufficient complexity to represent the key catchment processes, notably subsurface and overland flow pathways, and there is some evidence to link NFM measures to alteration of the parameters (e.g. transmissivity distribution) representing these processes. Its simple structure allows efficient operation, allowing thousands of model runs to be simulated to investigate uncertainties and sensitivities.

Overland flow velocity is fixed throughout each unit and can be changed to reflect changes in surface roughness introduced by, for example, peat restoration or riparian tree-planting. The maximum transmissivity at complete saturation T_0 $[L]^2/[T]$ is a measure of the local maximum saturated downslope transmissivity per unit hydraulic gradient (where transmissivity is the integral of the permeability to the saturated depth). This is a key parameter in identifying the onset of saturation overland flow (SOF). When downslope flows into lower slopes filling remaining storage capacity, return flow is produced. SOF is also generated when rain falls onto these areas of already saturated ground.

An exponential transmissivity profile is assumed. The use of such a form is supported by experimental evidence [71] and reproduces the typically higher values of permeability found near the ground surface. The recession parameter m [L] controls the rate of decline of transmissivity T as water table reduces. Small values of m lead to very rapid declines in transmissivity, suggesting shallower, faster responding streamflow generation systems. Deeper active hydrological systems are represented by a slower decline in transmissivity.

Dynamic TOPMODEL routes subsurface flow downslope between HRUs using a routing matrix derived from the local topography. It is assumed that the local slope is a reasonable approximation for the hydraulic gradient.

The root zone storage SRZ_{max} must be filled before any water table recharge begins through incident rainfall. Transpiration (and soil evaporation) is removed from this zone at a rate proportional to the actual storage. Direct observations of soil moisture content are unavailable for the catchments in the periods simulated, so a reasonable initial value was applied ($SRZ_0 = 95\%$).

Dynamic TOPMODEL represents an intermediate level of complexity that incorporates the key hydrological processes, but without imposing too many assumptions resulting in a large number of parameters (related to say soil properties), for which we do not have direct measurements. The model was also selected because of it:

- Makes use of standard data formats for catchment topography (elevations, channel network) and relevant spatial data such as land cover.
- Presents results back to the landscape as well as formats such as streamflow hydrographs that are easily understandable by partners.
- Uses real, spatially distributed rainfall data and can be rapidly calibrated against real event data (allowing for data quality).
- Incorporates spatial data overlays provided by the Rivers Trust (RT) and catchment partners of NFM interventions, for example, tree-planting, soil restoration and the addition of offline storage area (RAFs) and simulate the effect of these changes on streamflow response.
- Simulates a wide range of catchment scales (up to 223 km^2 in this study) and hydrological regimes such as the extreme flood event arising from Storm Desmond in December 2015.

- Allows for relatively quick application to future RT study catchments and different configurations of sub-catchments within an existing project.
- Uses a framework that allows assimilation of meteorological and streamflow data supplied in standard formats, such as those collected by the Environment Agency.
- Allows for free distribution of the source or compiled the code and, with suitable guidance and training, be operated by RT staff and catchment partners.

In conjunction with a hydraulic-routing scheme, also developed at Lancaster University, Dynamic TOPMODEL has been applied to the 28 km^2 Brompton, North Yorkshire, catchment in order to simulate the impact of up to 60 in-channel NFM interventions [5]. The model is written in the open source R language, distributed under the GNU Lesser Public Licence (GNU LGPL v2.1) and can be run on most common operating systems. The R implementation has been released as a package on the CRAN archive [72], passing the rigorous quality assurance and testing required by the submission process.

5.1.1. Representing NFM opportunities within HRUs for Dynamic TOPMODEL

A set of HRUs were developed based on local hydrological characteristics, the most important of which is the topographic wetness index (TWI). These were then split further so that individual NFM measures could be represented in the landscape. For example, the HRU representing the saturated area adjacent to the watercourse in the upper catchment becomes split into two, one where there is no change to the landscape and another where there is tree planting.

- HRU1–high SPR areas—mimicking tree planting effects.
  - Modification of T_0 (1.5–2.5 multiplicative factor of the un-logged value).
  - Decreased SOF velocity (reduced by 0.5 and 0.75–equivalent to a change in roughness for JFLOW).
  - Wet-canopy evaporation increases implemented as a loss to the gross rainfall input (using a similar strategy to Ref. [73]).
  - Modification of the initial root zone storage to mimic drier antecedent soil moisture conditions as a result of additional soil drying by enhanced wet-copy evaporation.
- HRU2–RAF features
  - Increased storage by introduction of RAFs is based upon modified JFLOW RAF opportunity maps
  - Implemented by a modification to the root zone storage to 1 m, making use of the existing store represented in the model
  - RAFs are 'leaky', draining at a rate q_{drain} [L]/[T] proportional to their specific storage S [L]: $q_{drain} = S/T$, where the constant of proportionality T [T] is the residence time

    - Assume all RAFs drain with the same time constant and 'piped' directly to a watercourse.
    - Implemented as the sensitivity of three pipe sizes to see differences in 'effective' size based on peak reduction.
- HRU3–Peat
    - Decreased SOF velocity (reduced by between 0.65 and 0.8–equivalent to a change in roughness implemented for JFLOW).

6. Uncertainty framework

The scale of the flooding caused by Storm Desmond is in part dependent on the catchment wetting from the storm events that occurred over the preceding weeks. In consequence, a 6-week period of rainfall and streamflow was selected in Nov-Dec 2015 (**Figures 4–6**) for the modelling, that included Storms Abigail (12–13th November), Barney (17–18th November), Clodagh (29th November) and Desmond (5–6th December). The first and last of these events had the highest impact on the streamflow. Clodagh was primarily a 'wind' storm and in Cumbria did not produce significant streamflow [11]. In November a south-westerly airflow described as an 'atmospheric river' became established bringing persistent warm moisture-laden air from subtropical regions resulting in persistent heavy rainfall. A 3-day total of 138 mm was recorded at the Shap automatic weather station in mid-November [11], compared to the total of 145–180 mm recorded at rain gauges around the Kent catchment in September and October. Two gauges lying within the Derwent catchment recorded new UK record rainfall totals: at Honister Pass, 341 mm fell within 24 h and 405 mm within 48 h at Thirlmere [11]. Dynamic TOPMODEL requires as input a time series of potential (or actual) evapotranspiration. We used the Calder et al. [74] approximation of a diurnal sinusoidal variation in potential evapotranspiration.

5000 Monte-Carlo simulations were undertaken before applying an acceptability criterion which sorts behavioural simulations, from those that are not. Within the Generalised Likelihood Uncertainty Estimation framework (GLUE [75–78]), the degree of acceptance of any simulation is weighted (or scored) quantitatively and is associated with the simulation during the entire analysis. Any simulations which are deemed physically unacceptable play no further part in the analysis and do not form part of the results.

The acceptance criteria were based on an overall performance measure over the whole modelled period (the Nash-Sutcliffe Efficiency statistic, NSE), the accuracy of the model prediction for the peak flow Q_{max} during Storm Desmond, and the maximum percentage of the catchment areas generating overland flow (by saturation through either of the processes described), SOF_{amax}. **Figure 10** illustrates the spread in model uncertainties through the calibration time series based on the resulting 'acceptable' parameter combinations.

Given the range of acceptable model predictions for each storm the problem arises of comparing the NFM interventions for all acceptable models. To compare the respective hydrographs contemporaneously is likely to be misleading, since each series' maxima may occur

http://dx.doi.org/10.5772/intechopen.68677

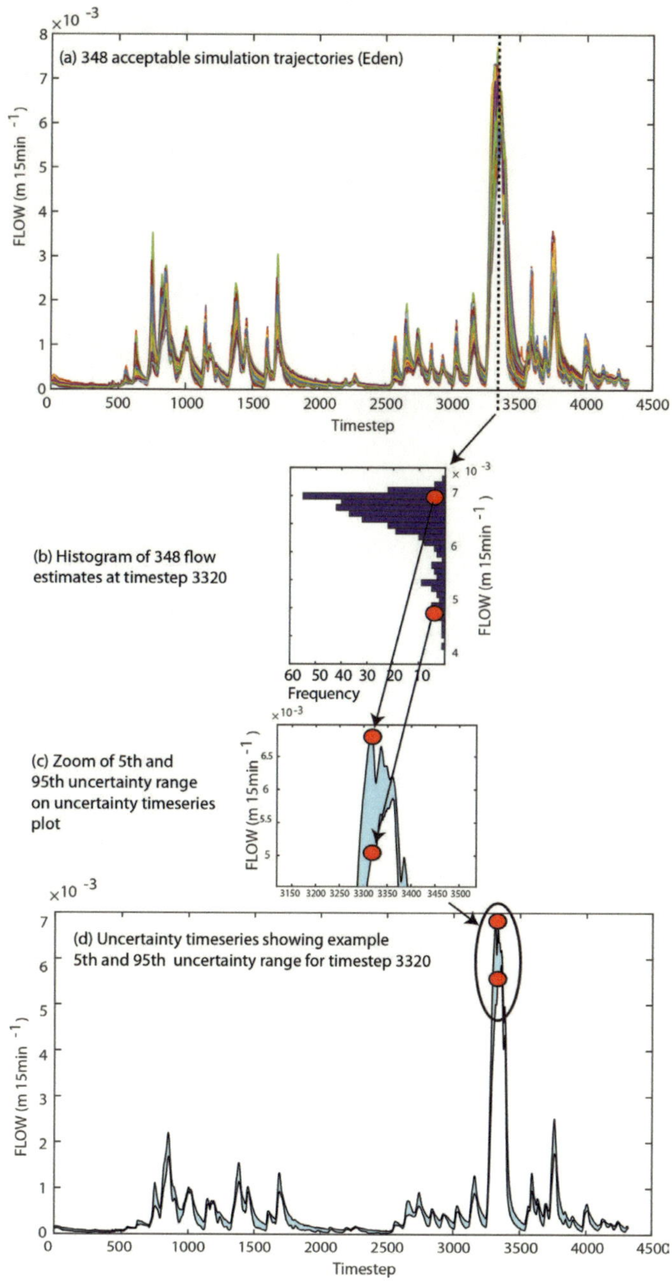

Figure 10. Translating uncertainty to model predictions (GLUE).

at different absolute times due, for example, to the retardation of the flood peak by the NFM measures. The distribution of flows for a window of time-steps around the peak has therefore been generated so that the shift in the distribution can be compared. This is illustrated in **Figure 11**.

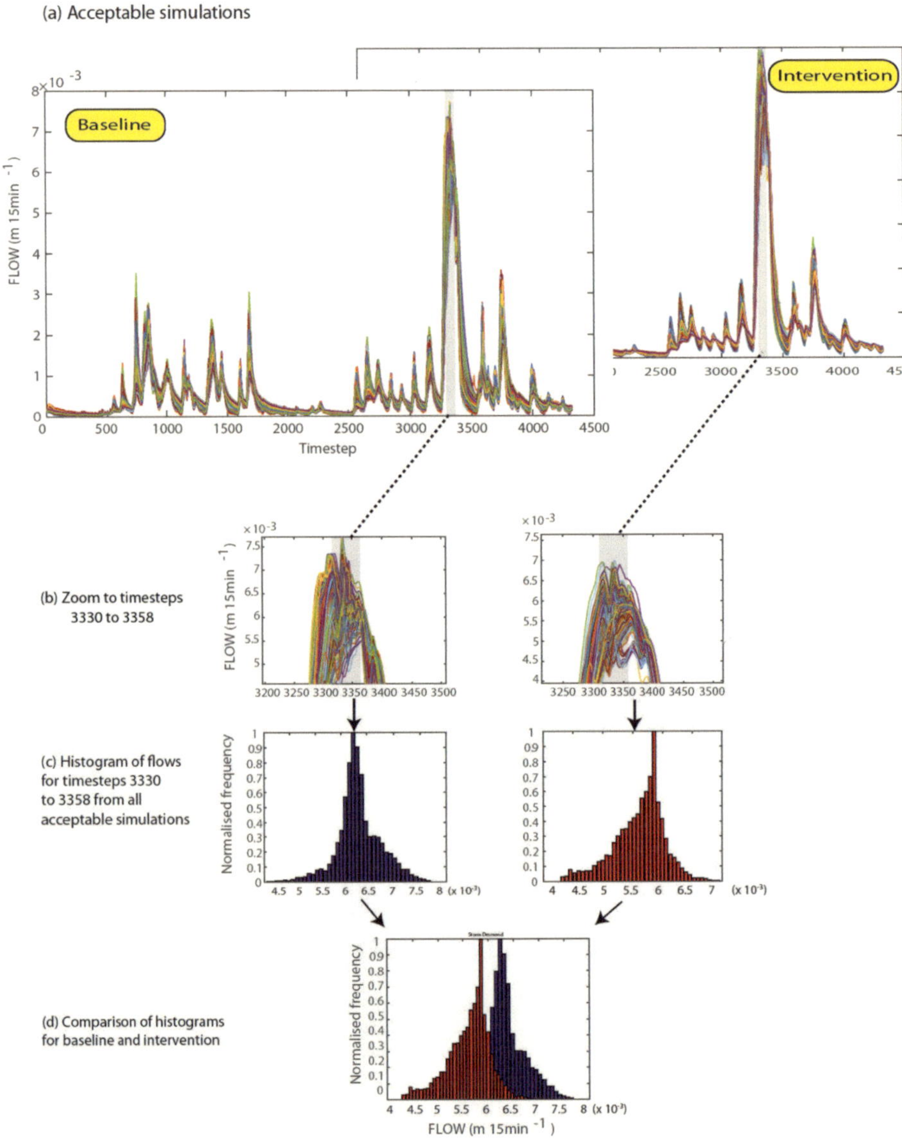

Figure 11. The change to predicted flow frequency distributions.

7. Results and discussion: comparison of models

We present here mainly the results for the Kent catchment, where the acceptance criteria discussed above were: NSE > 0.85; $6.9 < Q_{max} < 10.2$; $0.1 < SOF_{amax} < 0.95$, where Q_{max} is the peak flow per unit area for storm Desmond and SOF_{amax} was the very fuzzy range of acceptable fractional catchment areas producing SOF. This lead to 348 'acceptable' parameterisations, giving a range of predictions shown in **Figure 12**, for the RAF measures, designed to have a 1, 10 and 100 h retention time.

http://dx.doi.org/10.5772/intechopen.68677

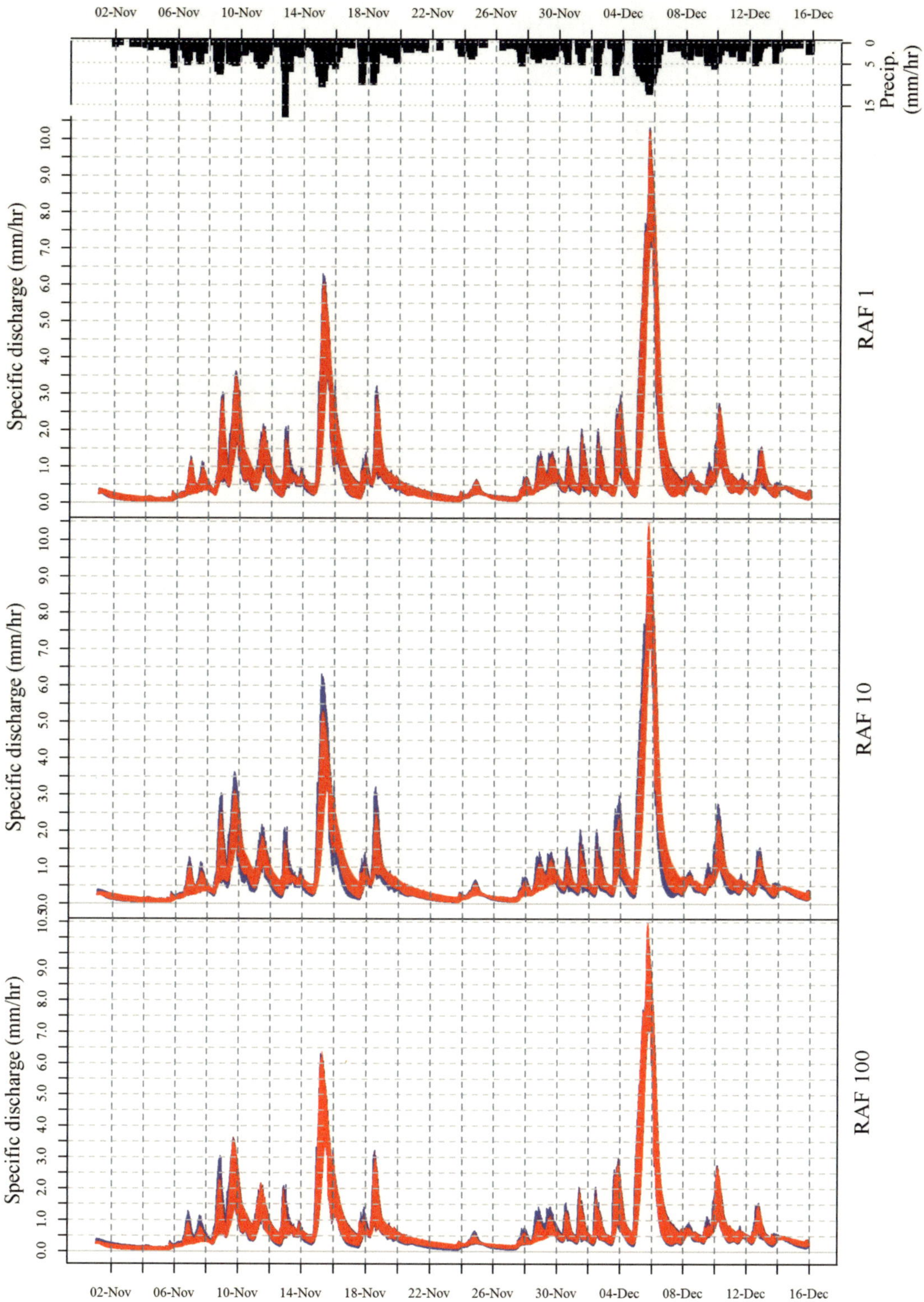

Figure 12. Range of predictions using acceptable combinations plus RAFs with a 1, 10 and 100 h retention time.

Figure 13 shows the range of potential changes to the storm profile when we apply two different levels of confidence (1 and 5, with 5 the least confident) in the evidence for the effect that tree-planting has on the different catchment processes.

It easier to consider the range of predicted peak flow reductions for each storm as a function of the confidence we place in the evidence (**Figure 14**) than by plotting the five different levels of confidence.

We can also make use of the generic approach from **Figure 11**, and plot the matrix of changes to the predicted peak flow distributions for a window around each named storm as a function of confidence (rows) in **Figure 15**, with a statistical significance of the changes highlighted in

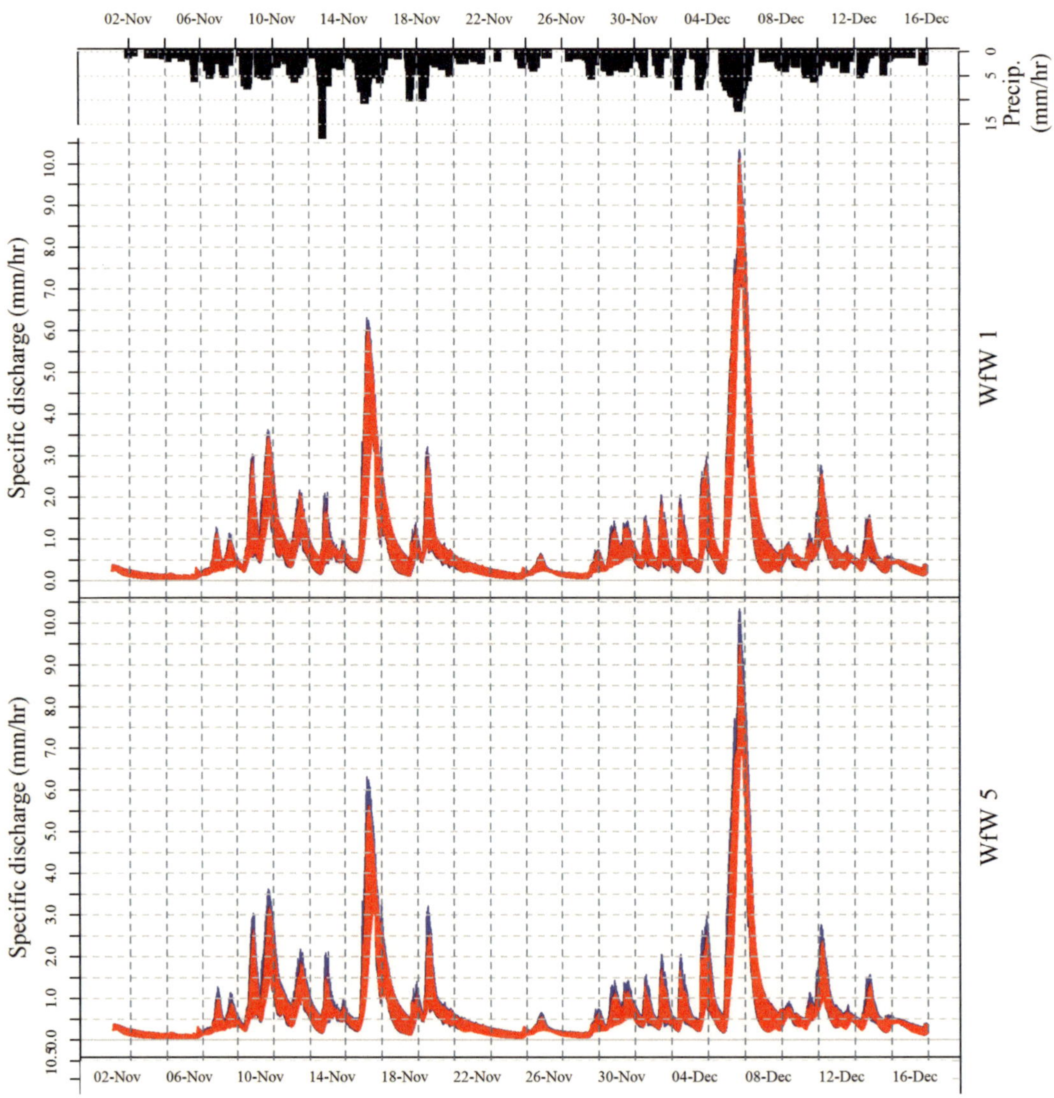

Figure 13. Effect of tree-planting opportunities with two levels of confidence in parameter changes.

http://dx.doi.org/10.5772/intechopen.68677

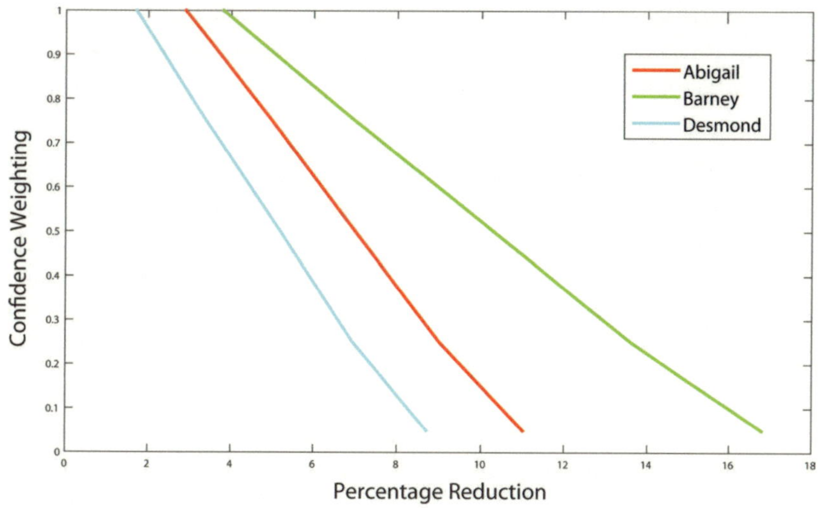

Figure 14. Confidence as a function of the percentage reduction for 3 of the named storms for tree-planting and roughening up.

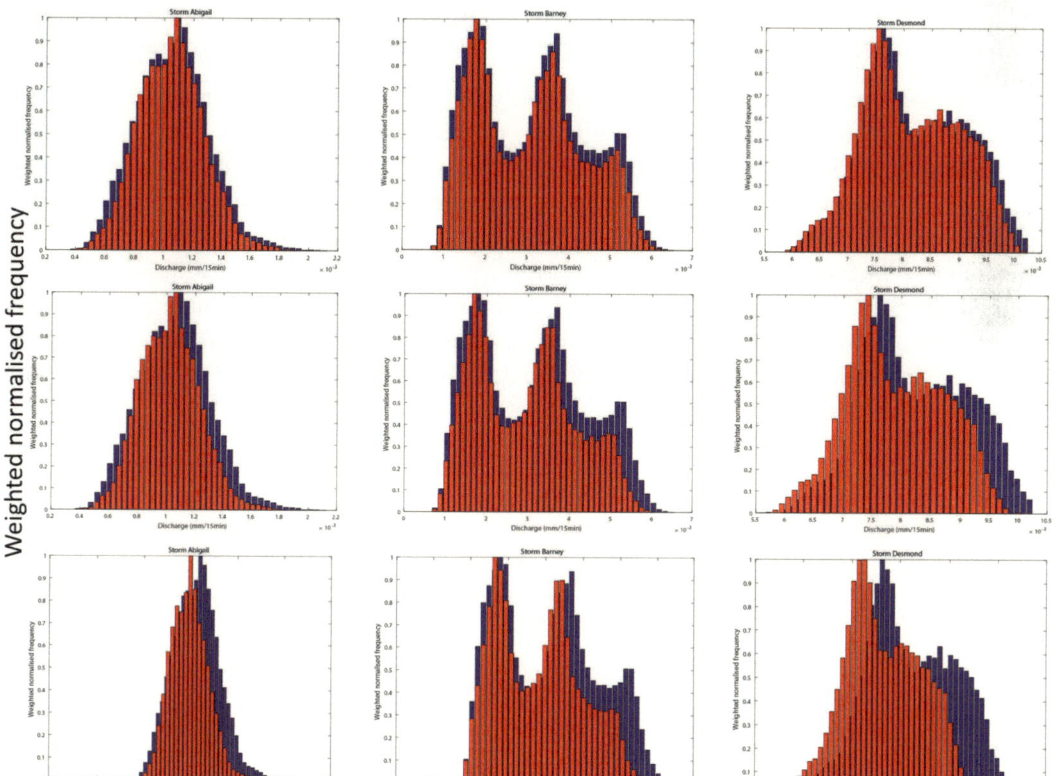

Figure 15. Matrix of shifts to the distribution of flows predicted around peaks for 3 storms (columns) for three confidence levels (rows).

Table 5. The WfW scenarios 1–5 were simulated, where there is more confidence in the evidence for scenario 1, and scenario 5 is the least confident. All the changes are more significant for storm Desmond, and the changes due to the RAF measures are generally more significant than the WfW.

Using Dynamic TOPMODEL, we have modelled drain-down of RAFs successfully with different time constants, and shown that for the Kent, RAFs designed with an intermediate residence time of around 10 h would be more effective for a series of flood events such as those in the period November through December 2015. The percentage reductions in peak flows are similar to the 2–5% peak runoff reduction predicted by JFLOW (30 year event) for the upper Kent for most of the period of modelling, apart from storm Desmond where **Figure 12** shows less reduction, potentially because the RAFs have not emptied.

It is less straightforward to compare the results for tree-planting between the modelling approaches, but for the Upper Kent, JFLOW predicts peak *streamflow* (without baseflow) reductions up to 30–40% for a 1 in 30 year design event (**Figure 7**), and based on **Figure 14**, the maximum reduction in *total flow* was half this, of the order 17% for storm Barney, but for simulations with the least confidence. However, these changes stem from very different physical processes, and further modelling has shown that although the perturbations applied to the wet canopy evaporation have the predominant impact, the *combined effect* was greater than the sum of the individual changes for evaporation, velocities and transmissivity.

The potential impact of large-scale NFM delivery on peak flow during extreme events such as storm Desmond is an important result. The evaluation of uncertainty enables us to use this finding appropriately and with greater confidence. The modelling allows us to see both the long-term potential of NFM and, critically, the model parameters and processes to which this prediction is most sensitive. These findings not only improve our understanding of the benefits of NFM but also guide future monitoring strategies that will be required to refine the modelling and adaptively manage NFM, and therefore flood risk, within a catchment. A consistent theme within the engagement was that it would be unlikely that advantage could be taken of every single potential opportunity, and so the results can seem overly optimistic but could be used in a relative sense.

	Abigail			Barney			Desmond		
	Δq_{max}	$\overline{\Delta q}$	K-S	Δq_{max}	$\overline{\Delta q}$	K-S	Δq_{max}	$\overline{\Delta q}$	K-S
WfW1	20.5	3.1	0.027	11.5	3.3	0.025	4.9	1.7	0.019
WfW5	30.9	9.6	0.139	26.9	12.9	0.124	14.2	7.9	0.076
RAF1	25.3	2.3	0.006	15.3	2.1	0.01	5.8	1.2	0.005
RAF10	28.6	7	0.067	22.5	13.7	0.118	4.5	0.1	0.055
RAF100	22.9	0.9	0.014	14.1	0.4	0.006	4.5	0.1	0.005

Δq_{max} = maximum relative reduction in peak (%); $\overline{\Delta q}$ = mean relative reduction in peak (%); K-S = Kolmogorov-Smirnov Statistic. The suffix to RAF corresponds to a retention time of 1, 10 and 100 h.

Table 5. Statistics for WfW and RAF interventions for each of the named storms in the simulation period.

http://dx.doi.org/10.5772/intechopen.68677

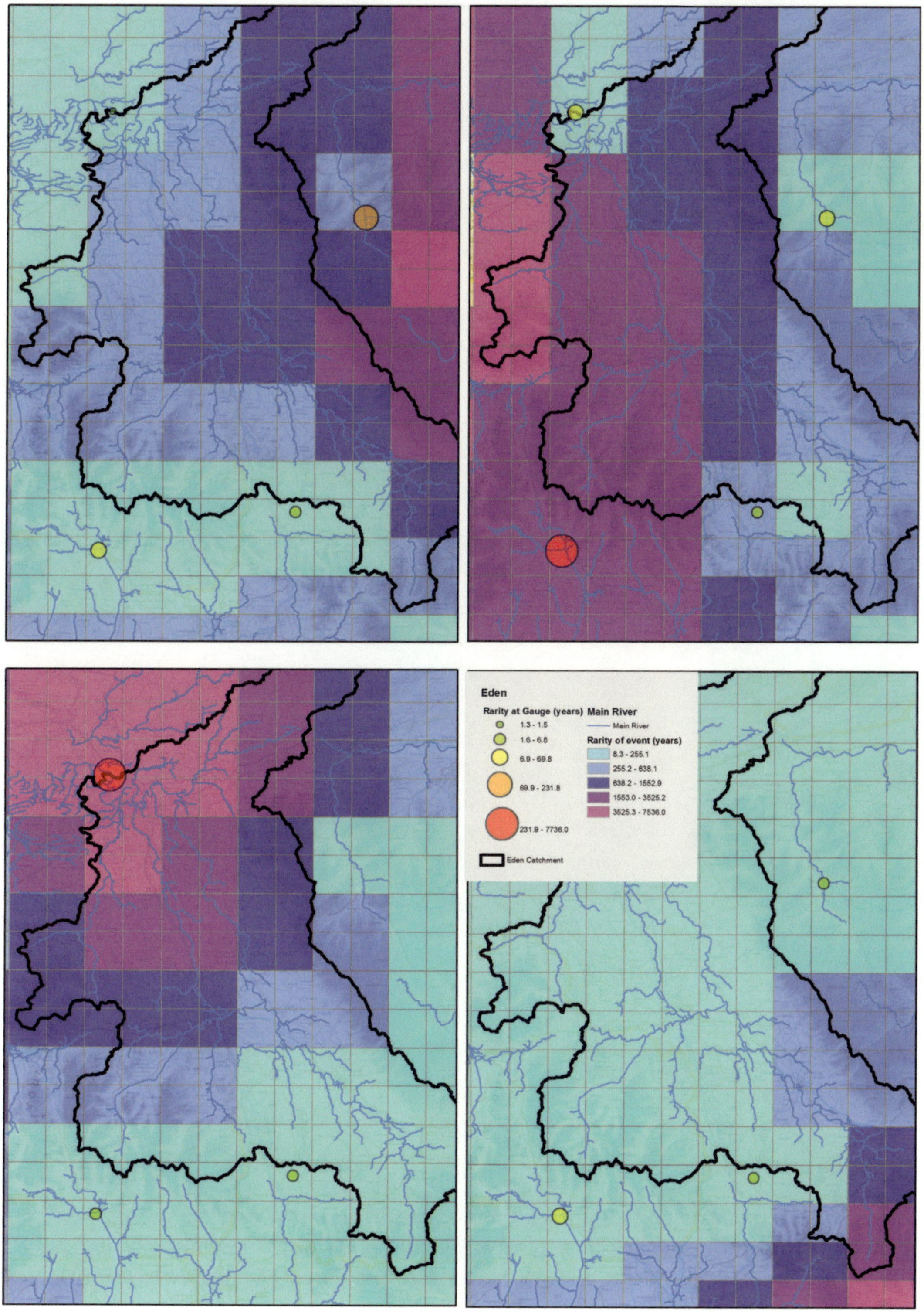

Figure 16. Spatial rainfall fields for plausible events.

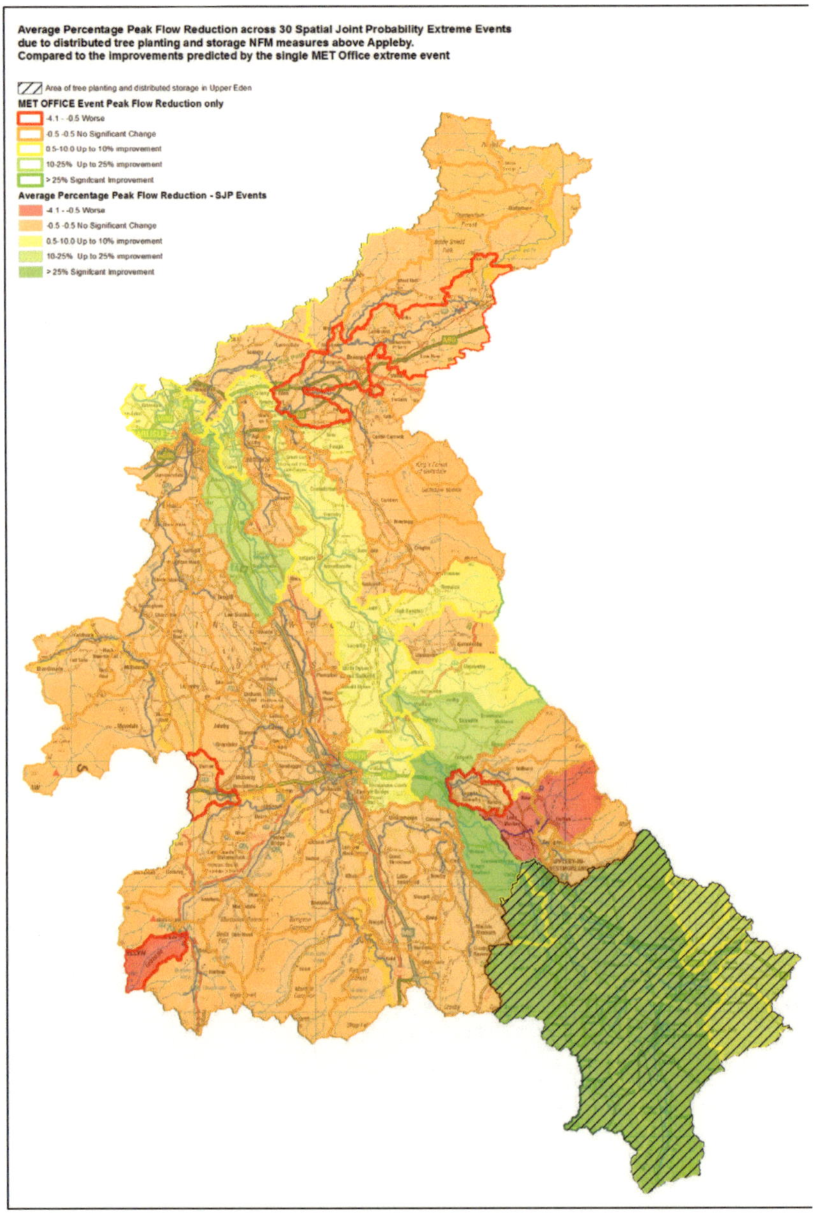

Figure 17. Average behaviour across 30 events with and without NFM.

7.1. Testing resilience

These types of analyses provide us with more confidence in the predicted response of the upper Kent to large-scale interventions of tree-planting and RAFs, yet we have not fully tested the long-term robustness nor the resilience in the face of different weather extremes. It would, in fact, be useful to test for a number of performance issues in order to gain greater confidence in the approach. These issues include:

- Synchronisation.
- Effect of sequences of events on antecedent conditions.
- Backwater effects.
- Sedimentation.
- Culvert or bridge blockage due to increased debris from tree-planting.

These can be examined and tested through modelling of extreme events with different spatial rainfall fields (especially for larger catchments) as performed for the Defra competition [6], for the Eden as shown in **Figure 16**.

Here the average beneficial effect of NFM across 30 extremes, but plausible, rainfall events was generated (**Figure 17**) in order to test for robustness in using such distributed events in the larger Eden catchment in Cumbria.

Ideally, the additional modes of failure of NFM should also be tested for through probabilistic modelling, as currently undertaken for established FRM in the UK. A systems-based approach could be applied [79] as discussed in relation to wider processes from synchronisation of flood peaks, backwater effects, sedimentation effects whereby RAFs fill in through time, and blockage of downstream features such as culverts near urban areas due to an increase in woody material in the longer term, although it is also important to re-think the implications of performance failure for very distributed small-scale measures, compared to established flood risk management approaches [80].

8. Conclusions

We have demonstrated a generic flood risk management framework that caters for the distinct differences between NFM and established approaches such as engineered downstream flood defences. NFM measures are characteristically small-scale and may need to be widely distributed to be effective at larger scales, and can influence a range of catchment processes. This combination means that modelling their effectiveness will be inherently uncertain, but we have demonstrated a framework that is tolerant of this uncertainty and the fuzziness in the evidence for how model 'effective' parameters can be plausibly changed to reflect their effect.

The tiered approach involves strategic modelling, mapping of opportunities and benefits, consultation and then prioritisation of areas for more detailed modelling and eventual implementation. The engagement phase is essential and ideally should be more of a continuous process through time with details of the model landscapes (topography, distributed roughness, storage, tree-cover, land-use), being revisited regularly. This is not new, but this chapter has hopefully shown how it can be achieved practically in the reasonable timescale of just 6 months.

However, differences in the model predictions from the different modelling approaches require careful interpretation and are only part of the picture set in the context of uncertainties on long-term ownership, maintenance and liabilities for NFM measures. The strategic overland flow process modelling is bound to lead to some different catchment responses than for a complete hydrological model, yet both shed light on different risk management issues. For example, the surface water modelling approach is useful for identifying key flow pathways and accumulations where partial blockage or storage can make a greater difference, but Dynamic TOPMODEL tells us more about the whole hydrograph and can integrate with more of the processes that tree planting is thought to influence. The two modelling strategies have helped to increase our knowledge of how and where in each catchment NFM measures can be more effective to reduce flood risk at large catchment scales. The models help shed more light on the complexities of parameterising change in model parameters, particularly at large catchment scales and for a range of flood events.

The potential for large-scale NFM delivery to provide significant flood risk benefits, even in extreme events, is based on a translation of the limited available evidence on the impacts of NFM between the 'real world' and the 'modelled world', using an uncertainty framework. If we believe that the model is a reasonable physical simulator of the whole catchment, then the potential benefits of following the risk management framework demonstrated are to help express model outputs in a language which highlights uncertainties and makes them more central to decision-making.

The models have helped to quantify by how much working with natural processes can improve flood regulation, depending on a fuzzy evidence base. The associated benefits can then be appraised alongside others, including carbon storage or reductions in diffuse pollution, sediment transport and improved community resilience through working together. However, we must stress the need for more information on the way in which model parameters might change as a result of land management changes as the information currently available is limited, in particular in respect of the type of flood event conditions relevant to NFM.

Acknowledgements

With thanks to the Rivers Trust for their guidance and engagement, and all the attendees of the workshop held on the 7th Oct 2016, and to the Environment Agency for the large quantities of data supplied.

Author details

Barry Hankin[1*], Peter Metcalfe[2], David Johnson[3], Nick A. Chappell[2], Trevor Page[2], Iain Craigen[1], Rob Lamb[4] and Keith Beven[2]

*Address all correspondence to: barry.hankin@jbaconsulting.com

1 JBA Consulting, Warrington, UK

2 Lancaster University, Lancaster, UK

3 The Rivers Trust, Malvern, UK

4 JBA Trust and Lancaster University, UK

References

[1] Pitt M. The Pitt Review, 2008. Learning lessons from the 2007 floods. June 2008. http://archive.cabinetoffice.gov.uk/pittreview/thepittreview/final_report.html

[2] JBA Trust. Working With natural Processes: A catalogue of nature-based flood risk management projects in the UK. 2016. http://naturalprocesses.jbahosting.com/

[3] Beven K. Environmental Modelling: An Uncertain Future? Routledge; London. 2009. ISBN10: 0-415-45759-9

[4] Lamb R, Crossley A, Waller S. A fast 2D floodplain inundation model. Proceedings of the Institution of Civil Engineers: Water Management. 2009;**162**: pp. 363-370

[5] Metcalfe P, Beven K, Hankin B, Lamb R. A modelling framework for evaluation of the hydrological impacts of nature-based approaches to flood risk management, with application to in-channel interventions across a 29 km^2 scale catchment in the United Kingdom. Hydrological Processes. 2017. https://doi.org/10.1002/hyp.11140

[6] Hankin B, Lamb R, Craigen I, Page T, Chappell N, Metcalfe P. Winning Flood Modelling Competition Entry [Internet]. 2017 [Updated: 19-01-2017]. Available from: https://consult.defra.gov.uk/water-and-flood-risk-management/flood-risk-management-modelling-competition/results/jba_defra_winning_entry_full_report.pdf [Accessed: 23 January 2017]

[7] Lamb R, Keef C, Tawn J, Laeger S, Meadowcroft I, Surendran S, Dunning P, Batstone C. A new method to assess the risk of local and widespread flooding on rivers and coasts. Journal of Flood Risk Management. 2010;**3**(4):323-336 (14 p). DOI: 10.1111/j.1753-318X.2010.01081.x

[8] Keef C, Tawn JA, Lamb R. Estimating the probability of widespread flood events. Environmetrics. 2013;**24**(1):13-21. DOI: 10.1002/env.2190

[9] Beven K, Freer J. A dynamic TOPMODEL. Hydrological Processes. 2001;**15**(10):1993-2011

[10] Metcalfe P, Beven K, Freer J. Dynamic TOPMODEL: A new implementation in R and its sensitivity to time and space steps. Environmental Modelling & Software. 2015;**72**:155-172

[11] Marsh TJ, Kirby C, Muchan K, Barker L, Henderson E, Hannaford J. The winter floods of 2015/2016 in the UK—A review. Wallingford, UK: Centre for Ecology & Hydrology; 2016

[12] Flooding after Storm Desmond. PERC UK 2015. http://www.jbatrust.org/wp-content/uploads/2016/08/flooding-after-storm-desmond-PUBLISHED-24-August-2016.pdf

[13] Medeiros SC, Hagen SC, Weishampel JF, Comparison of floodplain surface roughness parameters derived from land cover data and field measurements. Journal of Hydrology. 2012;**452-453**:139-149

[14] Chow VT. Open Channel Hydraulics. International Edition. McGraw-Hill; US. 1959. 680 p

[15] Holden J, Kirkby MJ, Lane SN, Milledge DG, Brookes CJ, Holden V, McDonald AT. Overland flow velocity and roughness properties in peatlands. Water Resources Research. 2008;**44**. DOI: 10.1029/2007wr006052

[16] Holden J, Burt TP, Evans MG, Horton M. Impact of land drainage on peatland hydrology. Journal of Environmental Quality. 2006;**35**:1764-1778

[17] Chappell NA, Tych W, Chotai A, Bidin K, Sinun, W, Thang HC. BARUMODEL: Combined Data Based Mechanistic models of runoff response in a managed rainforest catchment. Forest Ecology and Management. 2006;**224**:58-80

[18] Wheater H, Reynolds B, McIntyre N, Marshall M, JacksonB, Frogbrook Z, Chell J. Impacts of upland land management on flood risk: Multi-scale modelling methodology and results from the Pontbren Experiment. 2008. (http://nora.nerc.ac.uk/5890/1/ur16_impacts_upland_land_management_wp2_2_v1_0.pdf)

[19] Deasy C, Quinton JN, Silgram M, Stoate C, Jackson R, Stevens CJ, Bailey AP. Mitigation Options for Phosphorus and Sediment (MOPS): Reducing pollution in runoff from arable fields. The Environmentalist. 2010;**108**:12-17

[20] Nicholson AR. Quantifying and simulating the impact of flood mitigation features in as small rural catchment [PhD thesis]. UK: Newcastle University; 2013

[21] Nicholson AR, Wilkinson ME, O'Donnell GM, Quinn PF. Runoff attenuation features: a sustainable flood mitigation strategy in the Belford catchment, UK. 2012;**44**(4):463-469

[22] Vincke C, Granier A, Breda N, Devillez F. Evapotranspiration of a declining *Quercus robur* (L.) stand from 1999 to 2001. II. Daily actual evapotranspiration and soil water reserve. Annals of Forest Science. 2005;**62**:615-623

[23] Rychnovská M. Transpiration in wet meadows and some other types of grassland. Folia Geobotanica & Phytotaxonomica. 1976;**11**:427-432

[24] Black TA, Kelliher FM. Process controlling understory evapotranspiration. Philosophical transactions of the Royal Society of London. Series B. 1989;**324**:207-231

[25] Roberts JM, Rosier PTW. Comparative estimates of transpiration of ash and beech forests at a chalk site in southern Britain. Journal of Hydrology. 1994;**162**:229-245

[26] Verón SR, Paruelo JM, Oesterheld M. Grazing-induced losses of biodiversity affect the transpiration of an arid ecosystem. Oecologia. 2011;**165**:501-510

[27] Reynolds ERC, Henderson CS. Rainfall Interception by beech, larch and Norway Spruce. Forestry. 1967;**40**:165-184

[28] Kirby C, Newson M, Gilman K. Plynlimon research: The first two decades. IH Report No.109. Wallingford: Institute of Hydrology; 1991 188p

[29] Thurow TL, Blackburn WH, Taylor CA. Rainfall interception losses by midgrass, shortgrass, and live oak mottes. Journal of Range Management. 1987;**40**:455-460

[30] Herbst M, Rosier PTW, McNeil DD, Harding RJ, Gowing DJ. Seasonal variability of interception evaporation from the canopy of a mixed deciduous forest. Agricultural and Forest Meteorology. 2008;**148**:1655-1667

[31] Brown AE, Zhang L, McMahon TA, Western AW, Vertessy RA. A review of paired catchment studies for determining changes in water yield resulting from alterations in vegetation. Journal of Hydrology. 2005;**310**:28-61

[32] Chappell NA, Tych W, Chotai A, Bidin K, Sinun W, Thang HC. 2006. BARUMODEL: Combined Data Based Mechanistic models of runoff response in a managed rainforest catchment. Forest Ecology and Management. 2006;**224**:58-80

[33] Chappell NA, Jones TD, Tych W, Krishnaswamy J. Role of rainstorm intensity underestimated by data-derived flood models: emerging global evidence from subsurface dominated watersheds. Environmental Modelling and Software. Volume 88, February 2017, Pages 1-9

[34] Finch JW. Modelling the soil moisture deficits developed under grass and deciduous woodland: The implications for water resources. Journal of the Chartered Institution of Water and Environmental Management 2000;**14**:371-376

[35] Calder IR, Harding RJ, Rosier PTW. An objective assessment of soil-moisture deficit models. Journal of Hydrology. 1983;**60**(1-4):329-355

[36] Horton RE. The role of infiltration in the hydrologic cycle. Transactions, American Geophysical Union. 1933;**14**:446-460

[37] Dunne T, Black RD. Partial area contributions to storm runoff in a small New England watershed. Water Resources Research. 1970;**6**:1296-1311

[38] Cook HL. The infiltration approach to the calculation of surface runoff. Transactions, American Geophysical Union. 1946;**27**:726-743

[39] FAO-UNESCO. Soil map of the world. Revised legend. Soils Bulletin. 1990;**66** FAO, Rome

[40] Jarvis RA, Bendelow VC, Bradley RI, Carroll DM, Furness RR, Kilgour INL, King SJ. Soils and their use in Northern England. Soil Survey of England and Wales Bulletin No **10**. Harpenden; 1984

[41] Chappell NA, Ternan JL. Flow-path dimensionality and hydrological modelling. Hydrological Processes. 1992;**6**:327-345

[42] Chappell NA, Ternan JL. Ring permeametry: Design, operation and error analysis. Earth Surface Processes and Landforms. 1997;**22**:1197-1205

[43] Chappell NA, Lancaster JW. Comparison of methodological uncertainty within permeability measurements. Hydrological Processes. 2007;**21**:2504-2514

[44] Soil Survey of England and Wales. Soils of England and Wales, Sheet **1**: Northern England; 1983

[45] Boorman DB, Hollis JM, Lilly A. Hydrology of Soil Types: A hydrologically-based classification of the soils of the United Kingdom. IH Report No. 126. Wallingford: Institute of Hydrology; 1995

[46] Nisbet T, Silgram M, Shah N, Morrow K, Broadmeadow S. Woodland for water: Woodland measures for meeting WFD objectives. Forest Research Monograph, 4, Forest Research, Surrey; 2011. 156 pp

[47] Chandler KR, Chappell NA. Influence of individual oak (*Quercus robur*) trees on saturated hydraulic conductivity. Forest Ecology and Management. 2008;**256**:1222-1229

[48] Archer NAL, Bonell M, Coles N, MacDonald AM, Stevenson R, Hallett P. The relationship of forest and improved grassland to soil water storage and its implication on Natural Flood Management in the Scottish Borders. In: BHS 11th National Symposium, Hydrology for a Changing World; 9-11 July 2012; Dundee, Scotland. 2012. DOI: 10.7558/bhs.2012.ns04

[49] Archer NAL, Bonell M, Coles N, MacDonald AM, Auton CA, Stevenson R. Soil characteristics and land cover relationships on soil hydraulic conductivity at a hillslope scale: a view towards local flood management. Journal of Hydrology. 2013;**497**:208-222

[50] Chappell NA, McKenna P, Bidin K, Douglas I, Walsh RPD. Parsimonious modelling of water and suspended-sediment flux from nested-catchments affected by selective tropical forestry. Philosophical Transactions of the Royal Society of London. B. 1999;**354**:1831-1846

[51] McDonnell JJ, Beven KJ. Debates—The future of hydrological sciences: A (common) path forward? A call to action aimed at understanding velocities, celerities, and residence time distributions of the headwater hydrograph. Water Resources Research. 2014;**50**(6): 5351-5359. DOI: 10.1002/2013WR015141

[52] Ockenden MC, Chappell NA. Identification of the dominant runoff pathways from the data-based mechanistic modelling of nested catchments in temperate UK. Journal of Hydrology. 2011;**402**:71-79

[53] Jones TD, Chappell NA, Tych W. First dynamic model of dissolved organic carbon derived directly from high frequency observations through contiguous storms. Environmental Science and Technology. 2014;**48**:13289-13297

[54] Lane SN, Odoni N, Landstrom C, Whatmore SJ, Ward N, Bradley S, Doing flood risk science differently: An experiment in radical scientific method. Transactions of the Institute of British Geographers. 2011;**NS 36**:15-36

[55] Wilkinson ME, Quinn PF, Welton P, Runoff management during the September 2008 floods in the Belford catchment, Northumberland. Journal of Flood Risk Management. 2010;**3**(4):285-295

[56] Quinn P, O'Donnell G, Nicholson A, Wilkinson M, Owen G, Jonczyk J, Barber N, Hardwick M, Davies G. Potential use of Runoff Attenuation Features in small rural catchments for flood mitigation. England: Newcastle University, Environment Agency, Royal Haskoning DHV; 2013

[57] Odoni NA, Lane SN. Assessment of the impact of upstream land management measures on flood flows in Pickering using OVERFLOW. Contract report to Forest Research for the Slowing the Flow at Pickering Project. Durham: Durham University; 2010

[58] Moors for the Future Partnership. Peak District Moorland Gully Blocking in Deep Peat. Moors for the Future Research Note No 2; 2005

[59] Thomas H, Nisbet TR. An assessment of the impact of floodplain woodland on flood flows. Water and Environment Journal. June 2007; **21**(2): 114-126

[60] Nisbet TR, Thomas H. Restoring floodplain woodland for flood alleviation. Final report for the Department of environment, food and rural affairs (Defra), Project SLD2316. Defra, London; 2008

[61] Broadmeadow S, Thomas H, Nisbet T. Opportunity mapping for woodland creation to reduce diffuse pollution and flood risk for England and Wales. Forest Research. 2014

[62] Packman JC, Quinn PF, Hollis J, O'Connell PE. Review of impacts of rural land use and management on flood generation short-term improvement to the FEH rainfall-runoff model: Technical background. R&D Project Record FD2114/PR3; 2004

[63] Palmer RC, Smith RP. Soil structural degradation in SW England and its impact on surface-water runoff generation. Soil Use and Management. 2013;**29**(4)567-575 DOI: 10.1111/sum.12068

[64] Hunter et al. Benchmarking 2d hydraulic models for urban flooding. Proceedings of the Institute of Civil Engineers, Water Management. 2008;**161**(1): 13-30 (WM1)

[65] Hankin B, Waller S, Astle G, Kellagher R. Mapping space for water: Screening for urban flash flooding. Journal of Flood Risk Management. 2008;**1**(1):13-22

[66] Kjeldsen TR, Stewart EJ, Packman JC, Folwell S, Bayliss AC. Revitalisation of the FSR/FEH Rainfall-Runoff Method. Defra R&D Technical Report FD1913/TR; 2005

[67] Environment Agency. What is the updated Flood Map for Surface Water? 2013. https://www.gov.uk/government/uploads/system/uploads/attachment_data/file/297432/LIT_8988_0bf634.pdf

[68] Penning-Rowsell E, Priest S, Parker D, Morris Tunstall S, Viavattene C, Chatterton J, Owen D. Handbook for Economic Appraisal (Multi-Coloured-Handbook). Flood and Coastal Erosion Risk Management. 2013

[69] Beven K. TOPMODEL: A critique. Hydrological Processes. 1997;**11**(9):1069-1085

[70] Beven KJ. Rainfall Runoff Modelling—The Primer. John Wiley & Sons; Chichester. 2003. ISBN 0-470-86671-3

[71] Davies J, Beven K, Rodhe A, Nyberg L, Bishop K. Integrated modeling of flow and residence times at the catchment scale with multiple interacting pathways. Water Resources Research. 2013;**49**(8):4738-4750

[72] Metcalfe P, Beven K, Freer J. Dynatopmodel: Implementation of the Dynamic TOPMODEL Hydrological Model. R package version 1.1. 2016. https://CRAN.R-project.org/package=dynatopmodel

[73] Buytaert W, Beven K. Regionalization as a learning process. Water Resources Research. 2009;**45** ISSN: 0043-1397. doi:10.1029/2008WR007359.

[74] Calder IR, Reid I, Nisbet TR, Green JC. Impact of lowland forests in England on water resources: Application of the Hydrological Land Use Change (HYLUC) model. Water Resources Research. 2003;**39**:1319

[75] Beven KJ. A manifesto for the equifinality thesis. Journal of Hydrology. 2006;**320**(1-2):18-36

[76] Beven K, Binley A. The future of distributed models: model calibration and uncertainty prediction. Hydrological Processes. 1992;**6**(3):279-298

[77] Beven K, Binley A. GLUE: 20 years on. Hydrological Processes. 2014;**28**(24):5897-5918

[78] Blazkova S, Beven KJ. A limits of acceptability approach to model evaluation and uncertainty estimation in flood frequency estimation by continuous simulation: Skalka catchment, Czech Republic. Water Resources Research. 2009;**45**;W00B16

[79] Hall JW, Dawson RJ, Sayers P, Rosu C, Chatterton J, Deakin R. A methodology for national-scale flood risk assessment. Proceedings of the Institution of Civil Engineers—Water & Maritime Engineering. 2003;**156**(3): 235-247

[80] Hankin B, Burgess-Gamble L, Bentley S, Rose S. How to model and map catchment processes when flood risk management planning. Science Report SC120015/R1. Bristol: Environment Agency; 2016. Available at https://www.gov.uk/government/publications/how-to-model-and-map-catchment-processes-when-flood-risk-management.pdf

[81] Carroll, Z.L., Bird, S.B., Emmett, B.A., Reynolds, B., Sinclair, F.L. Can tree shelterbelts on agricultural land reduce flood risk? September 2004; **20**(3): 357-359

[82] Herbst, M., Roberts, J.M., Rosier, P.T.W and Gowing, D.J. 2006. Measuring and modelling the rainfall interception loss by hedgerows in southern England. Agric. For. Meteorol. 141: 244-256.

[83] Noirfalise, A. 1959. Sur l'interception de la pluie par le couvert dans quelques forêts belges. Bulletin de la Soc. Roy. Forest de Belgique 10: 433-9.

[84] Leyton, L., Reynolds, E.R.C. and Thompson, F.B. 1967). Rainfall interception in forest and moorland: In: Forest Hydrology, eds . Sopper, W.E. and Lull, H.W., Pergamon, New York, 163-178.

[85] Aussenac, G. 1968. Interception des préc ipitation par le couvert forestier. Annales des Sciences Forestieres 25: 135-156.

[86] White, E.J. and Carlisle, A. 1967. The interception of rainfall by mixed deciduous woodland. Quarterly Journal of Forestry 62: 310-20.

[87] Thompson, F.B. 1972. Rainfall intercept ion by oak coppice (Quercus robur L .). In: Research Papers in Forest Meteorology. Ed . J .A . Taylor, Aberystwyth. 59-74.

[88] Dolman, A.J. 1987. Summer and winter rainfall interception in an oak forest: Predictions with an analytical and a numerical simulation model. Journal of Hydrology 90: 1-9.

[89] Schnock, G. 1969. Le bilan de l'eau dans l'écosystème forêt. Application à une chênaie mélangée de haute Belgique. In: Productivity of forest ecosystems. Ed. P. Duvigneaiid. Proceedings of the Brussels syinposium organised by Unesco und the International Biological Programme, 27-31 October 1969. 41-47.

[90] Staelens, J., De Schrijver, A., Verheyen, K. and Verhoest, N.E.C. 2008. Rainfall partitioning into throughfall, stemflow, and interception within a single beech (Fagus sylvatica L.) canopy: influence of foliation, rain event characteristics, and meteorology. Hydrol. Process. 22: 33-45.

[91] Carlisle, A., Brown, A.H.F. and White, E.J. 1965. The interception of precipitation by oak (Quercus petraea) on a high rainfall site. Quarterly Journal of Forestry 592: 140-143.

[92] Gerrits, M. 2010. The role of interception in the hydrological cycle PhD thesis, Technische Universiteit Delft.

[93] Mawdsley, T., Chappell, N.A. and Swallow, E. 2016. Report on the saturated hydraulic conductivity dataset, Tebay Common, Cumbria. Lancaster University.

[94] Marshall, M.R., Francis, O.J., Frogbrook, Z.L., Jackson, B.M., Mcintyre, N., Medeiros, S., Hagen, S. and Weishampel, J. 2012. Comparison of floodplain surface roughness parameters derived from land cover data and field measurements. J Hydrol 452-453: 139-149.

Chapter 2

Flood Risk Mapping in the Amazon

Milena Marília Nogueira de Andrade,
Iris Celeste Nascimento Bandeira,
Dianne Danielle Farias Fonseca,
Paulo Eduardo Silva Bezerra,
Ádanna de Souza Andrade and
Rodrigo Silva de Oliveira

Additional information is available at the end of the chapter

http://dx.doi.org/10.5772/intechopen.68912

Abstract

Floods are part of the natural and cultural life in the Amazon. However, the issues and management of fluvial-disaster risks are poorly studied. Among the reasons for the lack of studies, the Amazon region has several gaps in information ranging from inadequate regional maps to spatially unsystematic local data. Flood patterns differ in urban and rural areas. Severe large-scale flooding took place during the previous and the current decades, such as those that occurred in 2009 and 2012. Between 1991 and 2010, official recorded data indicate about 3,292,888 people were affected in 6 regional states of the Amazon (Acre, Amapá, Amazonas, Pará, Rondônia, and Roraima) considering 7 different hazards. Because of the extensive damages, the national government started a mapping program for cities in Brazil that have a history of facing significant flood risks. The aim of this chapter is to analyse the flood-risk mapping conditions in the Amazon.

Keywords: Amazon region, disaster risk, flood hazard, vulnerability, risk mapping, risk management, urban planning

1. Introduction

Brazil recorded 4691 flood events between 1991 and 2012, which represent 12% of all natural hazards in the country. In the Amazon, floods affected around 2379 people [1]. However, rivers are a fundamental component of Amazon life. Many highly populated cities are located along the major rivers and floodplains on the Amazon and Tocantins river basins, which together house over five billion inhabitants [2] (**Figure 1**).

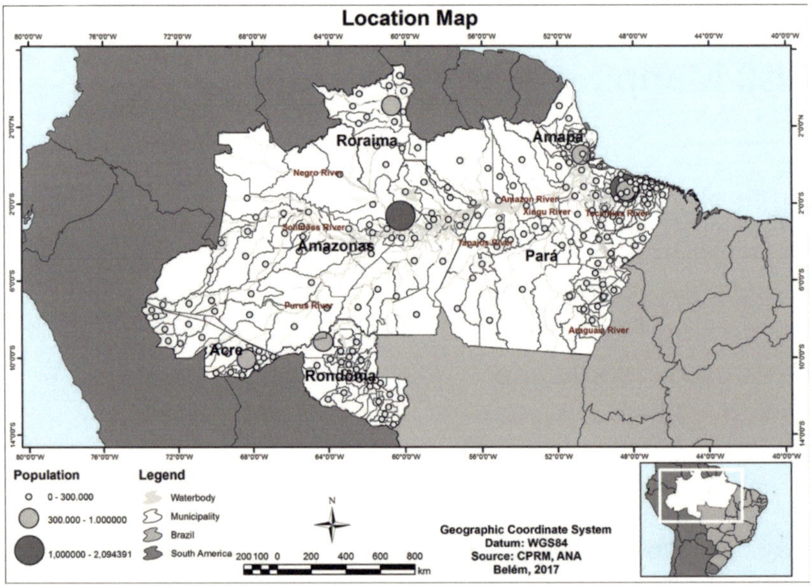

Figure 1. Study area showing the hydrographic networks and the pattern of the cities located nearby important rivers in the northern Amazon region.

Risk is a function of the probability that a particular hazard (such as flooding) might take place and the vulnerability of a particular location to being negatively affected by that hazard [3]. The likelihood of floods generating damage when river water overflows a river bank is a risk. Impacts are the generally negative effects of a hazard taking place in a given locale. Impacts vary from immaterial to material losses across cities such as individuals and families losing their homes and other dwellings, or losing access to such dwellings.

Included among material damages in urban areas are the destruction of public and private infrastructures, disruption of normal traffic flow, and reduction of accessibility to various locations and city spaces [4]. In rural areas, impacts include disruption of agricultural production, depopulation because of migration to the cities, and inflation of food prices at the local markets due to damage or destruction of crops near the floodplains. Small farming communities are particularly susceptible to extreme flood events because of the lack of adequate infrastructure for transporting people and their goods to industrial and commercial centers, the lack of information about extreme weather events at the community level, and the lack of sufficient economic resources to endure the effects of prolonged environmental catastrophes [5].

Since 2012, mapping flood risks has been one of the objectives of the National Program for Risk Management and Response to Disasters. Program priorities include investments in preventing, providing alerts for, monitoring, and responding to natural hazards. The program goal is to reduce the negative impacts of natural calamities on populations that live in risk-prone areas and guarantee the safety of communities from these calamities [6].

Therefore, response actions need precise physical aspects. Flood-hazard background information prioritizes the measures of frequency and magnitude of severe flooding, as well as the hydrodynamic and climatic scenarios [7, 8]. Response actions focus on the use of geoscience data and on structural engineering measures [9].

Another way to reduce risk is investments in prevention. Examples are the expansion of monitoring and warning systems, systematic mapping of high-risk areas [6], and preparation of vulnerability assessments. Vulnerability depends on the scale, time, and space units of analyses such as individuals, households, regional areas, and system-wide conditions [10]. Vulnerability definitions vary according to research approach and methodologies.

The International Strategy for Disaster Reduction (UN-ISDR) defines vulnerability as the combination of physical, social, economic, and environmental factors or processes that increase the susceptibility of a community to the impact of hazards [11]. The use of indicators is a common approach for vulnerability measurement [12, 13]. At a national scale, the social aspect of vulnerability displays higher levels of negative indicators for the Amazon region when compared to cities in the south-western region of Brazil [14].

2. Flood hazards in the Amazon region

Floods occur in the Amazon and Tocantins river basins in the northern region of Brazil. The Amazon river basin has flood water levels variability between 2 and 18 m [15]. The Amazon river receives the discharge of other big rivers (e.g., Negro and Tapajós), and floods occur periodically depending on the seasonal rainfall for each river. Floods usually occur in June for the Amazonas, Branco, and Negro rivers. At the Macapá station, tidal influences mean that river seasonality is barely noticeable throughout the year (**Figure 2**).

The historical data series varies according to the station. The water-level data from stream gauging stations cover 112 years at Manaus (1902–2014); 48 years at Boa Vista (1967–2015), Barcelos (1967–2015), Rio Branco (1967–2015), and Madeira (1967–2015); 39 years at Macapá (1976–2015) and São Felix (1977–2016); and 20 years at Baião (1971–1991). Some of the stations have been deactivated [16].

On many southern rivers and tributaries, water levels remain high during the first semester. Highest river stages occur during February and March, and there are more incidences of floods during this period (**Figure 3**). Flooding in the Tocantins river basin is limited to its eastern area with the Tocantins river being the main tributary that reaches a 10-m water level.

In 2009, the Amazon river basin experienced extreme flooding. At the Manaus station of the Negro river, the water level reached 29.75 m, the highest mark in 107 years since stage data became available [8]. At the reference point of the Obidos gauge station, the river reached 8.42 m. For the Tapajós river (Santarém station), the highest water-level mark was 8.31 m [4].

A complex combination of factors contributed to these recent extreme-hazard events; namely, large- and regional-scale climatic events, unusual flood mechanisms that produced complex interactions in time and space between the main system and its tributaries [7, 8], and recent urban growth without adequate planning. Long-term climate models show extreme precipitation events over the Amazon region [17].

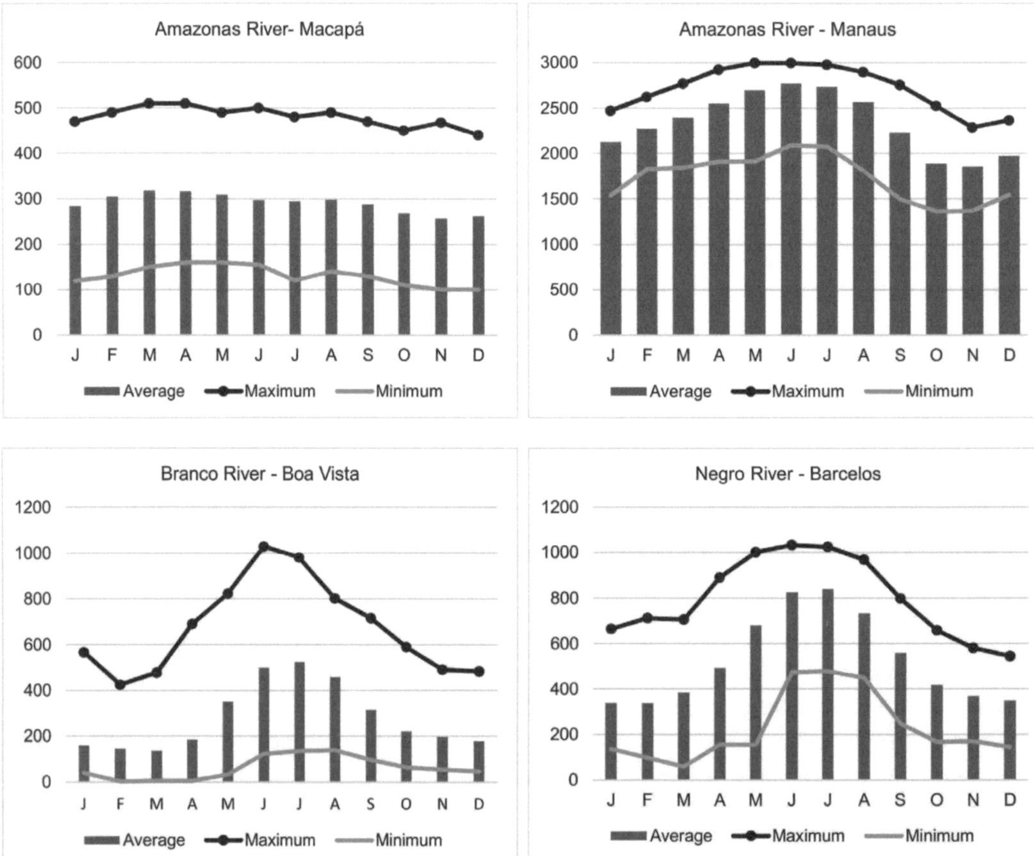

Figure 2. Historical river-level fluctuations (in cm) for the Amazon, Negro, and Branco rivers and their northern tributaries.

Landsat and shuttle radar topography mission (SRTM) data were used to define aspects of the morphologies of the fluvial systems in the Amazon and to construct the corresponding hydrodynamic models for these systems [18, 19]. Nevertheless, detailed geomorphologic maps that are fundamental to refining the fluvial models were not available [20]. For downscaling analysis, susceptibility maps for floods at 1:50,000 scale became available for some municipalities since 2013 [21]. The analysis covers 23 urban municipalities covering respective areas that have high, moderate, and low susceptibility to flooding. The methodological approach includes hypsometric, declivity, geomorphological, and drainage analyses [22–24]. Currently available status-of-susceptibility flood maps include 18 municipalities from Pará, 2 from Amapá, and 1 each from Acre, Rondônia, and Roraima.

The National Water Agency produced a flood-hazard atlas considering the frequency of floods and the impacts probability associated with each part of the river. The flood frequency corresponds to 5, 10, or more years of recurrence intervals, and the impact was measured according to the damage extension. The results aggregate the flood severity assessment according to international terminologies [25].

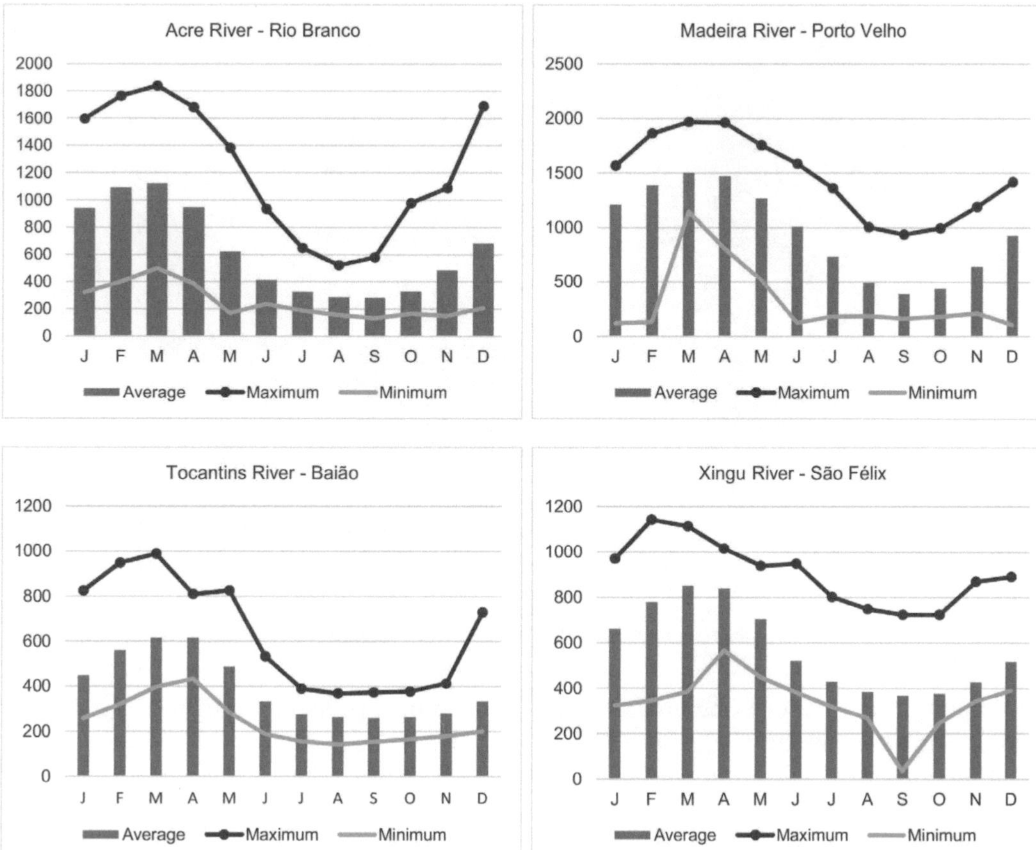

Figure 3. Historical river-level fluctuations (in cm) for the Acre, Madeira, Tocantins, and Xingu rivers.

In total, 68 different rivers were integrated in the analysis and 4756 km of river extension is considered with high impacts and frequencies to flooding (**Table 1**). The percentage for river high-hazard stretches are 27% for Acre, 30% for Amapá, 16% for Amazonas, 10% for Pará, 37%

State/flood-hazard stretch (km)	Low	Medium	High
Acre	3014	1241	1599
Amapá	7	819	362
Amazonas	690	7037	1555
Pará	1034	6304	813
Roraima	38	120	96
Rondônia	98	625	331
Total	4881	16146	4756

Table 1. Flood hazard stretches for rivers in the Amazon region grouped according to state.

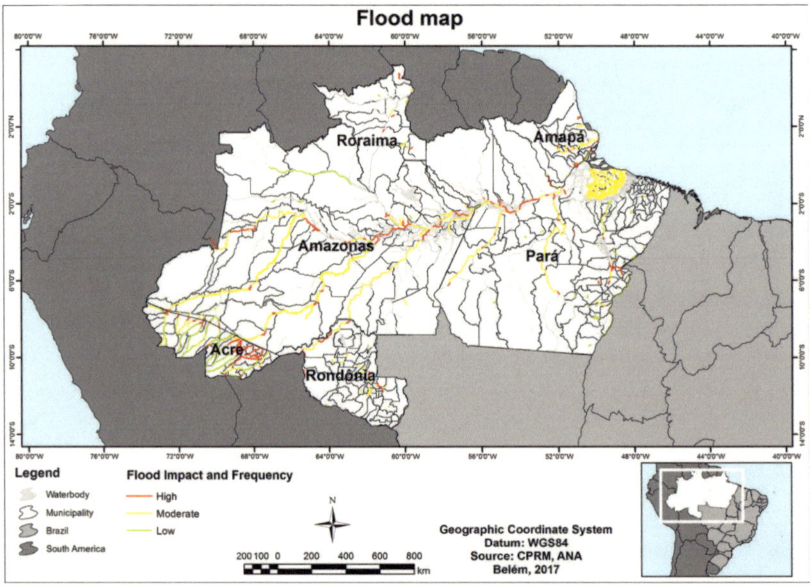

Figure 4. Flood map with parts of the river more susceptible to floods and impacts based on ANA data.

for Roraima, and 31% for Rondônia. The map is at national scale displaying areas with low, medium, and high impact frequencies [26] (**Figure 4**).

3. Vulnerability in the Amazon

Vulnerability is multidimensional and refers to social, economic, environmental, and physical categories [11]. Socio-economic vulnerability often describes the characteristics of people or groups of peoples [27]. Environmental vulnerability refers to natural-resources depletion and degradation [28]. Physical vulnerability refers to the susceptibility of particular locations to particular hazards and therefore represents the civilian infrastructure of a given place.

To deal with diverse research frameworks, the index of methodologies links natural-hazard risks and vulnerabilities. This means that when an extreme event leads to impacts, the latter is determined by the hazards aspect and by the respective vulnerabilities of communities, societal groups, or civilian infrastructures to such impacts [29].

Quantifying social vulnerability using various indicators can help identify which places are most vulnerable and which dimensions of social vulnerability are most relevant and therefore constitute the key drivers for change [30]. The Amazon region has 41% of its municipalities that are considered to have high social vulnerability. The analysis considered urban infrastructure, human capital, incomes, and work status as indicators [31]. This high percentage signifies a very low to medium-low Human Development Index (HDI) rating for rural and urban areas in the Amazon [32].

For a specific natural-hazard assessment, the Amazon region also has a very high social-vulnerability index. The indicators were (among others) gender, race and ethnicity, employment loss, social dependency, and migration rate [14]. Vulnerability to floods in rural communities is determined based on residency patterns, access to fishing and planting grounds, access to transportation and markets, water quantity and quality, and the prevalence of infectious diseases [33].

In large cities in the Amazon estuary, vulnerability was measured considering indicators of exposure (places and people located at hazardous areas), socio-economic situation, and condition of civilian infrastructure in terms of the availability and quality of sanitation services and housing structures (**Figure 5**). In these urban spaces, around 37,000 people live under very high-vulnerability conditions and 988,000 people live under high-vulnerability conditions [34].

Previous studies compiled for several locations along the Amazon coast in Pará state utilized vulnerability indices in addition to spatial information on floods and storm surges related to climate change and structural vulnerabilities [35]. Downscaling studies in vulnerability use methodological approaches based on the social-vulnerability index and an index on responses to hazards [36], on a methodological approach for community participation in flood mapping [4], and on a hazard-response identification scheme for urban planning for natural disasters issues [37].

Figure 5. The first and second photos showing Manaus (Amazonas state) infrastructure livelihood patterns and structural measures for accessibility in the central areas during the 2012 flood (photos in clockwise direction: CPRM, 2012). Santarem city (Pará state) during the 2012 flood showing wooden bridges and a view taken from the dike in front of the city during the 2009 flood (photos in clockwise direction: Milena Andrade 2012, Santarém Civil Defense 2009).

An effective vulnerability reduction plan should include adaptive capacity assessments. The human capacity to prepare for, respond to, and recover from natural disasters highlights local characteristics and situations [12, 38]. The structural and nonstructural measures for coping with hazards result from adaptive capacity. Pilot local-scale studies in the Amazon take into account qualitative variables and the influence of community participation in vulnerability assessments. These studies identify local knowledge as fundamentally important in coping and adaptation strategies and vulnerability reduction [4, 39].

4. Sectors of risk in the Amazon

The Geological Service of Brazil (CPRM) is responsible for mapping flood hazard-risk areas following the prevention criteria in the National Plan for Management and Response for Natural Hazards. The selection criteria for mapping emergency areas included the numbers of fatal victims and affected people, decreed state or level of emergency and calamity, and the official assistance requests from municipalities [40]. Between 2012 and 2016, CPRM mapped 1206 municipalities all over Brazil, of which 140 are located in the Amazon region.

Considering Pará, Amazonas, Acre, Rondônia, and Roraima states, around 180,000 people live within 418 risk sectors inside 99 municipalities with either high or very high risk of flooding [6] (**Figures 6** and 7).

To delineate the risk sectors, the methodology included previous classifications was proposed in Refs. [41–43]. Some adaptations were necessary for the risk sectors in the Amazon region that considered the following situations: a riverine community's location within a large floodplain, local water-level historical observations in existing infrastructures, lack of drainage and sanitation systems, household infrastructure fragility to flood effects (for wooden or brick houses),

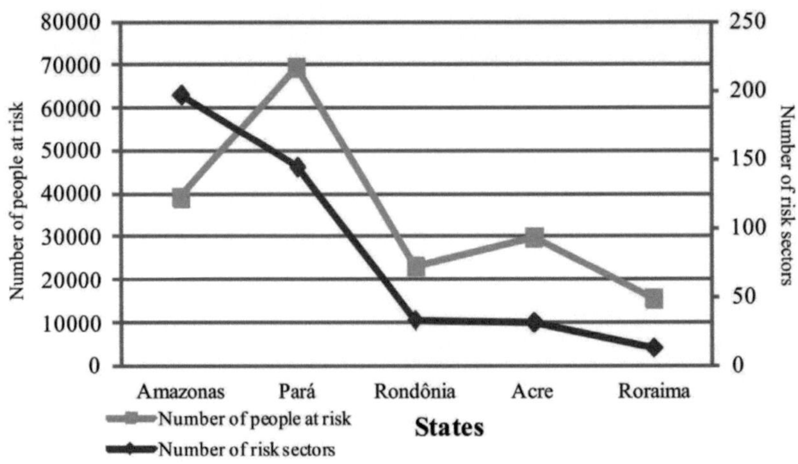

Figure 6. Graph of numbers of people in risk sectors of Amazonas, Pará, Rondônia, Acre, and Roraima states.

Figure 7. Sectors of risk in Epitaciolândia (Acre state) during the 2015 flood and Marabá City (Pará state) during the 2011 flood (photos in clockwise direction: Epitaciolândia and Marabá Civil Defense, respectively).

and measures of material and immaterial losses [6]. Very high-risk and high-risk areas were prioritized in the sector-mapping program because of the immediate need to prevent fatalities in these areas during hazard emergencies (**Figure 8**).

A very high-risk designation corresponds to a drainage with a high frequency (at least three events in the last 5 years) of flooding and a very high-damage probability because of the presence of vulnerable structures. High risk corresponds to a drainage with a moderate frequency (at least one event in the last 5 years) of flooding and a high-damage probability because of the presence of vulnerable household structures. Moderate risk corresponds to a moderate frequency (at least one event in the last 5 years) and low susceptibilities of structures to flood damage. Low risk corresponds to the absence of flood

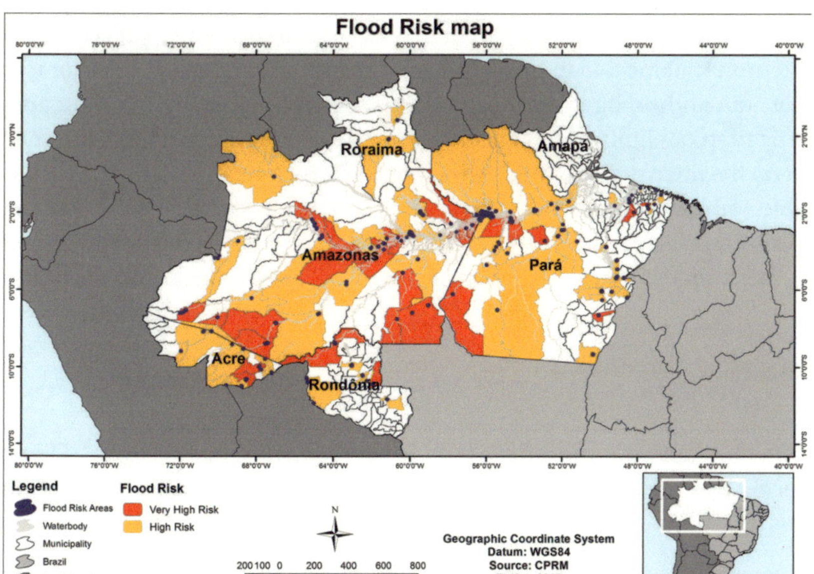

Figure 8. Map of municipalities in very high-risk and high-risk areas that are prone to flooding, according to CPRM data.

events in the last 5 years and the absence or presence of structures with low susceptibility to flood damage [43].

5. Conclusions

This study presented the developments in flood-risk mapping in the Amazon region. Flood-risk mapping projects are mostly by the national program for disaster reduction and counts on the support of national institutions. The effort to reach a nationwide methodology is a challenge due to the regional differences in Brazil.

Background information on flood hazards rely on fluvial water-level data taken along the river basins, on delineation of flood hazard-prone areas presented on 1:1,000,000-scale maps, and on semi-detail (1:50,000 scale) flood susceptibility maps for specific locations. These maps are the starting point for identifying critical areas for detailed hydrologic modeling. Extreme weather events have the potential to cause worse impacts on these high flood-risk areas.

The vulnerability studies in the Amazon region show some initial advances in understanding the effects of the various types of vulnerabilities on the impact of flood hazards. National hazard-risk studies provide the bases for national indicators. Therefore, some adaptations would be valuable for understanding the complex effects of vulnerabilities in the country. Some specific conditions in the Amazon should be taken into account. Infrastructure vulnerability is the main type of vulnerability that is most relevant in mapping flood-hazard risks.

Flood-risk mapping is a fundamental part of disaster-risk management. It is crucial in information-based decision-making and in providing guidance for defining priorities for risk reduction. In urban areas, the anthropogenic factors that influence or intensify flooding are considered for the minimization or elimination of risk sectors. Demographic and land-use changes have a direct impact on the intensity of floods. All sector risk maps generated by the CPRM have been made available for the civil defense activities of the municipalities, alert centers, ministers, and institutions associated with the National Program for Risk Management and Response to Disasters. This policy is a driver for transformative change that should continue over time.

Author details

Milena Marília Nogueira de Andrade[1*], Iris Celeste Nascimento Bandeira[2], Dianne Danielle Farias Fonseca[2], Paulo Eduardo Silva Bezerra[1], Ádanna de Souza Andrade[1] and Rodrigo Silva de Oliveira[1]

*Address all correspondence to: milenamarilia.andrade@gmail.com

1 Universidade Federal Rural da Amazônia, Belém, Pará, Brazil

2 Serviço Geológico do Brasil, Belém, Pará, PA, Brazil

References

[1] CEPED. Atlas Brasileiro de Desastres Naturais 1991-2012. Florianópolis: UFSC; 2013

[2] IBGE. Dados censitários das cidades [Internet]. 2016. http://www.ibge.gov.br/cidadesat/topwindow.htm?1 [Accessed: February 15, 2016]

[3] Wisner B, Blaikie P, Cannon T, Davis I. At Risk: Natural Hazards, People's Vulnerability and Disasters. 2nd ed. London: Routledge; 2004

[4] Andrade MMN de, Szlafsztein CF. Community participation in flood mapping in the Amazon through interdisciplinary methods. Natural Hazards. 2015;**78**(3):1491-1500

[5] Szlafsztein CF. Development projects for small rural communities in the Brazilian Amazon region as potential strategies and practices of climate change adaptation. Mitigation and Adaptation Strategies for Global Change. 2012;**19**:143-160. DOI 10.1007/s11027-012-9431-1

[6] CPRM Companhia de Pesquisa de Recursos Minerais. Setorização de Riscos Geológicos [Internet]. 2017. Available from: http://www.cprm.gov.br/publique/Gestao-Territorial/Geologia-de-Engenharia-e-Riscos-Geologicos/Setorizacao-de-Riscos-Geologicos-4138.html [Accessed: 10 February, 2017]

[7] Espinoza JC, Marengo JA, Ronchail J, Carpio JM, Flores LN, Guyot JL. The extreme 2014 flood in south-western Amazon basin: the role of tropicalsubtropical South Atlantic SST gradient. Environmental Research Letters. 2014;**9**:1-9

[8] Marengo JA, Tomasella J, Soares WR, Alves LM, Nobre CA. Extreme climatic events in the Amazon basin. Theoretical and Applied Climatology. 2012;**107**:73-85

[9] Rodrigues AC, Günther WMR, Vasconcellos MdaP, Di Giulio G, Boscov MEG. An overview of Brazilian scientific production on disasters in the beginning of this century. Desenvolvimento e Meio Ambiente. 2015;**34**:61-73

[10] Turner II B, Matson P, McCarthy J, Corell R, Christensen L, Eckley N, Hovelsrud-Broda G, Kasperson J, Kasperson R, Luers A, Martello M, Mathiesen S, Naylor R, Polsky C, Pulsipher A, Schiller A, Selin H, Tyler N. Illustrating the coupled human–environment system for vulnerability analysis: Three case studies. Proceedings of the National Academy of Sciences of the United States of America. 2003;**100**(14):8080-8085

[11] UN-ISRD (United Nations-International Strategy for Disaster Reduction). Living with Risk: A Global Review of Disaster Reduction Initiatives. Geneva: UN Publications; 2004. p. 429. ISBN/ISSN: 9211010640

[12] Adger N. Vulnerability. Global Environmental Change. 2006;**16**(3):268-281

[13] Kuhlicke C, Scolobig A, Tapsell S, Steinführer A, De Marchi B. Contextualizing social vulnerability: Findings from case studies across Europe. Natural Hazards. 2011;**52**(2):789-810

[14] Hummel BMdeL, Cutter SL, Emrich CT. Social vulnerability to natural hazards in Brazil. International Journal of Disaster Risk Science. 2016;**7**:111-122

[15] Filizola N, Guyot JL, Molinier M, Guimarães V, Oliveira Ede, Freitas MA de. Caracterização hidrológica da Bacia Amazônica. In: Rivas A, Freitas C, editors. Amazônia uma perspectiva interdisciplinar. Manaus: Manaus; 2002. pp. 33-53

[16] ANA Agência Nacional de Águas (Brasil). Dados sobre a Rede Hidrológica de Monitoramento em operação na bacia Amazônica [Internet]. 2016. Available from: http://www.ana.gov.br [Accessed: 10 November 2016]

[17] Marengo J, Jones R, Alves L, Valverde M. Future change of temperature and precipitation extremes in South America as derived from the PRECIS regional climate modeling system. International Journal of Climatology. 2009;**29**(15):2241-2255

[18] Hall A, Schumann G, Bamber J, Bates P. Tracking water levels changes of the Amazon Basin with space-borne remote sensing and integration with large scale hydrodynamic modeling: A review. Physics and Chemistry of the Earth. 2011;**36**:223-231

[19] Martinez J, Toan T. Mapping of flood dynamics and spatial distribution of vegetation in the Amazon floodplain using multitemporal SAR data. Remote Sensing of Environment. 2007;**108**:209-223

[20] Stevaux J, Latrubesse E, Hermann M, Aquino S. Floods in urban areas of Brazil. Development in Earth Surface Processes. 2009;**13**:245-266

[21] CPRM Companhia de Pesquisa de Recursos Minerais. Cartas de Suscetibilidade a Movimentos Gravitacionais de Massa e Inundações [Internet]. 2016. Available from: http://www.cprm.gov.br/publique/Gestao-Territorial/Geologia-de-Engenharia-e-Riscos-Geologicos/Cartas-de-Suscetibilidade-a-Movimentos-Gravitacionais-de-Massa-e-Inundacoes-3507.html [Accessed: December 27, 2016]

[22] Bandeira ICN, Conceição RAC, Mello LB, João XSJ. Carta de suscetibilidade a movimentos gravitacionais de massa e inundação do município de Itaituba—PA. In: Anais do 15° Congresso Brasileiro de Geologia de Engenharia e Ambiental; 18-21 October 2015; Bento Gonçalves, Rio Grande do Sul. 2015. ISBN: 978-8572700696

[23] Simões P, Conceição RAC, Vasconcelos K. Contribuição da carta de padrões de relevo para o mapeamento de suscetibilidade a inundação e movimentos de massa de Oriximiná-PA. In: Anais do 15° Congresso Brasileiro de Geologia de Engenharia e Ambiental; 18-21 October 2015; Bento Gonçalves, Rio Grande do Sul. 2015. ISBN: 978-8572700696

[24] Nogueira AC, Machado MF. Carta de suscetibilidade a movimentos gravitacionais de massa e inundação do município de Santarém—Pará. In: Anais do 15° Congresso Brasileiro de Geologia de Engenharia e Ambiental; 18-21 October 2015; Bento Gonçalves, Rio Grande do Sul. 2015. ISBN: 978-8572700696

[25] EEA (European Environmental Agency). Flood Risks and Environmental Vulnerability: Exploring the Synergies Between Floodplain Restoration, Water Policies and Thematic Policies. Luxemburg: EEA; 2016. p. 84

[26] ANA Agência Nacional de Águas (Brasil). Atlas de vulnerabilidade às inundações [Internet]. 2014. Available from: http://metadados.ana.gov.br/geonetwork/srv/pt/main.home [Accessed: November 15, 2016]

[27] Cutter S, Boruff B, Shirley W. Social vulnerability to environmental hazards. Social Science Quarterly. 2003;**84**(2):242-261

[28] Cardona, OD. The need for rethinking the concepts of vulnerability and risk from a holistic perspective: A necessary review and criticism for effective risk management. In: Bankoff G, Frerks G, Hilhorst D, editors. Mapping Vulnerability: Disasters, Development and People. London: Earthscan; 2004. pp. 37-51

[29] Birkmann J, Welle T. The WorldRiskIndex 2016: Reveals the necessity for regional cooperation in vulnerability reduction. Journal of Extreme Events. 2005;**3**:1650005-01-1650005-21. DOI: 10.1142/S2345737616500056

[30] Rufat S, Tate E, Burton CG, Marrof AS. Social vulnerability to floods: Review of case studies and implications for measurement. International Journal of Disaster Risk Reduction. 2015;**14**:470-486

[31] IPEA Instituto de Pesquisa Econômica Aplicada. Atlas da Vulnerabilidade Social nos municípios Brasileiros [Internet]. 2015. Available from: http://ivs.ipea.gov.br/ivs/data/rawData/publicacao_atlas_ivs.pdf [Accessed: October 9, 2016]

[32] PNUD Programa das Nações Unidas para o Desenvolvimento. Atlas do Desenvolvimento Humano no Brasil [Internet]. 2013. Available from: http://www.atlasbrasil.org.br/2013/ [Accessed: November 20, 2016]

[33] Brondízio ES, Moran E. Human dimensions of climate change: The vulnerability of small farmers in the Amazon. Philosophical Transactions of the Royal Society B. 2008; **363**:1803-1809. DOI: 10.1098/rstb.2007.0025

[34] Mansur AV, Brondízio ES, Roy S, Hetrick S, Vogt N, Newton A. An assessment of urban vulnerability in the Amazon Delta and Estuary: A multi-criterion index of flood exposure, socio-economic conditions and infrastructure. Sustainability Science. 2016;**11**:625-643

[35] Szlafsztein C, Sterr H. A GIS-based vulnerability assessment of coastal natural hazards, state of Pará, Brazil. Journal of Coastal Conservation. 2007;**11**:53-66

[36] Szlafsztein C, Marques O, Maia H, Prette M, Fischenich P, Altieri F. Referências Metodológicas para mapeamento de Riscos Naturais na Amazônia: Mapeando as vulnerabilidades. Brasília: MMA/GTZ; 2010. p. 59

[37] Silva Junior MO, Szlafsztein C. Análise de risco como critério a gestão do território: um estudo do plano diretor do município de Alenquer (PA). Olam: Ciência & Tecnologia. 2010;**10**:1-29

[38] Wiréhn L, Danielsson A, Neset T-S. Assessment of composite index methods for agricultural vulnerability to climate change. Journal of Environmental Management. 2015;**156**:70-80

[39] Vogt N, Pinedo-Vasquez M, Brondízio ES, Rabelo FG, Fernandes K, Almeida O, Riveiro S, Deadman PJ, Dou Y. Local ecological knowledge and incremental adaptation to changing flood patterns in the Amazon delta. Sustainability Science. 2016;**11**:611-623

[40] Fonseca D, Szlafsztein C. Riscos geológicos. In: João XS, Teixeira SG, Fonseca DDF, editors. Geodiversidade do Estado do Pará. Belém: CPRM; 2013. pp. 119-130

[41] PimenteL J, Ferreira CEO, Shinzato E, Maia MAM, Silva SF, Hoelzel M, Pfaltzgraff PA, Morais JM. de, Parisi GN, Silva CN, Filho IB. de O, Fregolentte A. Atuação do Serviço Geológico do Brasil no mapeamento de risco geológicos para a redução de impactos decorrentes de eventos extremos. In: Proceedings of Congresso Brasileiro de Geologia CD-Rom; 30 September–05 October 2012; Santos, São Paulo; 2012

[42] Bandeira ICN, Di Salvio LPP. Resultado da setorização das áreas de alto risco a inundações na zona urbana de Altamira no estado do Pará. In: Anais do 14° Congresso Brasileiro de Geologia de Engenharia e Ambiental; 1-6 December 2013; Rio de Janeiro. 2013. ISBN: 978-8572700696

[43] Ministério das Cidades. Capacitação em Mapeamento e Gerenciamento de Risco [Internet]. 2007. Available from: http://www.defesacivil.mg.gov.br/images/documentos/Defesa%20Civil/manuais/mapeamento/mapeamento-grafica.pdf [Accessed: December 12, 2016]

Chapter 3

Floods Forecast in the Caribbean

Luz Estella Torres Molina

Additional information is available at the end of the chapter

http://dx.doi.org/10.5772/intechopen.68783

Abstract

Floods are one of the most costly natural disasters in the world and represent a common hazard in Puerto Rico. Some floods develop slowly, sometimes over a period of a day or more. But flash floods can develop quickly, sometimes in just a few minutes. Puerto Rico, as well as other islands in the Caribbean, is subjected to flooding due to geographical location, topography, population distribution, and sudden rainfall events. The use of new technologies, such as apps and radars with higher spatial resolution that can cover areas missed by the NEXRAD radar, is important for flood forecasting efforts and for studying and predicting atmospheric phenomena. Recently, researchers in Puerto Rico initiated investigations using new technologies with high temporal and spatial resolution radars and systems that are able to monitor sudden floods in populated areas. These technologies will be used for hydrologic analyses and, specifically, for rainfall forecasting in Puerto Rico. A number of observational studies have shown that individual convective cells have mean lifetimes of about 20 min, with the best performance associated with a lead time of 10 min.

Keywords: floods, Caribbean, Puerto Rico western area, forecast, parameters

1. Introduction

Portions of western Puerto Rico are subject to flash flooding due to sudden, extreme rainfall events, some of which fail to be detected by NEXRAD radar located approximately 120 km away in the town of Cayey, Puerto Rico, and partially obstructed by topographic features. The use of new radars with higher spatial resolution and covering areas missed by the NEXRAD radar is important for flood forecasting efforts and for studying and predicting atmospheric phenomena.

The use of a radar network in Puerto Rico western area will monitor the lower atmosphere where the principal atmospheric phenomena occur. This work represents the first time that TropiNet radar technology will be used for hydrologic analyses and specifically for rainfall forecasting in western Puerto Rico.

Short-term rainfall forecasts have commonly been made using quantitative precipitation forecast (QPF). The introduction of quantitative precipitation forecasting (QPF) in flood warning systems has been recognized to play a fundamental role. QPF is not an easy task, with rainfall being one of the most difficult elements of the hydrological cycle to forecast [1], and great uncertainties still affect the performances of stochastic and deterministic rainfall prediction models [2].

Currently, this capability does not exist in western Puerto Rico, and it is needed because of the potential for flooding in certain areas (e.g., in flood plains near the principal rivers of the region). In this research, short-term rainfall forecast analysis is performed using nonlinear stochastic methods. Once obtained, the rainfall forecast is introduced into a hydrologic/inundation model Vflo and into the Inundation Animator configured for the Mayagüez Bay Drainage Basin (MBDB).

Specific components of the research are: the inclusion of calibration and validation of rainfall estimates produced by the TropiNet radar network and the development and validation of the stochastic rainfall prediction methodology.

2. Stochastic modeling and short-term rainfall forecasting

There are many approaches that can be used to predict the future direction and magnitude of a physical process, such as rainfall. Forecasting is a large and varied field having two predominant branches: qualitative forecasting and quantitative forecasting [3]. Quantitative forecasting should satisfy two conditions, the accessible numerical information about the past and assumptions that some aspects of the past patterns will continue into the future. Quantitative forecast can be divided into two classes: time series and explanatory models. Explanatory models assume that the variables to be forecasted exhibit an explanatory relationship with one or more other variables, in contrast, time series forecasting uses only information on the variable to be forecasted and makes no attempt to discover the factors affecting its behavior [3].

A number of researchers have developed hurricane prediction tracking models in Puerto Rico. For example, see Ref. [4] used historical data to develop a stochastic model to predict the behavior of hurricane tracks. The parameter estimation scheme, based on recursive and iterative algorithms, used historical records for hurricanes to fit vector autoregressive models. The identified models have been classified according to the order of the model. The first observations of a given hurricane are compared with historical hurricane tracks. The author [4] concluded that the vector ARMA model has excellent potential and may help reduce official forecasting error compared with a Statistical-Dynamical Hurricane Track Prediction Model (NHC90) from the National Meteorological Center [5]. Introduce a rainfall forecast methodology based on NEXRAD data, which was used as the basis to formulate the new rainfall methodology.

http://dx.doi.org/10.5772/intechopen.68783

Other examples of rainfall forecasting models were developed and available in the literature. Prediction of Rainfall Amount Inside Storm Events (PRAISE) is a stochastic model developed by [6] to forecast rainfall height at site. PRAISE is based on the assumption that the rainfall height accumulated on a delta time is correlated with a variable that represents antecedent precipitation. The mathematical background is given by a joined probability density function and by a bivariate probability distribution, which is referred to the random variable that represents rainfall in a generic site and antecedent precipitation in the same site. The peculiarity of PRAISE is the availability of the probabilistic distribution of rainfall heights for the forecasting hours, conditioned by the values of observed precipitation.

3. Methodology

The University of Puerto Rico at Mayagüez has a research weather radar network and a rain gauge network developed by Dr. Luz Estella Torres-Molina for research work. The radar network provides information with higher spatial and temporal precision. TropiNet (radar) has a 60 × 60 m spatial resolution at every pixel and temporal resolution of 1 min. A flood warning model must be operated based only on the data available at the time of forecast. Only the radar can display data in real time. This is not possible using rain gauges, but the rain gauges are used as data validation. Rain gauges-based systems must have a dependable and redundant telemetry system that will accurately and efficiently transmit data to a central location for processing. The data from TropiNet radar are used for rainfall prediction in the watershed, using stochastic methods. Once the rainfall forecast is obtained, the use of hydrologic models is necessary for analysis of flooding in this area. This research is focused on the western Puerto Rico and could be applied in general to other areas or regions with the same rainfall type with the corresponding hydrologic soil and coverage data.

The study area, which encompasses the watershed, is 819.1 km^2 in area [7] and is located in western Puerto Rico. The region has three important watersheds: Río Grande de Añasco, Río Guanajibo, and Río Yagüez. The area includes 12 municipalities: Mayagüez, Añasco, Las Marías, San Sebastián, Lares, Maricao, Yauco, Adjuntas, Sabana Grande, San Germán, Hormigueros, and part of Cabo Rojo.

Flood problems in this study area are serious and widespread. Periodic flood damage to pastureland, roads, and a number of residential areas is significant. Flood waters have inundated the main Río Grande de Añasco floodplain 17 times in a period of 31 years, an average of approximately once every 2 years. The floodplain of the lower Rio Grande de Añasco has been inundated extensively at least six times during the period 1899–1975. One of the greatest known floods occurred in September 1975. Another great flood was that of September 1928. Other major floods occurred in September 1932, September 1952, October 1970, August 1899, and September 1899 [8].

The soil textures present in this study as percent of area are clay with 62.49%, clay-loam 24.96%, rock 8.69%, loam 3.00%, sand 0.81%, and gravel 0.04%. A soil map describing the class distribution is necessary to assign the values of the Green-Ampt infiltration parameters (see **Figure 1**).

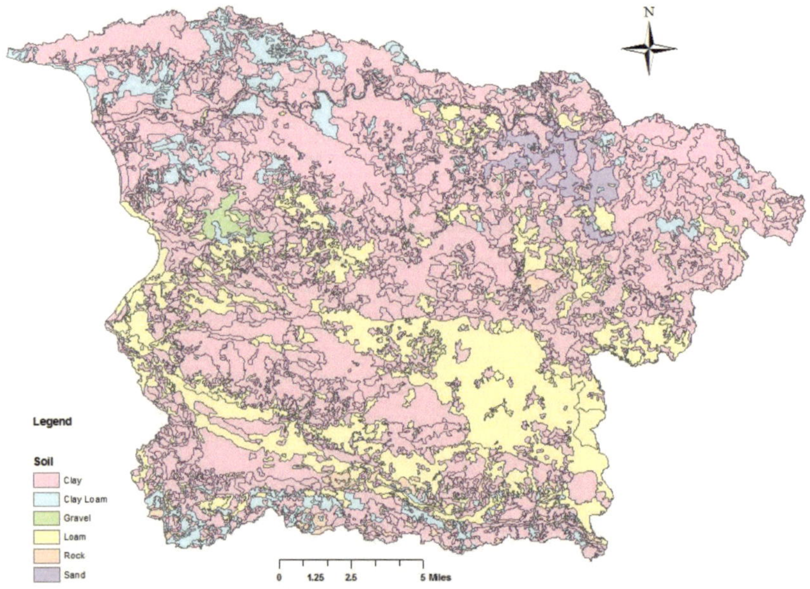

Figure 1. Soil present in the study area [Source: Soil Survey Geographic (SSURGO)].

Developed an algorithm of Water and Energy Balance for Puerto Rico using data from the geostationary operational environmental satellite (GOES) [9]. GOES-PRWEB utilizes an energy balance approach similar to [10]. The latent heat flux component of the algorithm is used to estimate actual evapotranspiration. The algorithm depends on solar radiation, which is determined using GOES satellite data. Ref. [11] were the first to propose a physical model for estimating the incident solar radiation at the surface from the Geostationary Operation Environmental Satellite.

Ref. [9] provided solar radiation data with spatial resolution of 1 km for Puerto Rico. In this research, we developed a subroutine in MatLab to convert the original 1 km resolution to 200 m resolution to obtain potential evapotranspiration (PET) estimation in a resolution compatible with our hydrologic model.

Twenty different classes of land cover and forest type are present over the study area corresponding to different kind of forest, woodland, and agriculture. The classification of land cover in this model is used to assign values for physical parameters which are important in the simulation with the hydrological model *Vflo*, and other important parameters with the land use are Manning's roughness coefficient, rainfall interception, evapotranspiration, crop coefficient, and other. Ref. [11] classified the land use for this watershed in six major categories, shrub land, woodland, and shade coffee with an area of 529.16 km^2, pastures with 172.84 km^2 of area, urban and barren area with 60.02 km^2, agriculture with 55.06 km^2, other emergent wetlands with 1.26 km^2, and quarries, sand, and rock with 0.75 km^2.

The climate of the study area is considered humid subtropical. The average temperature at the Mayagüez City, Puerto Rico station, is 70.7°F between the years 1971 and 2000, and the average maximum temperature in the Mayagüez city station between the years 1971 and 2000 is 88.7°F [12].

http://dx.doi.org/10.5772/intechopen.68783

Date	Duration (UTC)	Storm type	Storm impacts at western Puerto Rico
March 28, 2012	7 hr. 16:27–23:58	Stationary trough	Impacts rivers, water on the road, and significant rainfall accumulation
March 29, 2012	6 hr. 00:36–06:53	Stationary trough	Impacts rivers, water on the road, and significant rainfall accumulation
April 30, 2012	5 hr. 17:55–22:21	Convective storm	Numerous showers over western Puerto Rico at the afternoon
October 10, 2012	5 hr. 16:10–21:43	Convective storm	Some urban flooding
February 12, 2014	7 hr. 16:00–23:29	Heavy convective storm	Reduced visibilities and ponding of water on roadways and low-lying areas
May 6, 2014	7 hr. 16:45–23:59	Convective storm	Street flooding and reduced visibility on the highways.
May 21, 2014	7 hr. 16:46–23:00	Heavy convective storm	The water covers the roadway. Ponding of water on roadways
June 29, 2014	5 hr. 17:00–22:00	Convective storm	The shower activity produced periods of moderate to locally downpours
June 30, 2014	4 hr. 16:00–20:15	Thunderstorms associated to the leading edge of a tropical wave	Moderate to heavy rain, urban and small streamflood advisory
July 5, 2014	4 hr. 16:44–20:00	Convective storm	Heavy rain, urban flood

Table 1. Characteristics of studied storms.

The amount of rainfall varies considerably throughout the study area. Most of the rainfall occurs during the month of September with 10.62 inches on average. The months of January through April are considered the dry season with 1.60 inches in January, 2.59 inches in February, 3.35 inches in March, and 4.17 inches in April on average rainfall. In the west, the sea breeze effect carries wet air from the Mona Channel eastward, converging with the Trade Wind and resulting in intense convective rainstorms almost every afternoon within the watershed during the wet season. Rainfall and temperature data were obtained from the National Climatic data Center [12].

Table 1 includes the dates and specifications of every storm to the current study.

4. Radars

Radars are active sensors that emit electromagnetic pulses into the surroundings. A typical radar system consists of at least the following four components: a transmitter that generates high frequency signals, an antenna that sends the signal out and receives the echoes returned, a receiver that processes the returned signals, and a data display system [13]. Lower frequency and higher wavelength suggest that the radar has robust signal power and less attenuation, and the weather radar system discussed in the current research is based in X-band.

The TropiNet radars are Doppler polarimetric radars which allow the radar beam to measure reflectivity close to the ground, overcoming the shadow effect of the Earth's curvature, while maintaining high range and azimuth. The first TropiNet radar has been in operation since February 2012. TropiNet 1 is located in "Cerro Cornelia" Cabo Rojo, Puerto Rico, at 18.16°N, 67.17°W, and 200 ft elevation (msl), approximately. The radars, working with the X-band frequency, are about three times stronger than that of the traditional radar frequencies at S-band, making the measurements of rainfall more attractive. They have high space and time resolution for weather monitoring and detection and are capable of generating very high resolution data with a range of 40 km of radius or maximum radial distance (horizontal range) of 80 km of diameter.

Ref. [14] found a series of radars used for hydrological modeling. Additionally, they have indicated that in convective precipitation, when very steep horizontal gradients are observed, the information from a single rain gauge can be misleading. It must be stressed that radar and rain gauges are not competitive to each other because radars are used for spatial and temporal measurement, while the rain gauge only measures rainfall at a given single location [14]. If the comparison of a storm total is necessary, [15] explain that storm totals may be more accurately estimated by radar than any particular hourly accumulation when compared to rain gauges.

TropiNet radar being Doppler and Polarimetric can show velocity data of the cloud and reflectivity for every azimuth angle from 0 to 12°. TropiNet displays reflectivity logarithmically (10 log(Z)) or dBZ. The working frequency is 9.41 GHz ± 30 MHz, which corresponds to the X-band (in free space has a 3.19 cm wavelength). The TropiNet radar was designed and developed by Colorado State University (CSU) and UPRM to serve as the principal Internet-controllable node of the TropiNet radar network [16].

To analyze the data, it was necessary to develop a model to convert raw data into NetCDF data and then convert the reflectivity data in dBZ to rain rate in (mm/hr) using empirically derived *Z-R* relationships to transform reflectivity to rain rate. Ref. [17] equation is the default *Z*/*R* relationship employed by the WSR-88D and TropiNet.

NOAA-NWS [1995] report recommended that *Z-R* relationship in use at the time of the event be changed from $Z = 300R^{1.4}$ to a relationship more representative of raindrop distributions in a warm tropical storm. The *Z-R* relationship for warm tropical events recommended by the NWS Operational Support Facility since 1995 for all WSR-88D sites experiencing heavy rainfalls and now adopted by TropiNet is $Z = 250R^{1.2}$ [15].

The *Z-R* relationship used in Puerto Rico is the convective; furthermore, it was necessary to define a maximum precipitation rate threshold for decibels above 53 dBZ [15]. The convective rainfall is a type of precipitation with some characteristics like very high horizontal gradient and very large vertical depths. These characteristics mean that the weather radar is the best tool for detecting convective precipitation, but the presence of different types of hydrometeors, especially hail and storm dynamics yielding fast varying vertical profile reflectivity (VPR), usually results in considerable random error in quantitative precipitation estimates. Large differences can be found especially when comparing rain gauges and radar estimates because of the high temporal and spatial variability of the convective storm and related vertical profile of reflectivity [14].

A radar application in MatLab was developed to access the store of binary volume files that contain the respective information as determined by the operator like reflectivity, azimuth, velocity, beam width, range, elevation, and other radar products. The operator can apply one of several possible scan configurations. For instance, in the range height indicator (RHI), the radar holds its azimuth angle constant but varies its elevation angles. This is essential to provide vertical resolution where the radar continuously scans through elevation angles at a given azimuth angle. Another common scan configuration is the plan position indicator (PPI), where the radar holds its elevation angle constant but varies its azimuth angle, rotating through 360°.

For this research, it was necessary to hold the radar scan in PPI with a constant elevation angle of 3°. Every radar scan has two angles of 3° and 5° with a duration time of 30 seconds. The data information is saved in the server. The raw data files are stored by date every hour, minute, and second of scan in binary format. Each volume scan from radar has been interpolated to a fixed Polar grid and, after it is necessary, to convert to the fixed Cartesian grid. As part of the effort to further post-process the radar data, a model in MatLab was developed. This model performs the conversion from raw data polar coordinate system into ASCII data in Geographic coordinate system necessary for the hydrological software, *Vflo*.

5. Hydrologic model

The hydrologic model used in this research is *Vflo* [18]. *Vflo* is a fully distributed physically based hydrologic (PBD) model capable of utilizing geographic information and multi-sensory input to simulate rainfall runoff from major river basins to small catchments.

Vflo is a hydraulic approach to hydrologic analysis and prediction. Overland flow and channels are simulated using the Kinematic Wave Analogy (KWA). The model utilizes GIS grids to represent the spatial variability of factor controlling runoff. Runoff production is from infiltration excess and is routed downstream using Kinematic Wave Analogy. Computational efficiency of the fully distributed physics-based model is achieved using finite elements in space and finite difference in time. *Vflo* is suited for distributed hydrologic forecasting in post-analysis and in a continuous operation mode, derives its parameters from soil properties, land use, and topography, and in this case, the precipitation is obtained from radar TropiNet. The goal of distributed modeling is to better represent the spatial-temporal characteristics of a watershed governing the transformation of rainfall into runoff.

The hallmark of *Vflo* is prediction of flow rates and stages for every grid cell in a catchment, watershed, river basin, or region. *Vflo* provides high-resolution, physics-based distributed hydrologic modeling for managing water from catchment to river basin scale. Improved hydrologic modeling capitalizes on access to high-resolution quantitative precipitation estimates from model forecasts, radar, satellite, rain gauges, or combinations of multisensor products.

Model input consists of rain-rate maps at any time interval from radar or multisensor sources. Data input for this model (besides rainfall) is derived from various commonly available sources of digital data. Parameters include topography and drainage networks derived from a digital elevation model (DEM), infiltration derived from soils, and hydraulic roughness derived from land use/cover. These parameters may be input and edited manually or via ArcView grids.

For calibration performance, there is a sequence called the "ordered physics based parameters adjustment" (OPPA) method developed by [19]. The calibration process (OPPA) approach includes estimates of the spatially distributed parameters from physical properties, assigns channel hydraulic properties based on measured cross-sections where available, studies model sensitivity for the particular watershed, and identifies response sensitivity to each parameter. It furthermore runs the model for a range of storm from small, medium to large events. It observes the characteristics of the hydrograph over the range of storm size and any consistent volume bias. Then, it derives a range of response for a given change in a parameter and categorizes and ranks parameter sensitivity according to response magnitude.

The optimum parameter is that set which minimizes the respective objective function and matches volume by adjusting hydraulic conductivity. It can match the peak by adjusting overland flow roughness and re-adjust hydraulic conductivity and hydraulic roughness if necessary. The *Vflo* model does not simulate base flow, only direct runoff; it can be simulated assigning a fixed value to every channel cell for every event to simulate.

For a long-term analysis, it is necessary to quantify the base flow using known methodologies [20]. The OPPA procedure outlined above can be stated as: increasing the volume of the hydrograph is achieved by decreasing hydraulic conductivity, and similarly, increasing peak flow is achieved by decreasing hydraulic roughness.

Vflo model uses finite elements which can simulate streamflow based on geospatial data to simulate interior locations in the drainage network and determine channel flow and overland flow. It was fundamental to study the physical configuration of the watershed, such as a digital elevation model (DEM), the digitized topography, soils map, land use map, and information about the basin.

Some stations from the USGS were used to compare and validate the runoff with the results from the hydrological model using radar data. Ref. [8] implemented the most recent Flood Insurance Study (FIS) for the Commonwealth of Puerto Rico. Standard hydrologic and hydraulic study methods were used to determine the flood hazard data required for this countywide FIS. The flood events have magnitude of exceeding once at any given day during the recurrence period of 10, 50, 100, and 500 years. These events have a percent chance of 10, 2, 1, and 0.2%, respectively. The equation employed was mean annual rainfall (MAR) obtained from mean annual precipitation (MAP) developed by NOAA in 2006 [precipitation record 1971–2000]. The regression analysis was performed based on depth to rock (DR) and contributing drainage area (CDA) as variables that govern the peak streamflow.

Other product in this research was potential evapotranspiration. A GOES satellite-based potential evapotranspiration (*PET*) product, with resolution of 1 km over the entire island each day, was used in this research. One of the most used methods to calculate PET or reference evapotranspiration, and the method used in this study, is the FAO Penman-Monteith

http://dx.doi.org/10.5772/intechopen.68783

method. A large number of empirical methods have been developed over the last 50 years, and the Penman-Monteith method was considered to offer the best result with minimum possible error. The Penman-Monteith reference evapotranspiration equation is given by

$$ETo = \frac{0.408\Delta(R_n - G) + \gamma\frac{900}{T+273}u_2(e_s - e_a)}{\Delta + \gamma(1 + 0.34u_2)} \tag{1}$$

where ETo is reference evapotranspiration (mm day^{-1}), R_n is net radiation at the crop surface (MJm^{-2}day^{-1}), G is soil heat flux density (MJm^{-2}day^{-1}), T is mean daily air temperature at 2 m height (°C), u_2 is wind speed at 2 m height (ms^{-1}), e_s is saturation vapor pressure (kPa), e_a is actual vapor pressure (kPa), $e_s - e_a$ is saturation vapor pressure deficit (kPa), Δ is slope vapor pressure curve (kPa°C^{-1}), and γ is psychrometric constant (kPa°C^{-1}). The equation uses standard climatological records of solar radiation, air temperature (°C), humidity, and wind speed (ms^{-1}). The weather measurement should be made at 2 m (or converted to that height) above an extensive surface of an hypothetical green grass with an assumed height of 0.12 m, a fixed surface resistance of 70 sec m^{-1}, and an albedo of 0.23.

Also a slope map was developed using the digital elevation map (DEM) at 200 and 10 m resolution from USGS. The digital elevation model (DEM) data consist of a sampled array of regularly spaced elevation values referenced horizontally either to a Universal Transverse Mercator (UTM) projection or to a geographic coordinate system. The grid cells are spaced at regular intervals along south to north profiles which are ordered from west to east. **Figure 2** presents the slope map for the basin derived from the DEM at 200 m resolution.

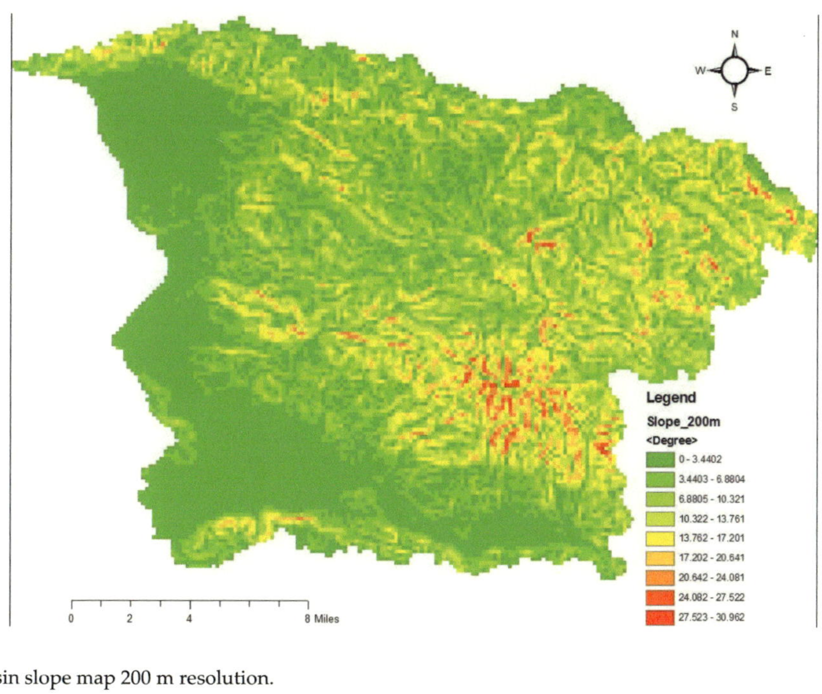

Figure 2. Basin slope map 200 m resolution.

The watershed presents the six basic textures into the basin area, where a large amount area of clay is observed. The soil names present into the clay area are: Alluvial land, Aguilita, Aibonito, Bajura, Consumo, Daguey, Delicias, Humatas, Lares, Jacana, Los Guineos, Malay, Mabi, Mariana, Mariaco, Montegrande, Mucara, Nipe, and other. For the clay-loam texture, the soil names present are Anones, Caguabo, Descalabrado, and Morado. For the loam texture, the soils are Coloso, Corcega, Dique, Guainabo, Mani, Maresua, Palmarejo, Reilly, Talante, Toa, and other. Soils that correspond to the rock texture are including Limestone, Serpentine, and Volcanic rock land, for the sand texture was found in the soils: Cataño, Leveled and River wash and the last texture is Gravel which only has one soil with the same texture name (**Figure 3**).

Figure 4 presents the percent's of textures into the watershed, in which the clay encompasses most of the study area with 63% of total area.

On the other hand, the minimum texture present in the basin is the Gravel with a value approximate to 0.02%.

The hydrologic group is a parameter that affects the infiltration and runoff. **Figure 5** presents the basin area with the hydrologic groups A, B, C, and D. The most representative groups are C and D: C with a 32% total area and D with a 40% total area.

Other parameters such as hydraulic conductivity, wetting front, and effective porosity were assigned from literature [20, 21]. The hydraulic conductivity (*K*) may especially control the infiltration process when rainfall occurs over already saturated soil; the hydraulic conductivity was specified for a single layer soil profile for this study area.

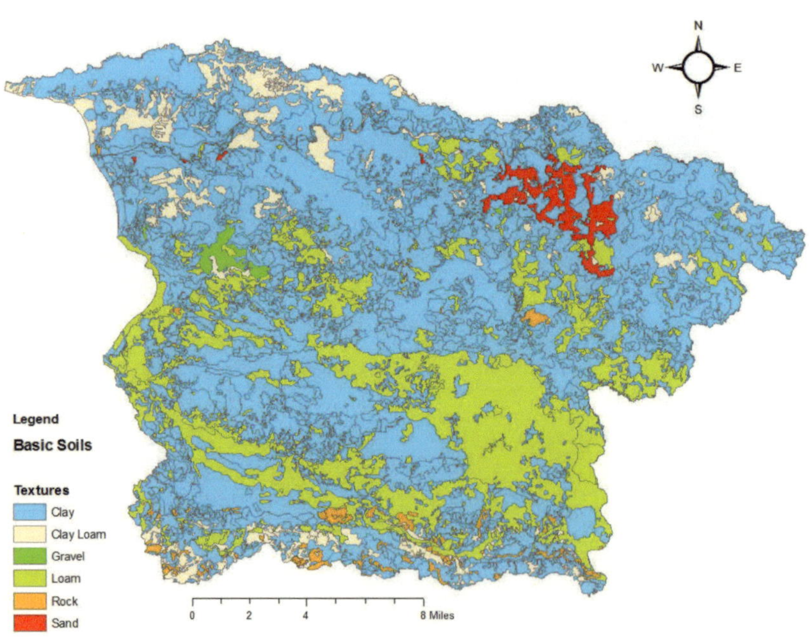

Figure 3. Basic soils textures in the basin area.

http://dx.doi.org/10.5772/intechopen.68783

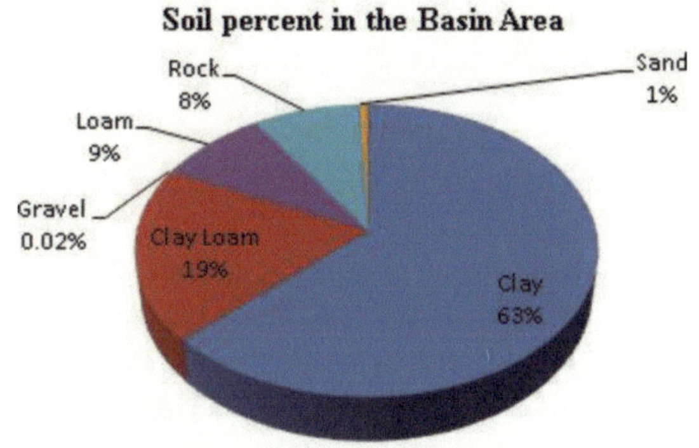

Figure 4. Soil texture percent in the basin area.

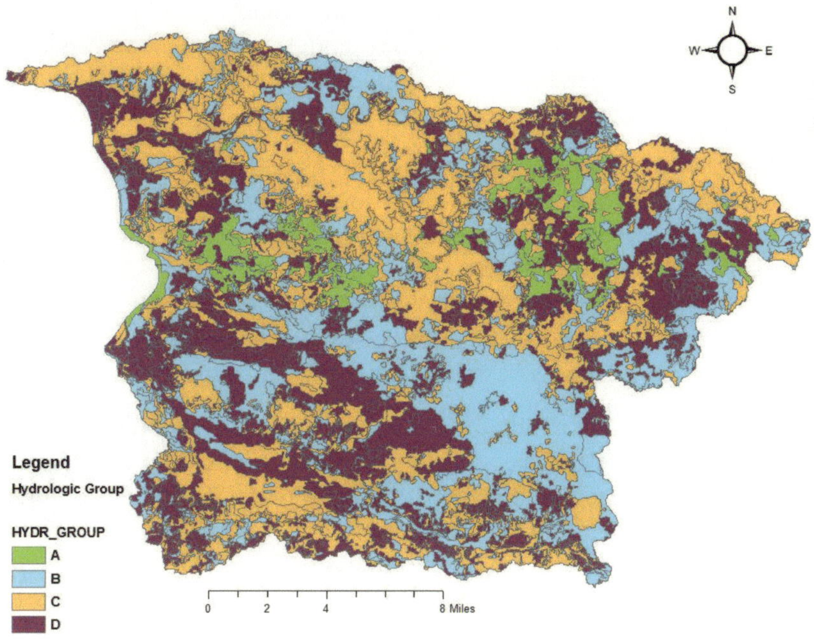

Figure 5. Hydrologic group with basin map areas.

The wetting front is the average capillary potential of the Green-Ampt infiltration routine, and this parameter is important because it can calculate infiltration under unsaturated conditions and its value is independent of soil moisture at any particular time. The effective porosity is the difference between total porosity and residual soil moisture content. This property is independent of soil moisture at any time, its range is between 1 and 0, with complete porosity being a value of one (1), and the value zero (0) is for the zero porosity. The soil depth is the depth to which the infiltration can occur in the soil, and if the wetting front is obstructed by a perched water table, then the depth to the water table is the limiting depth. If the soil profile is limited by an impermeable layer, then the depth to that layer is the limiting depth. Soil depth can be modified through the calibration of simulations to observed streamflow [21]. The soil depth data were obtained using USDA [22–25].

6. Types of forecasts

There are some properties needed to distinguish between different types of forecast. Forecast can extend to different scales in space and time; the spatial is doing reference in a fixed location in a specific area of city, e.g., the precipitation on a grid from TropiNet radar over Mayagüez city. The temporal range of a forecast is furthermore called lead-time. Short-range forecast covers very close events, like the next few hours or next minutes as in our case, and the long-range forecast considers the mean value of a meteorological parameter over a few days or months.

In this research, the data are correlated in space and time, where the strength in general decreases with spatial and temporal distance. Our models are designed to do forecast in time and space. This increases the difficulty as compared with prediction models that only use the forecast in time at a given place (e.g., forecast in rain gauges).

Other types of forecast are deterministic. In this case, a single forecast value is issued at each occasion, pretending a confidence that hides the forecaster's uncertainty about the outcome. They are easy to interpret even for user without stochastic background knowledge. The simplest case is a deterministic binary forecast. This area decision, like yes or no, and additionally a generalization in the forecast if necessary, distinguishes between types of variables to be forecasted. The variable of interest can be ordinal, which can be expressed by a number and can be defined by an appropriate number of threshold values (e.g., light rain, middle rain, or heavy rain).

Other variable of interest is the nominal, where there is no natural ordering, like qualitative observation of the kind of precipitation (e.g., snow, rain, ice, or other). A deterministic evaluation is furthermore named quantitative precipitation forecast (QPF), which induces the user to suppress information and judgment about uncertainty. In fact, it may create the illusion of certainty, while a probabilistic forecast is indicated as probabilistic quantitative precipitation forecast (PQPF). In order to reflect the uncertainty of the future outcome, probabilistic statements are more appropriate.

For this research, a methodology that embraces a space-time stochastic model is used and is considered a "discrete time-series model" that includes a special kind of nonlinear model with stochastic and deterministic components. Here, the rainfall process is described at a discrete time steps, is not intermittent, and, therefore, can be applied for describing the forecast within storm rainfall.

The other case is the use of meteorological model. These are useful qualitative and quantitative rainfall forecasting tools on 24–72-h interval and on a large spatial scale. In such cases, indeed absolute precision is not required for practical application. In meteorological models when the forecasting lag time and spatial scale decrease, the effectiveness and the precision of kind of model additionally decrease [26].

Autoregressive-moving-average models (ARMA) are mathematical models of autocorrelation in a time series. ARMA models are widely used in hydrology and were popularized by [27] who elaborated a comprehensive theoretical and practical development of time series models. There are several possible reasons for fitting ARMA models to data. ARMA modeling can contribute to understanding the physical system by revealing something about the physical process that builds persistence into the series. ARMA models can additionally be used to predict behavior of a time series from past values alone. Such a prediction can be used as a baseline to evaluate possible importance of other variables to the system.

The model consists of two parts: an autoregressive (AR) part and a moving average (MA) part. The AR model expresses a time series as a linear function of its past values. The order of the AR model indicates how many lagged values are included. The moving average (MA) model is a form of ARMA model in which the time series is regarded as a moving average of a random shock. The model is usually then referred to as the ARMA(p,q) model where p is the order of the autoregressive part and q is the order of the moving average part. ARMA models in general, after choosing p and q, are fitted by iterative procedure of a nonlinear least squares regression to find the values of the parameters which minimize the error term. The ARMA modeling process is commonly an iterative, trial, and error process. Thus, it is necessary to use the least possible number of parameters that will adequately produce forecasted values with similar statics of the historical data [28].

ARMA is a methodology widely used to do predictions of all types, for economy as well as for the weather predictions. In any case, it is necessary to have a long historical data. In the literature, ARMA model has been used to predict at one or two rain gauges at a single point but not at radar field. Since the ARMA model predicts at a single point, this is an important reason to avoid the use of ARMA methods in this research. This principle was applied to this thesis or this model and at the same time to the principle of parsimony to obtain results in the model with small possible error.

The Point process is a type of random process for which any action consists of a set of isolated points in time or in space. The example more global in point process model is the Poisson process that counts the number of events (storm) and the time that these events occur in a given time interval. Usually, the time between each event's development has an exponential distribution and the numbers of occurrences are independent of each event (storm).

The Point process model has been used commonly to forecast rainfall in which storm origins occur in a Poisson process. The Point process model is applied at a single site or fixed point where the storms arrive in a Poisson process. Each storm incorporates a group of random number of rain cell, where each cell has a random duration or lifetime and depth. The total rate of precipitation at time (t) is the sum of contributions from all active cells at (t) [29]. This type of model uses complex equations, and the analysis of precipitation is in time at a fixed point in space and the properties of the natural process can be deduced via the mathematical model.

Ref. [30] have a model for daily rainfall in which wet and dry days occur in a Markov chain with seasonally dependent transition probabilities. In it, the amounts of rain per wet day have a gamma distribution with seasonally dependent parameters.

An algorithm for predicting 10, 20, and 30 min in advance the spatial distribution of rainfall rate is introduced in this work. The algorithm is based on the assumption that TropiNet radar rainfall rate data provide estimations of the rainfall with high spatial and temporal resolution. Some researchers have compared radar rainfall data with rain gauge measurements [31–33]. These comparisons may not be useful since a rain gauge measures precipitation at a single point located at the surface level, whereas the weather radar measures the average of reflectivity at certain elevation and over a much larger area. A stochastic function is used to estimate the rainfall rate based on reflectivity.

When a rain gauge is compared with radar, it is expected that the average rainfall will behave as an individual point. It is known that the average will behave differently than that of an individual observation; therefore, these quantities should not be expected to be equal. When several rain gauges are averaged and compared with the radar measurements, the average of the rain gauges is inconsistent because it was developed with few points, whereas the average of the radar was developed with a much larger number of points. The rainfall modeled over a watershed shows that the peak flow measurements and overall runoff from radar performed better than the estimated peak flow using rain gauges [34]. Additional studies have concluded that the peak discharge of streamflow computed with radar data was more accurate than those computed from rain gauges alone [35]. Thus, there is no instrument that precisely measures the amount of rainfall over a large area. The weather radar provides an estimation of the rainfall rate over larger areas.

The suggested algorithm uses TropiNet (RXM-25) data to predict the variability of the rainfall field in time and space. It is assumed that for a short time period (10, 20, and 30 min), a rain cloud behave as a rigid object, with all pixels moving in the same direction at a constant speed. Thus, the most likely future rainfall areas are estimated by tracking rain cell centroid advection in consecutive radar images. The suggested algorithm is a special kind of nonlinear model with stochastic and deterministic components. The rainfall process exhibits significant changes in time and space, and it can be characterized as a nonstationary stochastic process. To face the nonstationary characteristic of the process, parameters are estimated at every time and spatial domain.

The model consists in considering the rainfall shape data as a rectangular grid with 940 columns and 740 rows of pixels for a total of 695,600 pixels, with every pixel size is 0.06 km wide and 0.06 km long. From the grid data, select a zone of 81 pixels that was divided in squares of $\Delta x \times \Delta y$ pixels, where (Δx) is referenced to columns of 9 pixels and (Δy) to rows of 9 pixels with total zones of 8,528 (82 x 104) in every window, as shown in **Figure 6**. Several zone sizes were explored for Δx and $\Delta y = \{7, 9, 11, \ldots, 25\}$, and it was found that the larger the zone size, the larger the number of degree of freedom. However, resolution was degraded with increased zone size.

In the model, the use of the same zone in the previous window ($t - 1$) and ($t - 2$) is necessary. Every zone (9×9) should have a minimum of 24 rain pixels with 20 degrees of freedom. Zones with less pixel of rain could not be selected to forecast analysis. In zones where the prediction movement suggests that there is a rainfall cell but the zone has no necessary pixels required (24 pixels), an interpolation was applied. The interpolation used was "*Kriging*" between the 25 pixels nearest to pixel that has no prognostic.

http://dx.doi.org/10.5772/intechopen.68783

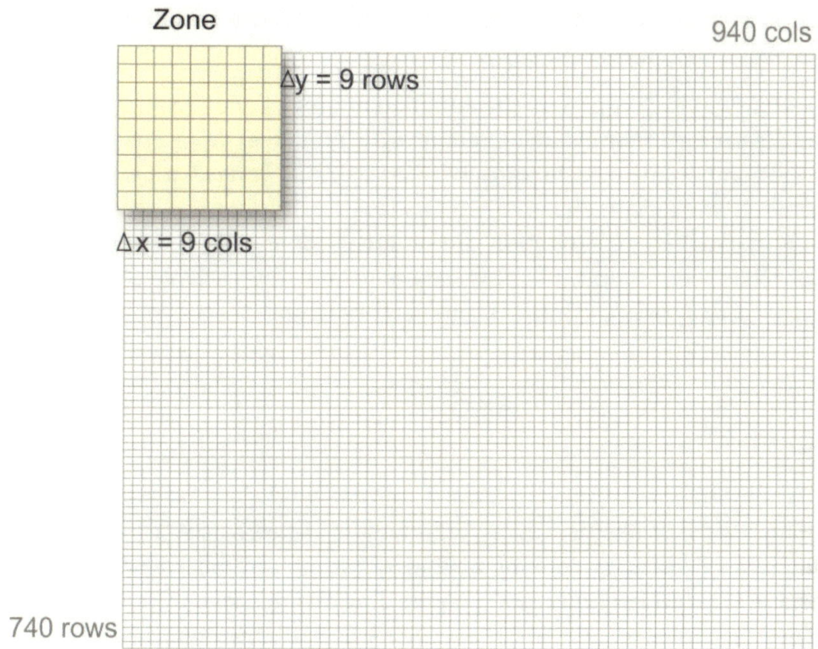

Figure 6. Rectangular grid of rainfall data.

The model is defined by the equation

$$h_{t,k(i,j)} = \alpha_{t,k} + \left(\beta_{t,k} - \alpha_{t,k}\right) \quad (2)$$

$$\Phi_{t,k}\left[1\text{-}e^{-\Sigma\left(\delta 1_{t,k}\bar{h}_{t-1,k(i,j)}+\delta 2_{t,k}\bar{h}_{t-2,k(i,j)}+\delta 3_{t,k}Z_{t-1,k(i,j)}\right)}\right] + \varepsilon_{t,k(i,j)} \quad (3)$$

where (i, j) represents the geographic position or coordinates, latitude, and longitude of every pixel in the grid, and k is the zone. This process starts in pixel 1 until pixel 8,528. In every zone, unknown parameters should be determined (α, β, Φ, $\delta 1$, $\delta 2$, $\delta 3$): α is the minimum value found between previous values of $h_{t-1,k(i,j)}$ and $h_{t-2,k(i,j)}$ in their respective zones, and (k), β is the reflectivity maximum value found between previous values of $h_{t-1,k(i,j)}$ and $h_{t-2,k(i,j)}$ in the specific zone (k). The mathematical structure of the model was inspired on a previous work [36]. In the current work, this model was used because this scheme ensured that rainfall forecast will fall inside of the most likely rainfall intensity domain [α, β], which was derived by the observed local rainfall distribution.

$\bar{h}_{t-1,k(i,j)}$ is the reflectivity average value in the time $(t-1)$. The average value was determined in every pixel into each zone. It was obtained averaging the eight pixels closest to the pixel under study. Similarly, $\bar{h}_{t-2,k(i,j)}$ is the average reflectivity value in the time $(t-2)$.

The variable $Z_{t-1,k(i,j)}$ is the ratio between the pixels with maximum reflectivity. $Z_{\max(t-1),k(i,j)}$ in every cloud or cell and the nearby pixels $Z_{i(t-1),k(i,j)}$ forming the cloud or cell, and the random variable $\varepsilon_{t,k(i,j)}$ is a sequence of an unobserved random variable with mean zero and constant variance associated with the pixel (i, j)

The variable Phi ($\Phi_{t,k}$) is changing in the equation at every zone (9 × 9) in each window. This variable was determined first by linearization of the nonlinear equation (Phi-initial) and after using optimization nonlinear techniques with constrains "*Sequential Quadratic Programming*" (SQP). The initial coefficient deltas ($\delta 1$, $\delta 2$ *and* $\delta 3$) were obtained through the estimation method "*least squares*" by linearization of nonlinear equation (exponential). Once the variable's initial deltas were found, the next step is to find the variable Phi ($\Phi_{t,k}$) initial. These values were used to forecast rainfall at one lead-time and successively with the following forecasts.

7. Results

There are many methods for forecasting with long lead-times as 8, 24, and 36 hr. or weekly, using autoregressive methods, moving averages, and others. However, the current study is a special kind of method for nowcasting (short time as minutes). In the western Puerto Rico occur sudden precipitations with short durations due to atmospheric conditions and topographic features at a given locations. Precipitation events may develop, occur, and dissipate immediately and the duration will be about 1, 2, or 3 hr.

Knowing the precipitation characteristics, the nowcasting model developed in the current research only needs two lag times for prediction. This means that the model has the capacity to forecast the rainfall even if the duration is very short. The developed model is presenting the best prediction when the lead-time is 10 min. The postulated rainfall nowcasting algorithm involves two major tasks: a) predicting the future location of the rain pixels and b) predicting rainfall at each pixel.

Figure 7 shows the cloud motion comparison between observed (right) movement and estimated (left) movement at storm date (March 28, 2012; 17:10 hr.).

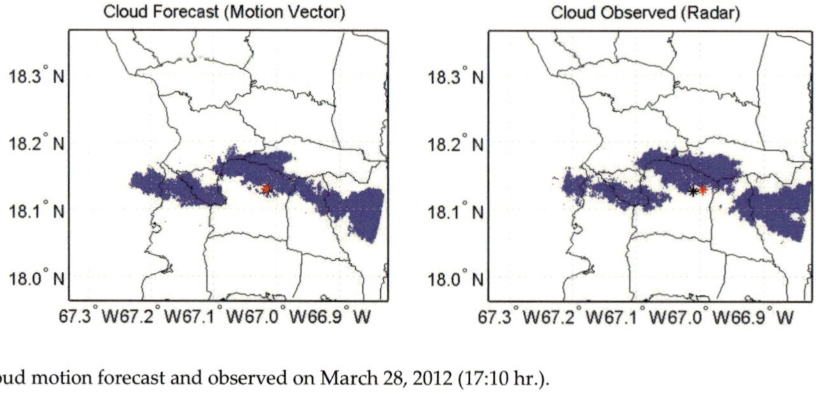

Figure 7. Cloud motion forecast and observed on March 28, 2012 (17:10 hr.).

The point black is the centroid at initial time, and the red point is the centroid at the final time. In some cases, there is more than one clouds centroid, and so these are more than one black and red points in the figure. This happens when the division cloud method has detected more than one cloud system within the area.

For all events, the best results were presented with a prediction of 10 min (see **Figure 8**). The western Puerto Rico area geographical position makes it susceptible to sudden rainfalls that are changing rapidly in time and space. Due to this change, a lead-time of 10 min is the time prediction more adequate to this precipitation class. A larger lead-time results in greater statistical errors. Contrarily, using a lead-time smaller than 10 min, the purpose of flood alert system will be annulled by the absence of time to evacuation.

It is important to mention that the algorithm to forecast precipitation uses a sequence of the observed rainfall data to estimate the movement direction and size of the cloud or cell, and then using the main equation, rainfall is estimated in each pixel within every zone. Thereby, the suggested regression model was developed under the following assumption: It is expected that in a short time (10 min) period, a rain cloud behaves approximately as a rigid object and the cloud rain pixels move in a constant speed and direction. Thus, the most likely future rainfall areas can be estimated by using the advection of the centroids of the rain cells in consecutive images. The current estimation reflectivity is a function of the previous reflectivity images observed. Rainfall nowcasting algorithm task is predicting rainfall rate at each pixel.

This methodology was applied to four unknown parameters ($\delta 1, \delta 2, \delta 3$ *and* Φ) to find the optimum values with a bounded constraint: first, linearize the main equation; second, identify the initial point through a nonlinear regression model where the phi (Φ) is temporarily ignored and the delta values initially are obtained by solving the linear regression; and third, find the optimum values using a constrained nonlinear optimization technique to estimate the final parameter set for each zone (9×9) and every window where the phi (Φ) parameter is a bias correction factor introduced in the optimization. The optimum parameters for the nonlinear regression model were estimated by solving a constrained nonlinear optimization problem (*fmincon*). The derived initial point was ingested into the constrained nonlinear subroutine to facilitate convergence, the delta parameters were restricted to be positives, and phi parameter was restricted to be in the range of 0–1.1 values.

An analysis for the nowcasting requires a combination of meteorological and hydrological statistics, which permits a better understanding of behavior of the spatial and temporal accuracy of storm prediction. A good nowcasting includes accuracy of the spatial and the temporal levels and accuracy of the predicted rainfall intensity. Model performance criteria for the prediction required quantitative comparison measures; these measures include ten storms.

The accuracy of rainfall prediction of each pixel can be measured by decomposing the rainfall process into sequences of discrete and continuous random variables, i.e., the presence or absence of rainfall events and rainfall intensity. Examples of quantitative measures used in the current research include Contingency table, mean square error (MSE), root mean square error (RMSE), bias ratio (BR), and mean absolute error (MAE). These measures will be discussed in detail below.

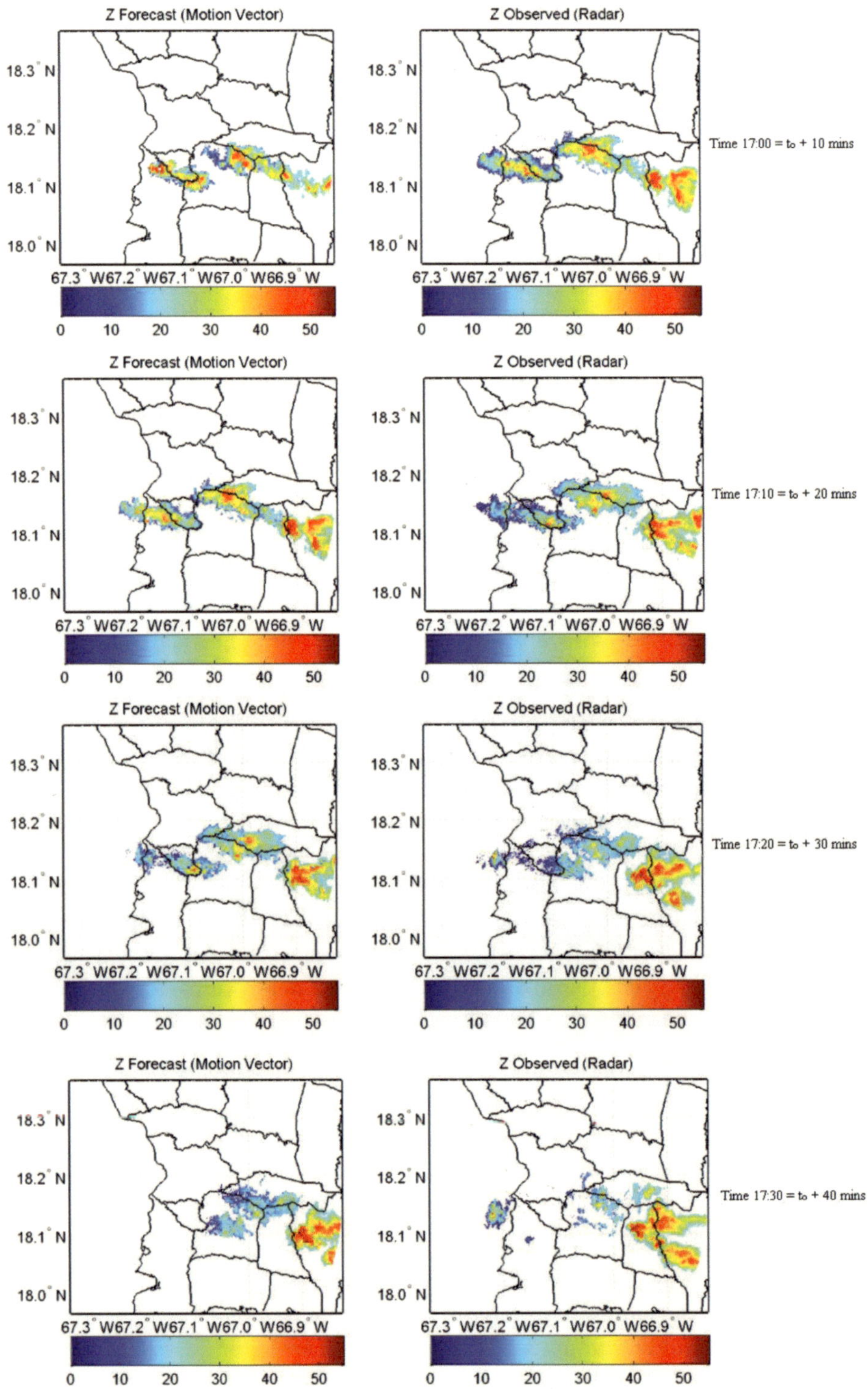

Figure 8. Reflectivity sequence with a lead-time of 10 min, t_o = 16:50 hr. on March 28, 2012.

The joint distribution of the forecast and observations has fundamental interest with respect to the verification of forecasts. In the most practical setting, both the forecast and observations are discrete variables, even if the forecasts and observations are not already discrete quantities.

For lead-times of 10, 20, and 30 min, the storms provide an average hit rate (*HR*) of 0.90, 0.86, and 0.84, respectively. The hit rate score is the fraction of observed events that is forecast correctly. It ranges from 0 at the poor end to 1 at the good end.

The probability of detection (*POD*) of storms varies from 0.61, 0.50, and 0.41, while the false alarm rates (*FAR*) are 0.27, 0.38, and 0.46 for lead-times of 10, 20, and 30 min, respectively.

Figure 9 shows *POD* values and *FAR* values for the complete set of storms. In the ideal situation, *POD* should approach to 1, while the *FAR* results should approach to 0.

The performance indices of 0.74, 0.66, and 0.60 for 10, 20, and 30 min, respectively, for the model, are shown in **Figure 10**.

Similarly, the hit rates (*HR*) of the model for the all storms were 0.90, 0.86, and 0.84 for the 10, 20, and 30 min. The root mean square error (*RMSE*) and bias ratio (*BR*) measure the accuracy

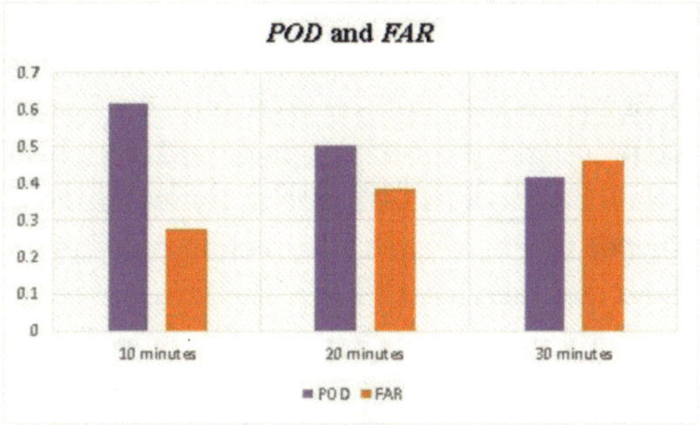

Figure 9. Probability of detection and false alarm for the all storms.

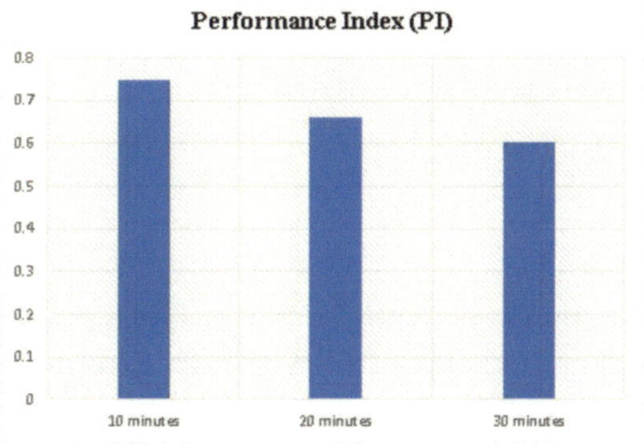

Figure 10. Performance index for the all storms.

of the simulation for all 10 studied events, which furthermore shows the corresponding average values for each lead-times of 10, 20, and 30 min, respectively. The *RMSE* average values are 0.026, 0.077, and 0.144 mm, and the Bias average values are 0.97, 0.98, and 1.04 for lead-times of 10, 20, and 30 min, respectively.

The estimation of bias ratio for a lead-time of 30 min presents an average overestimation prediction, while the estimation of bias ratio for a lead-times of 10 and 20 min shows sub-estimation.

The forecast results present the same tendency of the observed data where the peaks with more precipitation in TropiNet events are coinciding with the forecasted data. They are in good agreement considering that the prediction is in short time and space.

The hydrological model *Vflo* required the ensemble of various layers that perform the physical and topographic characteristics of the basin area. These layers are formed by parameters that were previously presented as effective porosity, hydraulic conductivity, wetting front, roughness, soil depth, and initial saturation, which can be most sensitive in the watershed. Spatially distributed parameter and input from radar rainfall require new methods for adjustment in order to minimize differences between simulated and observed hydrographs. The hydraulic roughness (n), hydraulic conductivity (K), and initial saturation (θ) are the most sensitive parameters of the hydrological model. These values are estimated from physical properties of the watershed adjusted to reproduce system behavior [19]. The hydraulic conductivity controls the total amount of water that will be split into the surface runoff. The hydraulic roughness affects the peak flow, and the time to peak and initial saturation is related to the existing humidity into the soil. Study of model sensitivity was done for the watershed to identify response sensitivity for peak flow to each storm changing the multiplicative factor in the parameters. When the adjustment factor is above 1, the range of change in peak flow falls and tends to remain constant or with a minimum change in the peak flow, just below the referenced value. The analysis suggests that the initial saturation is the parameter with the highest sensitivity in the peak flow for different storms with short duration. Initial saturation is a parameter that depends on how many storms have occurred previously to the studied storm (antecedent soil moisture). Different results are possible to obtain with a sample of continuous storms.

Similar results were founded in peak flow with variations of roughness and hydraulic conductivity for all events. Low variations were founded in peak flow when the adjustment factor takes values greater than one.

8. Conclusions

The best statistical results were found in the rainfall nowcasting model with a lead-time of 10 min as expected. It is well known that prediction of sudden storms using rainfall nowcasting models represents the category that is the most difficult to predict, and consequently, providing accurate flash flood warnings from these types of storms is a major challenge.

The nowcasting model has a limitation in the timeshift because we are assuming that the cloud is a rigid object and that the cloud speed is constant, when in reality these parameters could vary. To find the actual weather conditions, more atmospheric parameters would need to be taken into account. In fact, cloud speed depends on its formation and other physical parameters that are constantly changing [37]. These factors should be taken into account in future works.

Author details

Luz Estella Torres Molina

Address all correspondence to: torresl6@suagm.edu

Universidad del Turabo, Gurabo, Puerto Rico

References

[1] French MN, Krajewski WF, Cuykendall RR. Rainfall forecasting in space and time using a neural network. Journal of Hydrology. 1992;**137**:1-31

[2] Toth E, Brath A, Montanari A. Comparison of short-term rainfall prediction models for real-time flood forecasting. Journal of Hydrology. 2000;**239**:132-147

[3] Hyndman, R. J. (2010). "Forecasting overview". In: International Encyclopedia of Statistical Science. Ed. by M. Lovric. Springer, pp.536-539

[4] Ramirez-Beltran ND. A vector autoregressive model to predict hurricane tracks. International Journal of Systems Science. 1996;**27**(1):1-10

[5] Ramirez-Beltran, N.D., J. M. Castro, and J. Gonzalez. (2005). An algorithm for predicting the spatial and temporal distribution of rainfall rate, An International Journal of Water. In press

[6] Sirangelo R, Versace P, De Luca DL. Rainfall nowcasting, by at site stochastic model P.R.A.I.S.E. Hydrology and Earth System Sciences. 2007;**11**:1341-1351

[7] Rojas AM. Flood Prediction Limitations in small watersheds with mountainous terrain and high rainfall variability. Doctor of Philosophy in Civil Engineering, University of Puerto Rico at Mayagüez, PR; 2012

[8] FEMA. Flood Insurance Study. Commonwealth of Puerto Rico, Volume 1 of 5, Preliminary: June. 22, 2012. Federal Emergency Management Agency

[9] Harmsen EW, Mecikalski J, Mercado A, Tosado Cruz P. Estimating evapotranspiration in the Caribbean Region using satellite remote sensing. Proceedings of the AWRA

Summer Specialty Conference, Tropical Hydrology and Sustainable Water Resources in a Changing Climate. San Juan, Puerto Rico. August 30–September 1. 2010

[10] Yunhao C, Xiaobing L, Peijun S. Estimation of regional evapotranspiration over Northwest China by using remotely sensed data. Journal of Geophysical Sciences. 2001;**11**(2):pp 140-148

[11] Gautier C, Diak GR, Masse S. A simple physical model to estimate incident solar radiation at the surface from GOES satellite data. Journal of Applied Meteorlogy. 1980;**19**:1007-1012

[12] NCDC. National Climatic Data Center – National Oceanic and Atmospheric Administration (NOAA). (http://www.ncdc.noaa.gov/). 2013

[13] Rinehart RE. Radar for Meteorologists. 3rd Ed; 1997, Rinehart Publications

[14] Rossa A, Bruen M, Frühwald D, Macpherson B, Holleman I, Michelson D, Michaelides S. Use of radar observations in hydrological and NWP models. COST 717 Final Report. 2005

[15] Vieux BE, Bedient PB. Estimation of rainfall for flood prediction from WSR-88D reflectivity: A case study, 17-18 October 1994. American Meteorological Society. 1998 June:407-415

[16] Galvez MB, Colom JG, Chandrasekar V, Junyent F, Cruz-Pol S, Rodríguez-Solís RA, León L, Rosario-Colon JJ, De Jesús B, Ortiz JA, Mora Navarro KM. First Observations of the Initial Radar Node in the Puerto Rico Tropinet X-Band Polarimetric Doppler Weather Testbed, IGARSS, 2013, 2337-2340

[17] Marshall JS, Palmer W. The distribution of raindrops with size. Journal of Meteorology. 1948;**5**:165-166

[18] Vieux BE, Vieux JE. *Vflo* ™: A real-time distributed hydrologic model. *Proceedings of the Second Federal Interagency Hydrologic Modeling Conference*, July 28–August 1, Las Vegas, NV 2002. Abstract and paper in CD-ROM. 2002

[19] Vieux BE, Moreda FG. Ordered physics-based parameter adjustment of a distributed model In: Duan Q, Sorooshian S, Grupta HV, Rousseau AN, Turcotte R, (Eds.). Water Science and Application Series, vol. 6. American Geophysical Union; 2003. pp. 267-281, ISBN 0-87590-355-X (Chapter 20)

[20] Gupta RS. Hydrology and Hydraulic Systems. Illinois: Waveland Press, Inc; 1989. pp. 302-307. ISBN 0-88133-865-6

[21] Vieux BE. Vieux, INC. GIS, Hydrology and Radar Rainfall, viewed 12 January 2013. 2013<bodyhttp://www.vieuxinc.com>

[22] USDA. U.S. Department of Agriculture, Natural Resources Conservation Service. Soil Survey Geographic (SSURGO) database for Mayagüez area. Puerto Rico Western Part, pr684, Fort Worth, Texas, Publication date: Dec. 26. 2006a URL:http://SoilDataMart.nrcs.usda.gov/

[23] USDA. U.S. Department of Agriculture, Natural Resources Conservation Service. Soil Survey Geographic (SSURGO) database for Lajas Valley Area. Puerto Rico Western Part, pr687 Fort Worth, Texas, Publication date: Dec. 26. 2006b URL:http://SoilDataMart.nrcs.usda.gov/

[24] USDA. U.S. Department of Agriculture, Natural Resources Conservation Service. Soil Survey Geographic (SSURGO) database for Arecibo area. Puerto Rico Western Part, pr682, Fort Worth, Texas, Publication date: Dec. 26. 2006c URL:http://SoilDataMart.nrcs.usda.gov/

[25] USDA. U.S. Department of Agriculture, Natural Resources Conservation Service. Soil Survey Geographic (SSURGO) database for Ponce area. Puerto Rico Western Part, pr688, Fort Worth, Texas, Publication date: Dec. 26. 2006d URL:http://SoilDataMart.nrcs.usda.gov/

[26] Koussis AD, Lagouvardos K, Mazi K, Kotroni V, Sitzmann D, Lang J, Zaiss H, Buzzi A, Malguzzi P. Flood forecasts for urban basin with integrated hydro-meteorological model, Journal of Hydrological Engineering. 2003;**8**(1):1-11

[27] Box G, Jenkins G. Time Series Analysis, Forecasting and Control. Rev. Ed. Holden-Day; 1976. 575

[28] Dizon CQ. ARMA modeling of a stochastic process appropriate for the angat reservoir. Philippine Engineering Journal (accepted for publication, March 2010). 2010;**XXVIII**(1, June 2007)

[29] Rodriguez-Iturbe I, Cox DR, Isham V. Some models for rainfall based on stochastic point process, Proceedings of Royal Society London, Series A. 1987;**410:**269-288

[30] Stern RD, Coe R. A model fitting analysis of daily rainfall data (with discussion). Journal of the Royal Statistical Society, Series A. 1984;**147**;1-34

[31] Seo DJ, Breidenbach JP. Real-time correction of spatially non-uniform bias in radar rainfall data using rain gauge measurements. Journal of Hydrometeorology. 2002;**3**:93-111

[32] Yilmaz K, Hogue TS, Hsu K, Sorooshian S, Gupta HV, Wagener T. Intercomparison of rain gauge, radar and satellite-based precipitation estimates with emphasis on hydrologic forecasting. Journal of Hydrometeorology. 2005;**6**(4):497-517

[33] Rossa A, Laudanna Del Guerra F, Borga M, Zanon F, Settin T, Leuenberger D. Radar-driven high-resolution hydrometeorological forecasts of the 26 September 2007 Venice flash flood. Journal of Hydrology, in revision. 2010

[34] Schell GS, Madramootoo CA, Austin GL, Broughton RS. Use of radar measured rainfall for hydrologic modelling. Canadian Agricultural Engineering. 1992;**34**(1):41-48

[35] Robinson JS, Sivapalan M, Snell JD. On the relative roles of hillslope processes, channel routing, and network geomorphology in the hydrological response of natural catchments, Water Resources Research. 1995;**31**:3089-3101

[36] Ramírez-Beltran ND, Castro JM, Harmsen E, Vasquez R. Stochastic transfer function models and neural networks to estimate soil moisture. Journal of the American Water Resources Association. 2008;**44**(4):pp 847-865

[37] Corfidi SF, Merritt JH, Fritsch JM. Predicting the movement of mesoscale convective complexes. Weathre Forecasting. 1996;**11**:41-46

Chapter 4

Economic Growth and Employment Effects as a Result of the Upper Austrian Flood Protection Building Program

Sebastian Goers, Friedrich Schneider,
Horst Steinmüller and Andreas Zauner

Additional information is available at the end of the chapter

http://dx.doi.org/10.5772/intechopen.68889

Abstract

Starting in 2002, a continuous building of flood protection infrastructure aiming at preventing high monetary damages has been taking place in Upper Austria. As a result of investments in these flood protection measures between 2002 and 2015, significant positive effects on the gross regional product as well as on the employment level have been generated. However, the macroeconomic effects are found to depend substantially on the import quota for required materials.

Keywords: flood protection infrastructure, investment, macroeconometric simulation, economic effects, Upper Austria

1. Introduction

Austria has always been confronted with flood situations, which pose a threat to the inhabitants of the particular areas and cause enormous monetary damages. Since the turn of the century, there have been 200-year floods just 11 years apart, one in August 2002 and the other one in June 2013. As a consequence, the state and federal Governments of Austria and Upper Austria, respectively, have been continuously investing into flood protection measures. The present study aims to quantify the value added and the effects on employment as a consequence of implementing the flood protection program and the measures therein. The main focus is on an ex-post analysis of the Upper Austrian economy between 2002 and 2015.

An assessment of the flood damages which were avoided due to making these investments is not the focus of this study. The damages caused by floods occurring in recent years are only rudimentarily documented. It is also noted that there is scarce or insufficient information on the economic costs of flood damage corresponding to social opportunity costs as these are not directly apparent from existing statistics [1].

Section 2 provides an overview of flood damages in Upper Austria between 2002 and 2013. In Section 3, the effects of the investments on the regional value added and the rate of employment are quantified on the basis of ex-post analyses. These are based on a comparative static assessment and a macroeconometric simulation by means of a dynamic time-series model. The conclusions of this study are drawn in Section 4.

2. Flood damages in Upper Austria

2.1. Flood of 2002

The flood of August 2002 exceeded all floods in Austria since 1965/1966 [2]. It claimed nine lives and resulted in monetary damages of € 3 bn [3]. € 420 m of those monetary damages were covered by insurances, another € 414 m were covered by the disaster fund, resulting in a deficit of approximately € 2.2 bn [4, 5]. Waterways Authority conducted a damage survey in their jurisdiction (e.g., damages to towpaths, the riverside and approach piers, silting up of harbor basins, aggradations, and suspended loads) along the Danube, March, and Thaya, estimating the damages caused by the flooding to amount to approximately € 3 m [6].

There have been different statements about the amount of damages in Upper Austria. According to Ref. [7], the flood in August 2002 caused total monetary damages of approximately € 464 m. 23.9% of the damages are apportioned to households, 49.6% to companies, 7.8% to agriculture and forestry, and 18.7% to the public sector. More than twice as much, approximately € 1.1 bn of monetary damages were reported by a territorial council at a press conference on March 9, 2013. € 500 m thereof are apportioned to the Upper Austrian Machland, a region and a cultural landscape along the Danube. The Upper Austrian agriculture faced monetary damages of € 11.6 m. A total of approximately 10,500 ha of agricultural land were affected [8].

2.2. Flood of 2002

The flood in June 2013, the second 100-year flood within 11 years, led to overall monetary damages of approximately € 870 m. The EU's disaster fund covered just under € 22 m [9]. Insurances paid out € 250 m. In Upper Austria, the total damages of the flood of 2013 amount to approximately € 220 m. The EU granted € 2.9 m from their solidarity fund as flood aid for Upper Austria [10]. The Upper Austrian agriculture was faced with estimated damages of € 15 m as a result of the flood. 247 farms were damaged throughout the state, as well as approximately 10,500 ha (as in 2002) of agricultural land. Of this, 7400 ha are forage crops,

2400 are grasslands, and 700 are vegetables and strawberries. However, the damages of the flood in June 2013 exceeded those of the flood in 2002. The reasons for this are the flooding of higher quality goods on the one hand and on the other hand, the fact that in August 2002, many fields had already been harvested. Additionally, approximately 4000 hectares of forest areas were flooded in 2013 [8].

2.3. Flood protection

In general, an increase in severe flood disasters can be observed [7]. Due to those flood events, efforts in the area of flood protection have been intensified. In 2005, the largest flood protection program to date was launched in Upper Austria. Thus, the damages of the flood in 2013 could be reduced from € 1.1 bn to one-fifth to a quarter, despite being more severe than the flood in 2002 [11]. Data about the amount of investment into the flood protection program vary between € 690 and 700 m over the period from 2002 to 2015 [12, 13]. The average infrastructure investment costs per year of approximately € 50 m are similar to those in the assessment of alternative flood control policies in the Netherlands [14]. For another flood protection project in the Eferdinger basin, a budget of € 250 m is available for the implementation period from 2014 to 2022 [15].

3. Economic ex-post analysis of the Upper Austrian flood protection program

This section investigates the effects of the investments into flood protection in Upper Austria in the period from 2002 to 2015. As shown in Ref. [16], the effects on the regional economy (e.g., employment) play an important role within the decision-making of environmental adaption measures. In the present study, the economic effects are calculated by two methods, applying comparative static analysis on one hand (see Section 3.1) and a dynamic simulation (see Section 3.2) based on the macroeconomic simulation model MOVE2 [17] on the other hand. Since this is an ex-post analysis, the primary research question is defined as follows: what economic contribution with regard to employment and value added has been created in the past (2005–2015) by the Upper Austrian flood protection construction program?

3.1. Comparative static analysis

The data of investments made between 2002 and 2015 within the context of flood protection are used as an input for the dynamic simulation of the economic effects in Upper Austria (see Section 3.2). For the period of time between 2002 and 2015, overall investments of approximately € 690 m were made (see **Figure 1** and **Tables 1** and **2** for a temporal aggregation and arrangement according to specific measures). All data were provided by the Upper Austrian Government, Environment and Water Management and are identical to the data in Ref. [1] for the time period 2002–2012.

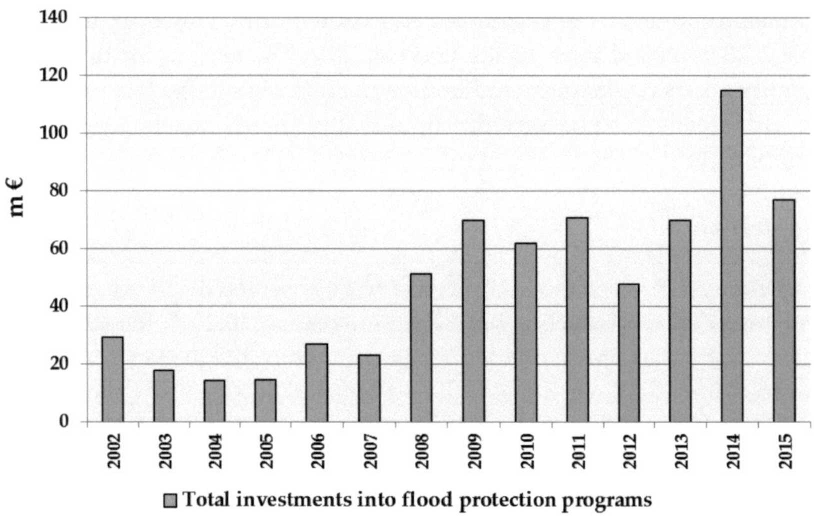

Figure 1. Total investments into flood protection programs in Upper Austria, 2002–2015.

In general, investments into the flood protection are grouped as follows:

a. Immediate measures (€ 27 m/approximately 4% of overall investments).

b. Flood protection measures (€ 137 m/approximately 20% of overall investments).

c. Maintenance (€ 44 m/approximately 6% of overall investments).

d. Costs of planning (€ 5 m/approximately 1% of overall investments).

e. Torrent control and immediate measures (€ 140 m/approximately 20% of overall investments).

f. Flood protection projects along the Danube[1] (€ 337 m/approximately 49% of overall investments).

The types of financing also are considered (see **Table 3**). Analogous to Ref. [1], there are three different types of financing: (1) federal funds, (2) state funds, and (3) funds from interested parties. As shown in **Table 3**, approximately 39% (€ 265 m) are accounted for by federal funds, approximately 52% (€ 361 m) by state funds, and approximately 9% (€ 64 m) by interested parties.

As imports may lead to outflows of added value, parts of the investments do not take effect in Upper Austria. Analogous to Ref. [1], three scenarios with different import quotas (0, 10, and 20%) are defined for construction services within the scope of immediate measures, flood

[1]Measures of the Austrian Ministry for Transport, Innovation and Technology along the Danube in Upper Austria.

	2002	2003	2004	2005	2006	2007	2008
	(m €)	(m €)	(m €)	(m €)	(m €)	(m €)	(m €)
Immediate measures	16.6	0.0	0.0	1.0	0.1	0.0	0.0
Flood protection measures	2.3	5.1	3.7	0.1	12.7	11.0	11.7
Maintenance	2.1	2.0	2.8	3.4	1.9	1.5	5.5
Planning	0.4	0.0	0.0	0.7	0.4	0.2	0.7
Torrent control and immediate measures	7.5	10.2	5.1	6.5	6.2	7.4	16.8
Danube projects	0.4	0.4	2.9	2.9	5.6	3.1	16.7
SUM	29.2	17.7	14.4	14.5	26.8	23.1	51.3

Table 1. Investments by the type of flood protection measures, 2002–2008.

	2009	2010	2011	2012	2013	2014	2015
	(m €)	(m €)	(m €)	(m €)	(m €)	(m €)	(m €)
Immediate measures	0.1	0.1	0.2	0.1	3.9	3.1	2.3
Flood protection measures	27.3	9.2	3.0	4.3	19.4	15.5	11.6
Maintenance	4.4	3.6	1.7	0.4	6.3	5.0	3.8
Planning	0.3	0.2	0.2	0.0	0.6	0.5	0.4
Torrent control and immediate measures	10.0	9.0	8.5	5.1	19.8	15.8	11.9
Danube projects	27.8	39.9	57.3	38.2	20.0	75.0	47.0
SUM	70.0	62.0	70.9	48.0	70.0	115.0	77.0

Table 2. Investments by type of flood protection measures, 2009–2015.

protection measures, torrent control and immediate measures, and maintenance as well as projects related to the Danube (see **Table 4**).

3.2. Dynamic simulation analysis

The input for the dynamic analysis of economic effects in Upper Austria with the simulation model MOVE2 [17] is provided by the comparative statistic preparation of the data (see Section 3.1) of the investments which were made between 2002 and 2015 within the scope of the flood protection program. The focus is on macroeconomic effects due to the investments, whereas economic optimality, as for example examined in Refs. [18, 19], is not assessed. The dynamic economic impacts of the investments into the flood protection program are described

	Total investment	Federal funds	State funds	Interest parties' funds
	(m €)	(m €)	(m €)	(m €)
Immediate measures	27	15	9	3
Thereof public waters	2	2	0	0
Thereof interest parties' waters	26	14	9	3
Flood protection measures	137	72	48	17
Thereof public waters	28	24	1	3
Thereof interest parties' waters	109	48	47	14
Maintenance	44	19	12	13
Thereof public waters	7	6	0	1
Thereof interest parties' waters	37	13	12	12
Planning	5	3	2	0
Thereof public waters	1	1	0	0
Thereof interest parties' waters	4	2	2	0
Torrent control and immediate measures	140	83	26	31
Danube projects	337	74	264	0
SUM	690	265	361	64

Table 3. Investments by the type of flood protection measures and the method of funding, 2009–2015.

	Import quota: 0%	Import quota: 10%	Import quota: 20%
Year	(m €)	(m €)	(m €)
2002	29.2	26.3	23.4
2003	17.7	15.9	14.1
2004	14.4	12.9	11.5
2005	14.5	13.1	11.8
2006	26.8	24.2	21.5
2007	23.1	20.8	18.6
2008	51.3	46.2	41.2
2009	70.0	63.0	56.0
2010	62.0	55.8	49.6
2011	70.9	63.8	56.7
2012	48.0	43.2	38.4
2013	70.0	63.1	56.1

	Import quota: 0%	Import quota: 10%	Import quota: 20%
Year	(m €)	(m €)	(m €)
2014	115.0	103.6	92.1
2015	77.0	69.3	61.7
SUM	690	621	553
Ø	49.3	44.4	39.5

Note: Rounded values. Outflows of value added are taken into account for construction services within the scope of immediate measures, flood protection measures, torrent control and immediate measures, and maintenance as well as projects related to the Danube. Planning costs are not affected by outflows of added values

Table 4. Investments affecting the value added.

below. In particular, the progression and interactions of the macroeconomic parameters such as gross regional product (GRP), investments, employment, and private consumption are observed and elucidated. For the general interpretation of the results, there has to be emphasized that the simulation result shows the difference of the two development paths of the model—the difference between the business-as-usual and the simulation scenario of each endogenous variable of the model—and not the absolute values of both scenarios as in the case of a prognosis model [17].

Over the period of 2002 until 2015, positive effects of the investments in flood protection on the Upper Austrian GRP can be observed at large (see **Table 5** and **Figure 2**). For the time period from 2002 to 2015, the annual average increase of the GRP is approximately € 83, 78, or 73 m higher at import rates of 0, 10, or 20%, respectively.

	Gross regional product		
	Import quota: 0%	Import quota: 10%	Import quota: 20%
Year	(m €)	(m €)	(m €)
2002	+31	+29	+27
2003	+29	+27	+25
2004	+28	+26	+24
2005	+28	+26	+24
2006	+41	+39	+36
2007	+42	+39	+37
2008	+73	+68	+64
2009	+104	+97	+90
2010	+108	+101	+94
2011	+121	+114	+106
2012	+104	+97	+91

	Gross regional product		
	Import quota: 0%	Import quota: 10%	Import quota: 20%
2013	+123	+116	+108
2014	+176	+165	+154
2015	+155	+145	+136
Ø	+83	+78	+73

Note: Displaying of direct, indirect, and induced effects. Own calculations with the simulation model MOVE2 [17].

Table 5. Impacts on Upper Austria's gross regional product as a result of investments in flood protection measures, taking into account different import quotas, 2002–2015.

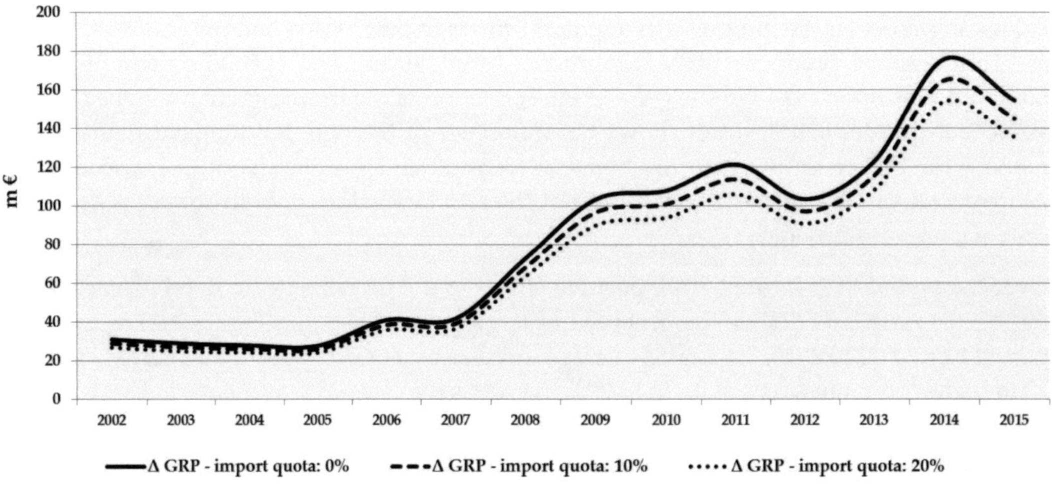

Figure 2. Impacts on Upper Austria's gross regional product as a result of investments in flood protection measures, taking into account different import quotas, 2002–2015. Note: Displaying of direct, indirect, and induced effects. Own calculations with the simulation model MOVE2 [17].

Investments into flood protection as well as the increase of private consumption of households between € 25 and 26 m per year (depending on the import quota) as a result of the induced effects of the investment impulses are the cause for this development (see **Tables 6** and **7**). The outflow of value added caused by imports of foreign materials or technologies is superimposed and thus weakened by the increase in exports as a result of the positive development of the economy. This results in net exports (difference between exports and imports) of approximately € 15, 10, or 6 m at import rates of 0, 10, or 20%, respectively (see **Table 8**).

As for the rate of employment (see **Figure 3** and **Table 9**), as a result of the increase in GRP over the time period 2002–2015, an increase of 590, 520, or 460 persons can be observed at import rates of 0, 10, or 20, respectively. 60% of the additional employees originate from the construction sector.

	Investments		
	Import quota: 0%	**Import quota: 10%**	**Import quota: 20%**
Year	(m €)	(m €)	(m €)
2002	+20	+20	+20
2003	+15	+15	+15
2004	+13	+13	+13
2005	+13	+13	+13
2006	+22	+22	+21
2007	+20	+20	+20
2008	+40	+40	+40
2009	+56	+56	+55
2010	+54	+53	+53
2011	+61	+61	+60
2012	+47	+46	+46
2013	+61	+61	+60
2014	+95	+94	+93
2015	+72	+71	+70
Ø	+42	+42	+41

Note: Displaying of direct, indirect, and induced effects. Own calculations with the simulation model MOVE2 [17].

Table 6. Impacts on Upper Austria's investments as a result of investments in flood protection measures, taking into account different import quotas, 2002–2015.

	Private consumption		
	Import quota: 0%	**Import quota: 10%**	**Import quota: 20%**
Year	**(m €)**	**(m €)**	**(m €)**
2002	+10	+10	+10
2003	+7	+7	+7
2004	+9	+9	+8
2005	+9	+8	+8
2006	+13	+13	+13
2007	+13	+12	+12
2008	+24	+23	+23
2009	+31	+31	+31
2010	+32	+32	+31

	Private consumption		
	Import quota: 0%	**Import quota: 10%**	**Import quota: 20%**
Year	**(m €)**	**(m €)**	**(m €)**
2011	+39	+38	+37
2012	+33	+32	+31
2013	+42	+41	+41
2014	+58	+57	+56
2015	+48	+47	+46
Ø	+26	+26	+25

Note: Displaying of direct, indirect, and induced effects. Own calculations with the simulation model MOVE2 [17].

Table 7. Impacts on Upper Austria's private consumption as a result of investments in flood protection measures, taking into account different import quotas, 2002–2015.

	Net exports		
	Import quota: 0%	**Import quota: 10%**	**Import quota: 20%**
Year	**(m €)**	**(m €)**	**(m €)**
2002	+1	−1	−3
2003	+7	+5	+3
2004	+6	+5	+3
2005	+6	+5	+3
2006	+6	+4	+2
2007	+9	+7	+4
2008	+9	+5	+1
2009	+16	+10	+4
2010	+22	+16	+10
2011	+21	+15	+9
2012	+24	+19	+14
2013	+20	+14	+7
2014	+24	+14	+4
2015	+34	+27	+19
Ø	+15	+10	+6

Note: Displaying of direct, indirect, and induced effects. Own calculations with the simulation model MOVE2 [17].

Table 8. Impacts on Upper Austria's net exports as a result of investments in flood protection measures, taking into account different import quotas, 2002–2015.

http://dx.doi.org/10.5772/intechopen.68889

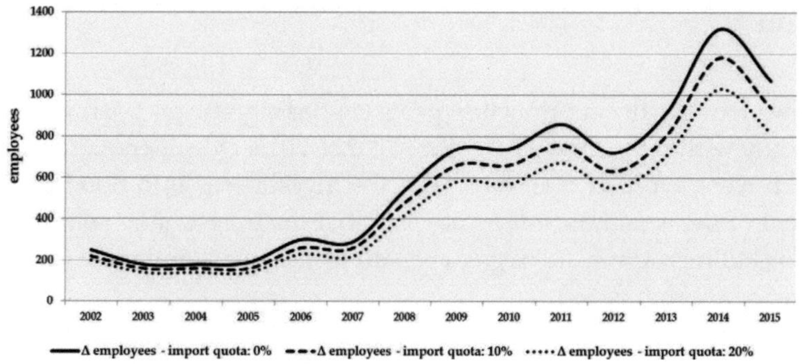

Figure 3. Impacts on Upper Austria's employment as a result of investments in flood protection measures, taking into account different import quotas, 2002–2015. Note: Displaying of direct, indirect, and induced effects. Own calculations with the simulation model MOVE2 [17].

	Employment		
	Import quota: 0%	**Import quota: 10%**	**Import quota: 20%**
Year	**(employees)**	**(employees)**	**(employees)**
2002	+250	+220	+200
2003	+180	+160	+140
2004	+180	+160	+140
2005	+190	+160	+140
2006	+300	+260	+230
2007	+290	+260	+220
2008	+540	+480	+420
2009	+740	+660	+580
2010	+740	+660	+570
2011	+860	+760	+670
2012	+720	+630	+550
2013	+920	+810	+710
2014	+1.320	+1.180	+1.030
2015	+1.070	+940	+820
Ø	+590	+520	+460

Note: Displaying of direct, indirect, and induced effects. Own calculations with the simulation model MOVE2 [17].

Table 9. Impacts on Upper Austria's employment as a result of investments in flood protection measures, taking into account different import quotas, 2002–2015.

4. Conclusion

The study shows that the flood protection program has significant positive effects on Upper Austria's economy within the time frame from 2002 to 2015. Despite not quantifying the flood damages, which were avoided due to making the investments into flood protection, in this study, the extent of the economic effects suggest that the investments within the framework of the flood protection program are of great benefit from an economic perspective. The following points have to be emphasized.

- Positive effects of the investments into flood protection on the GRP can be obtained. Overall, the annual average increase of the GRP is approximately € 83, 78, or 73 m higher at import rates of 0, 10, or 20%, respectively.
- As a result of the increase in GRP over the time period, from 2002 to 2015, an increase in employment of 590, 520, or 460 persons can be observed at import rates of 0, 10, or 20%, respectively. 60% of the additional employees originate from the construction sector.
- As a result of the increase in employment, there is an increase of the gross income and thus private consumption, which generate positive effects on the economy. Investments into flood protection, as well as the increase of private consumption of households between € 25 and 26 m a year (depending on the import quota) as a result of the induced effects of the investment impulses, are the cause for this development.
- The outflow of value added caused by imports of foreign materials or technologies is superimposed and thus weakened by the increase in exports as a result of the positive development of the economy.
- Higher import quotas cause a smaller amount of the investments being effective on the value added and the rate of employment in Upper Austria.

Acknowledgements

The authors would like to thank Julia Mayrhofer for drafting this chapter. Further, the support of this work by the Upper Austrian Government is gratefully acknowledged.

Author details

Sebastian Goers[1*], Friedrich Schneider[1,2], Horst Steinmüller[1] and Andreas Zauner[1]

*Address all correspondence to: goers@energieinstitut-linz.at

1 Energieinstitut at the Johannes Kepler University, Linz, Austria

2 Department of Economics, Johannes Kepler University, Linz, Austria

References

[1] Schneider F, Jenewein S, Haigner S. Nutzen-Kosten-Analyse der Investitionen zur Schadensminimierung von potenziellen Hochwasserschäden. Innsbruck: Gesellschaft für Angewandte Wirtschaftsforschung mbH; 2012

[2] Godina R, Lalk P, Lorenz P, Müller G, Weilguni V. Das August-Hochwasser 2002. WasserWirtschaft. 2005;**95**(9):14-18

[3] Stiefelmeyer H, Sattler J. Bundesministerium für Land- und Forstwirtschaft, Umwelt und Wasserwirtschaft, editors. Schutz vor Naturgefahren in Österreich. Vienna; 2012

[4] Zens A, Neumüller A. Hochwasser: Versicherungen decken nur einen Bruchteil des Schadens [Internet]. 2013. Available from: http://www.nachrichten.at/nachrichten/wirtschaft/Hochwasser-Versicherungen-decken-nur-einen-Bruchteil-des-Schadens;art15,1132844,PRINT?_FRAME=33 [Accessed: March 6, 2017]

[5] Austria Presse Organisation. Allianz Risikoforschung: Deckungslücke von 80 Prozent bei Erdbeben und Hochwasser [Internet]. 2014. Available from: http://www.ots.at/presseaussendung/OTS_20140402_OTS0035/allianz-risikoforschung-deckungsluecke-von-80-prozent-bei-erdbeben-und-hochwasser [Accessed: March 6, 2017]

[6] Habersack H, Bürgel J, Petraschek A. Bundesministerium für Land- und Forstwirtschaft, Umwelt und Wasserwirtschaft, editors. Analyse der Hochwasserereignisse vom August 2002—FloodRisk-Synthesebericht. Vienna; 2004

[7] Habersack H, Fuchs H. Hochwasser 2002: Datenbasis für Schadensbilanz 2002. Vienna: Zentrum für Naturgefahren und Risikomanagement (ZENAR); 2003

[8] ORF. Hochwasser: 15 Mio. Euro Schaden in Landwirtschaft [Internet]. 2013. Available from: http://ooe.orf.at/news/stories/2588212/ [Accessed: March 6, 2017]

[9] Austria Presse Organisation. EU-Hochwasserhilfe für Oberösterreich: BM Mikl-Leitner und LH Pühringer unterzeichnen Vereinbarung [Internet]. 2014. Available from: http://www.ots.at/presseaussendung/OTS_20140228_OTS0097/eu-hochwasserhilfe-fuer-oberoesterreich-bm-mikl-leitner-und-lh-puehringer-unterzeichnen-vereinbarung [Accessed: March 6, 2017]

[10] Lichtenberger P. Jahrhundert-Hochwasser 2013 in Österreich: ein Jahr danach [Internet]. 2014. Available from: https://www.gmx.at/magazine/panorama/jahrhundert-hochwasser-2013-oesterreich-jahr-18981542 [Accessed: March 6, 2017]

[11] Anschober R. Hochwasserkatastrophe 2013: aufarbeiten—verbessern—schützen: Zwischenbericht und die weiteren Pläne der Aufarbeitungsarbeit [Internet]. 2013. Available from: http://www.hochwasser2013.at/images/Dokumente/PKAnschober.pdf [Accessed: March 6, 2017]

[12] OÖNachrichten. Hochwasser 2013: Schutzprojekte in Oberösterreich um 700 Millionen Euro nötig [Internet]. 2013. Available from: http://www.nachrichten.at/nachrichten/

ticker/Hochwasser-2013-Schutzprojekte-in-Oberoesterreich-um-700-Millionen-Euro-noetig;art449,1137631 [Accessed: March 6, 2017]

[13] Pühringer J, Anschober R. 2 Jahre nach der Hochwasserkatastrophe 2013: Eine Zwischenbilanz über die bisher gesetzten und geplanten Verbesserungsmaßnahmen beim Hochwasserschutz OÖ [Internet]. 2015. Available from: https://www.land-oberoesterreich.gv.at/Mediendateien/LK/PKLHPuehringerLRAnschober02062015_Internet.pdf. [Accessed: March 6, 2017]

[14] Brouwer R, van Ek R. Integrated ecological, economic and social impact assessment of alternative flood control policies in the Netherlands. Ecological Economics. 2004;**50**:1-21. DOI: 10.1016/j.ecolecon.2004.01.020

[15] Anschober R, Theobald S. Aufarbeitung der Hochwasserkatastrophe 2013 an der OÖ Donau: Kassel-Studie über Wehrbetriebsordnungen und Sedimentmanagement liegt vor [Internet]. 2015. Available from: https://www.land-oberoesterreich.gv.at/Mediendateien/LK/PKAnschober30042015_Internet.pdf [Accessed: March 6, 2017]

[16] Marttunen M, Hämäläinen R. Decision analysis interviews in environmental impact assessment. European Journal of Operation Research. 1995;**87**:551-563

[17] Goers S, Baresch M, Tichler T, Schneider F. MOVE2—Simulation model of the (Upper) Austrian economy with a special focus on energy including the socio-economic module MOVE2 social: integration of income, age and gender. Linz: Energieinstitut at the Johannes Kepler University; 2015

[18] de Bruin K, Ansink E. Investment in flood protection measures under climate change uncertainty. Climate Change Economics. 2011;**2**(4):321-339. DOI: 10.1142/S2010007811000334

[19] Kousky C, Luttmer E, Zeckhauser R. Private investment and government protection. Journal of Risk Uncertainty. 2006;**33**:73-100. DOI: 10.1007/s11166-006-0172-y

Chapter 5

Geodesign a Tool for Redefining Flood Risk Disaster in Developing Countries: A Case Study of Southern Catchment of Ankobra Basin, Ghana

Adams Osman and Benjamin Nyarko

Additional information is available at the end of the chapter

http://dx.doi.org/10.5772/intechopen.68782

Abstract

Flood is a hazard with increasing number of events and damages in all parts of the worlds. Many studies and researches have been conducted in understanding the causes of flood, effects and finding solutions to reduce its occurrences. Most of the solutions proposed have failed in the light of changing climate, river morphology and human modifications of watersheds in the world. In this regard, this research adapted geodesign as a flood reduction technique in stimulating flood risk within the Ankobra lower basin. Geodesign framework by Steinitz was adopted where various physical characteristics of the study area were remodelled. The results indicated that increasing building foundation above 88 m reduced flood risk. Also, further simulations by increasing drainage and connectivity reduced flood area size of extreme flood risk zone and high risk zones.

Keywords: geodesign, spatial multi-criteria evaluation, flood risk

1. Introduction

Man's life through centuries has been engulfed within diverse risk. Manifestation of these risks to reality is term disaster and it brings untold hardship. Flood-related disasters are one of the oldest disasters experienced by man since the dawn of time. A translation of cuneiform symbols of the Weld-Blundell Prism shows that flood occurrences go way back in antiquity during the Sumerian civilisation with both negative and positive effects [1]. In modern times, flood disasters are increasing with huge tolls on people and governments. Flood disasters are working against the attainment of Sustainable Development Goals as destructions levels are increasing human insecurity and poverty levels. According to the International Strategy for Disaster Reduction [2], there has been an increase in flood disasters with over 3455 reported cases between 1980 and 2011. Increase in flood disasters have puts over 24 million people and an estimated $2203.97 billion worth of assets at risk in the world [3]. In Ghana, it is estimated that floods have affected over 4 million people and brought about damages to the tune of US$ 78 hundred million between 1990 and 2014 [4].

Based on the destructive effects of flood, greater attention has been devoted to studying the causes and finding possible solutions to flood. Diverse research has been able to document numerous causes of flooding for all terrains in the world. But, solutions for curbing the occurrences and effects of flood have been limiting. Most proposed flood solutions have been inadequate in reducing risk to flood because of changing nature of river flow characteristics, climate change and rapid land cover/use changes [5]. Structural (building dykes, floodwalls and widening river channels) and non-structural (land use planning and flood warning systems) flood control measures are failing [6]. In the Ankobra Basin, numerous measures have been adopted by local and disaster managements to prevent flood and reduce its impact. These measures over the decades have ranged from dredging the Ankobra river bed, creating new drains, retrofitting of buildings. All these measures have succumbed to increasing flood intensity. This problem is not peculiar to the study area, but rather an emerging trend in the world. In 2005, over 80% of New Orleans was submerged in flood waters as dykes and flood walls were unable to flood waters from the Hurricane Katrina [7]. Also, comprehensive land use plan and dykes were unable to shield the Hague-Netherlands in 2012 from inundation [8]. The inadequacy of current flood control techniques calls for newer techniques [9].

Geodesign is a new intervention that can help in the fight of reducing flood events and decrease risk levels. This is mainly because it gives planners and disaster managers the ability to stimulate risk and disaster variables to ascertain which variable when changed can have the optimal effect on reducing flood risk. Also with geodesign, geospatial technicians have the capability to control and manage floods by undertaking numerous flow stimulations to ensure that drains maximise flow of flood waters [10]. Geodesign combines the age-old practice of planning, designing, implementing and evaluating changes to our built and physical environment with modern tools, including digital databases and representational and analysis software tools [11]. Implementing geodesign can ensure a win-win situation for both man and the natural environment by taking into consideration the full spectrum of the earth's life support including everything that lies below, on and above the surface system [12].

The potential of geodesign has been tested by some planners and disaster managers since its inception in 2010. Planners in Asheville and Cap Cod (United States of America), Sabah Al-Salem (Kuwait) and Bodegraven (Netherlands) have been able to redesign their landscapes using geodesign concepts. The results have been a remarkable decrease in flood occurrences and a friendly coexistence of man and nature (flood) [13]. It is based on these success stories that this study seek to assess the ability of geodesign as a flood reduction technique in the Southern Basin of Ankobra.

2. Literature on geodesign

In reducing flood risk, most measures have centred on storage dams, barriers, land use plans, flood proofing of buildings, land reclamation and flood forecasting and warning [14]. McMillan and Brasington [5] assert that these measures are inadequate considering the constant changes in river dynamics and climate. As such, there is a need for new approaches to solve flood issues [9]. Geodesign is a new approach, which is gaining momentum in the field of geospatial technology with capabilities to solve complex environmental issues by finding the right balance between settlements and nature [11]. The use of the concept of geodesign is within the context that spatiotemporal (geographic) dynamics of events conditions what and how we design to tweak and adapt to our environment [12]. The effectiveness of any form of geodesign is conditioned on having sufficient knowledge of the relevant spatio-temporal characteristics of the area under study.

Geodesign originates from the merging of two words geography (geo) and design [15]. Mathur [16] states that geodesign is the intersection of geography and design. Flaxman [17] defined geodesign as a design and planning method, which combines the creation of design proposals with impact simulations informed by geographic contexts. Impact simulation abilities of geodesign are what most flood reduction measures lack. But geodesign helps in averting these problems by envisioning possible future scenarios with predictive alternatives whose consequence can be evaluated before implementation [18]. Geodesign is an interventionist approach in contrast to the more detached and dispassionate approaches [19].

Geodesign made the world stage in 2010 at the ESRI conferences. As a new field, various practitioners have diverse concepts about what geodesign is or should be. Others have even criticised the concept of being and old age technique of multi-criteria evaluation, as such it is not a new concept. But, Carl Stenitz, a pioneer and advocate of geodesign, developed a framework, which reduces the arguments around the concept. In Stenitz framework, geodesign involves specific actions or activities (**Figure 1**).

According to Steinitz [15], the assessment phase deals with the modelling of the environment, understanding it and the assessing of the elements in the environment, whereas the intervention phase looks at changing the modelled environment, analysing its impact and making a decision. By this, you sketch an idea, find out its implications, make adjustments and try again until a suitable alternative is achieved [20].

2.1. Geodesign as a structural and non-structural measure

Practitioners in the field of geographic information systems have been able to use geodesign to solve diverse environmental problems since it provides an excellent concept for proposing change to the geographical area [11]. Geodesign can be used as structural or non-structural measure in flood risk reduction. This is mainly because some disaster managers concern themselves on modifying only the physical characteristics of a landscape. But others rather introduce new physical elements into a landscape when alteration of already elements is not sufficient to solve flood occurrence.

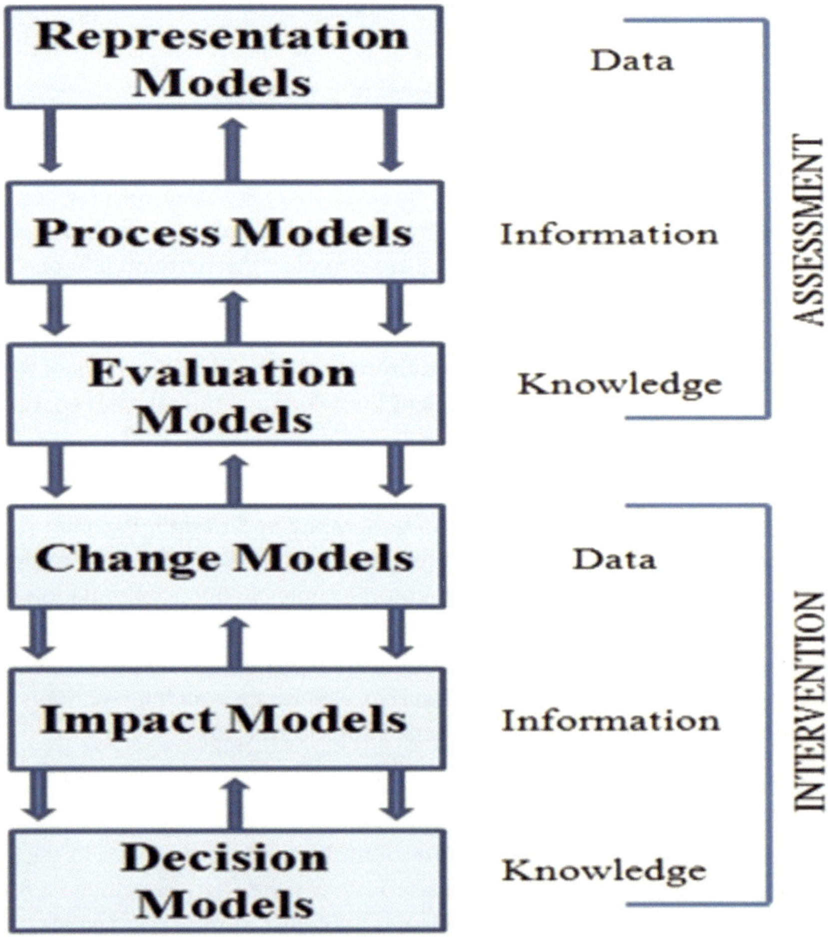

Figure 1. Geodesign framework. Source: Steinitz [15].

In 2008, geodesign was employed by the town Charleville, Queensland in Australia to help reduce flood risk [21]. Geodesigners were able to remodel the town's landscape and the impact of their new model which informed them to construct a 375 m of geodesign pallet barrier serving as a flood defence wall. Since the flood defence wall was constructed in January 28, the town has been safe from spillages coming from Waitaki River. In February 2004, River Severn caused havoc when torrential rain raised its level in Ironbridge town, United Kingdom. In less than 5 hours, a 550 m geodesign steel barrier, which was 1.8 m high, was erected by the United Kingdom Environment Agency along the Wharfage in Ironbridge Gorge. Geodesign barriers have a standard protection height of 0.65, 1.25 and 1.8 m with the ability to interlock, making it easy to superimpose them to increase their height as against flood walls that are static [21]. Also, geodesign barriers were constructed in the River Calder at Hebden Bridge, West Yorkshire, United Kingdom. This diverts water from an old riverside wall reducing and preventing flooding. These examples show the ability of geodesign to aid conventional structural flood reduction approaches.

http://dx.doi.org/10.5772/intechopen.68782

As a non-structural measure was employed in Cape Cod, Massachusetts in the United States of America, when the town threatened by sea level rise and coastal flood [22]. Through alternative scenarios modelling, Snyder and Lally [22] were able to find zones fit for human developments free from coastal and sea level rise in the future. China, a country with flood problems costing billions of dollars yearly have resorted to geodesign by remodelling its urban landscape ecologically to help reduce flood. Mainly because the cost to be incurred from geodesign simulations will be lower than normal structural measures [23].

Geodesign flood reduction strategy has challenges like any other intervention. Field experience shows that when geodesign steels are not firmly installed and the plates properly locked high pressure waters can topple over them [24]. On non-structural usage, Ervin [25] argues that there are some ethical issues, which will emerge in the future about geodesign since it does not have a set of ethics.

3. Methodology

The Ankobra coastal estuary lies within 4054'55"N and 2017'44"W to the upper left, 4054'55"N and 2015'58"W to upper right, 4053'41"N and 2015'58"N to lower right and 4053'41"N and 2017'44"W to the lower left. The study area is drained by the Ankobra River, which increases the areas vulnerability to flooding during rainy season (**Figure 2**). Rainfall in the zone is mostly in the ranges of 1500–2000 mm [26]. The area is relatively flat with most part well below 10 m above sea level. There are about 4069 people living within the estuary [27] who are mostly affected during flood.

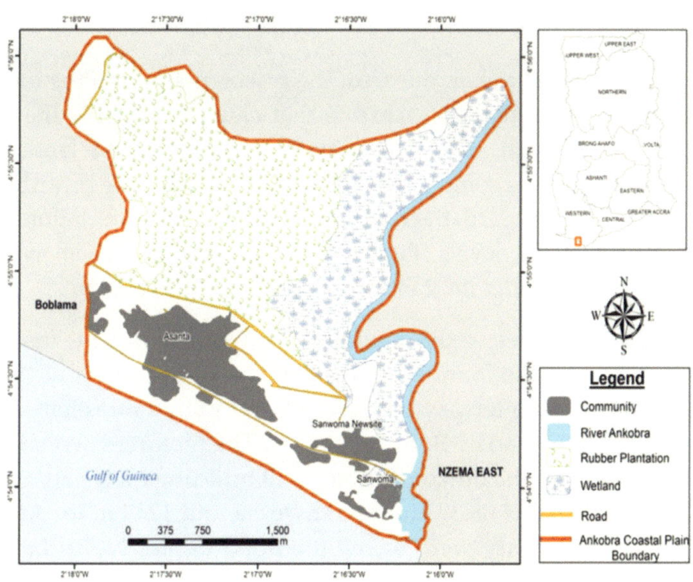

Figure 2. Map of study area.

3.1. Data processing and analysis

The flood risk map was used as the basis for remodelling the landscape to reduce flood risk in the study area. The intervention phase of [15], geodesign framework was adopted. The first part of the intervention phase of geodesign deals with change models. Steinitz [15] explains that change models require remodelling of the landscape. That is, the geodesigner changes some or social features, which have contributed to improper functioning of a zone. In meeting this requirement, some landscape features in the Ankobra estuary had to be changed or remodelled. The first change model of the geodesign undertaken was increasing the foundation heights of buildings in the study area above the worst flood depth experienced in the communities. The second stage of geodesign model was also undertaken by remodeling flow channels (direction, length) of drains. In this regard, a hydrological model was run for the Ankobra estuary from the digital elevation model. The hydrological model tool in ArcMap 10.1 ESRI software was used.

The processes for generating the hydrological model were checking for sinks in the elevation data, filling these sinks and running a flow direction function as well as a flow accumulation function. Finally, the hydrological tool (flow accumulation algorithm) was run to determine where runoffs are likely to move downslope in the Ankobra estuary. This helped in generating a drainage network of the Ankobra estuary. The drainage network of the landscape, the researcher remodelled some part of the land use in Ankobra estuary, channelling away from the communities the excess water, which mostly causes floods. After the remodelling or the change model process, the impact model stage was reached. This stage ascertained whether the landscape model of the estuary designed has the ability to reduce risk.

4. Results and discussions

In an effort to reduce the flood risk in the area, the research adopted the interventionist part of the geodesign concept. Steinitz [15] explains that change model is the first stage of the intervention phase of geodesign, which deals with remodelling the landscape. That is the geodesigner changes or modifies some physical features, which have contributed to improper functioning of an area. In this regard, the drainage and building foundations were remodeled to ascertain their impact on risk levels (**Figure 3**). Risk levels for [28] analysis were extreme: 46,725 m^2, high: 701,525 m^2, moderate: 248,150 m^2 and low: 9,167,758 m^2.

This function of change model is the hallmark of geodesign. The first model element changed for evaluation was the foundation parameters of buildings in the Ankobra estuary. This is because risk is an interplay of a hazard and vulnerable elements together with their coping capacity as indicated by Bollin et al. [29]. The researcher considered a "what if" scenario such as what will be the outcome (risk) if all buildings were above the worst flood risk depth in the estuary, that is 88.39 cm for Sanwoma and 121 cm for Asanta. Because all buildings in the Ankobra estuary were below the flood depth. Nevertheless, if the buildings were constructed above a flood depth, then flood risk levels might differ. Evidence of this is seen in Nzulezu where houses are less risky to floods due to their construction above

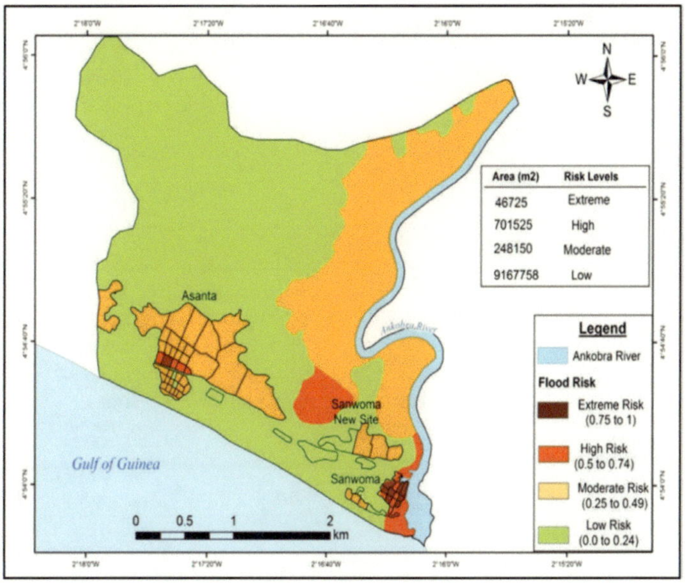

Figure 3. Flood risk of the Ankobra estuary. Source: Osman et al. [28].

the Amansuri water and wetland. Based on this, the researcher held all the variables of risk constant and changed the building foundation height of structures in the area. The second part of the geodesign interventionist phase (impact model) was then considered as it looks at assessing the effect of the change model. The spatial criteria evaluation (processes, problem definition, standardisation, weighting and slicing were considered) was therefore used to run the impact again. The output is depicted in **Figure 4**.

Extreme risk zone completely diminished to zero when compared with the original flood risk map, the high risk zone decreased by 53.27%, moderate risk zone gained by 169.42% and low risk remained the same. The changes in the new flood risk as against the original (flood risk) can be attributed to the fact that a building foundation above flood depth will prevent flood waters causing direct damages to buildings and their contents. This will have an enormous effect on physical vulnerability and therefore reduce flood risk areas.

A second change model undertaken was remodelling of the flow channel in the Ankobra estuary. The flow channels were identified as one of the factors causing floods in the study area. In Asanta community, the main cause of flood is blockage of excess rain water by the Axim-Elubo road, hence no proper drainage to carry excess run-off. In Sanwoma, a stream west of the community mostly causes floods as there are no flow outlets which would divert the water from the community. It was, therefore, important to generate flow channels in the study area (**Figure 5**) to know how they interconnect.

The result showed improper functioning of the water flow system in the Ankobra estuary, the flow in Asanta and Sanwoma. As such, the flow channels were modelled again by adjusting the land use and introducing a water outflow channel to connect the channels which were not

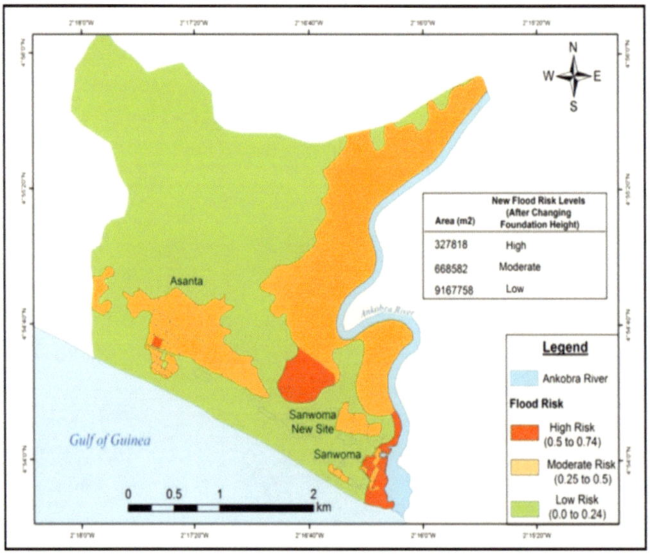

Figure 4. Flood risk reduced map after foundation parameters were increased above the worst flood depths.

linked and also to channel water from the settlements to the sea. In Sanwoma, a channel was constructed to link the Ankobra River to the stream west of the community, which sometimes causes flood during the rainy season. The length of the newly modelled channel was 98.23 m from the stream to the Ankobra River. Also an artificially closed lagoon was constructed to serve as the drain point for the two channels joined at the south west of Sanwoma. This is because a closed lagoon will prevent direct contact between the sea (waves and surges) and the channel water which can lead to flooding. The area of the closed lagoon modelled was

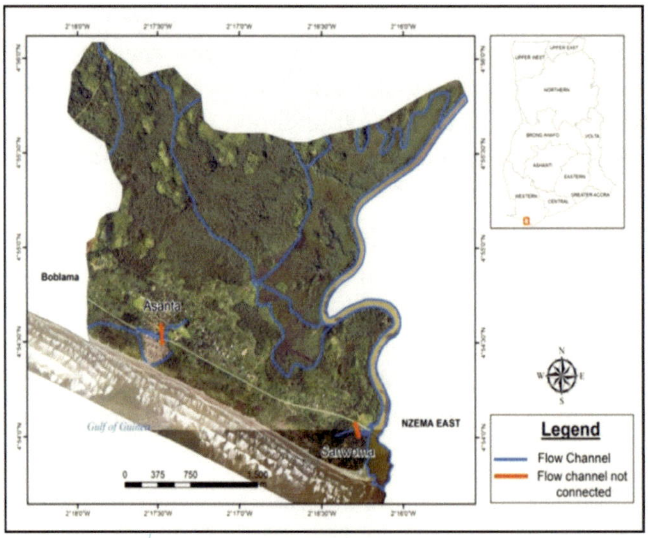

Figure 5. Flow channel of Ankobra estuary generated from DEM.

1347.85 m^2 with a depth of 242 twice the depth of the worst flood in Asanta. **Figure 5** shows the new modelled landscape with the channel. In Asanta, the main cause of flood is blockage of rain water by the Axim-Elubo road. In order to solve the problem, the researcher created a new channel within the community (**Figure 6**) from the northern part of the community to the southern part of the community. The length of the new channel in Asanta is 103.24 m.

The second stage of the intervention phase of geodesign was again employed where the new landscape was fed into the spatial multi-criteria evaluation to ascertain its impact. The impact assessment of the newly modelled landscape shows a new flood risk map with flood levels reduced (**Figure 7**).

The new flood risk zones from the remodelled landscape and drainage were high, moderate and low risk zones without extreme risk zones. Also, the area covered by these flood risk levels changed, the low risk zone remained the same in area size and moderate risk zone increased while high and extreme risk zones decreased. **Table 1** shows differences in area sizes of the risk levels before and after the geodesign application. The extreme flood risk area diminished from 46,725 m^2 to zero (0), whereas the high risk zone decreased by 72.23%. The moderate risk zone also gained 223.03%, whereas the low risk zone remained the same. In the outputs of the first change model (foundation) and this new change model (remodelled landscape and drainage system), the low risk zones remained the same because the researcher did not apply the interventions in these areas as they were already safe from flood risk.

It can be concluded that the impact assessment of the change models (foundations parameters and remodelled landscape and drainage) reduced flood risk levels in all the various risk levels. But as Fisher [18] remarks, after the last part of geodesign is the decision making stage where whether the impact was desirable and should be accepted or rejected by the geodesigner

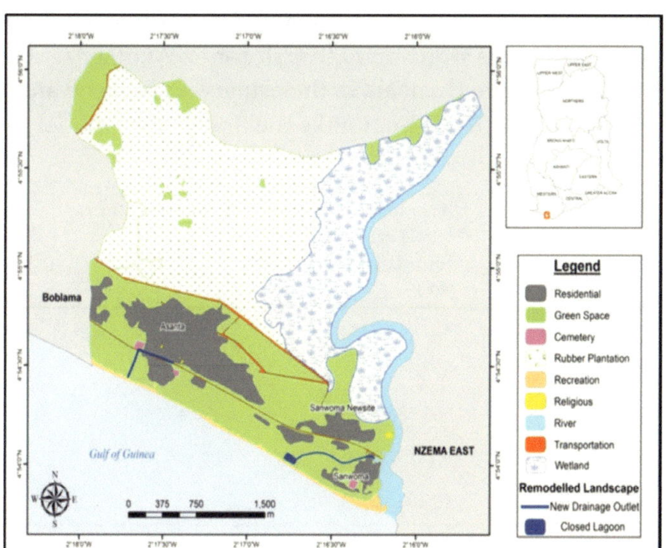

Figure 6. Remodelled landscape and the new channels.

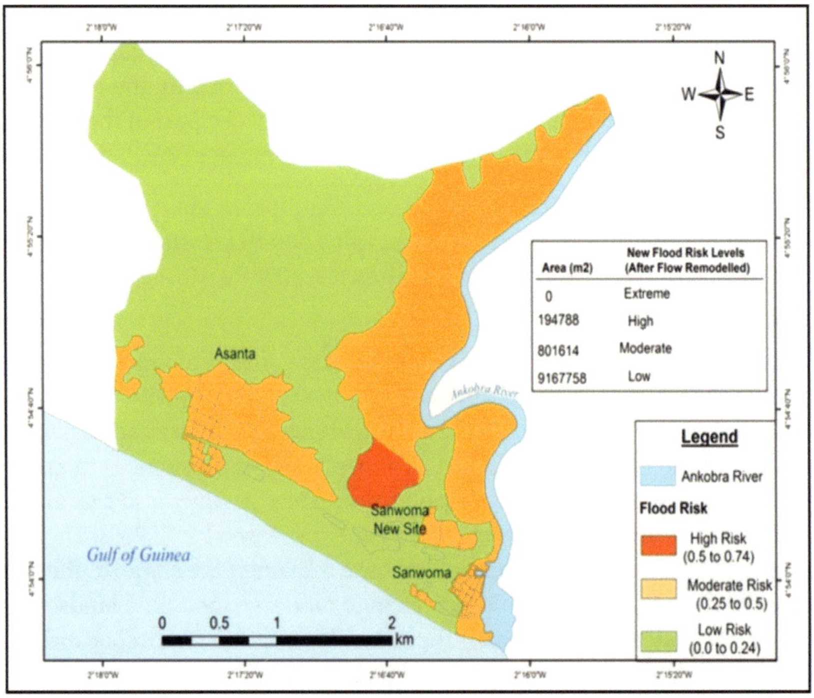

Figure 7. Flood risk reduced map after remodelled landscape and drainage.

is determined. Taking into account the two results (statistics of flood risk reduced map) in **Tables 1** and **2**, the researcher decided to adopt the results (flood reduced map) of the remodelled landscape and drainage system as the best intervention compared with remodelled foundation of buildings for the estuary since the flood risk reduction map from the change of building heights still had some wards in high-risk zones. Another reason for this decision is that it would be easier to create channels in the estuary since those areas where the channels to be created are not developed. Also, unlike the first change model (change of building

Flood risk levels	Area of flood risk levels before geodesign (m^2)	Area of flood risk levels after changed foundation parameter (m^2)	Differences (m^2)	Percentage change
Extreme	46,725	-	−46,725	−100
High	701,525	327,818	−373,707	−53.27
Moderate	248,150	668,582	+420,432	169.42
Low	9,167,758	9,167,758	-	-
Total	10,164,158	10,164,158		

Source: Osman et al. [28].

Table 1. Flood risk levels after change in foundation parameters.

Flood risk levels	Area of flood risk level before geodesign (m^2)	Area of flood risk level after geodesign (m^2)	Differences (m^2)	Percentage change
Extreme	46,725	-	−46,725	−100.00
High	701,525	194,788	−506,739	−72.30
Moderate	248,150	801,614	+2,927,044	223.03
Low	9,167,758	9,167,758	-	-
Total	10,164,158	10,164,158		

Table 2. Flood risk levels after change in drainage channels.

parameters), physically changing the building foundations laterally will mean demolishing buildings for new buildings with higher foundations to be put up. Most of the buildings had their foundations in the ground. Furthermore, this activity (physically changing the building foundations) will put much financial burden on inhabitants in the Ankobra estuary considering the fact that their socio-economic vulnerability is high. Analysis and results from the geodesign confirm Fisher's [18] assertion about geodesign capabilities and Dangermond [11] believe that geodesign can help men live in harmony with nature.

5. Conclusion and Recommendations

Geodesign is an innovative way of solving modern environmental problems in the face of climate change. This study adopted geodesign approach in testing its effectiveness as a flood risk reduction measure. Data employed were drainage channels, digital elevation model, land use, building parameters. Analysis was based on changing parameters of building foundation and drainage channels. Results showed, all change models generated better results and were effective in reducing flood risk levels. The downside of geodesign is where physical developments have fully taken place as implementing results might be difficult. In this research, changing building foundations provided the best option to reducing flood risk. However, buildings are structural entities which will be difficult to change in the shortest time to get the results of the model. The study recommends that planners should be made to adopt geodesign frameworks and models before developing physical plans. This is to give them a better understanding of the likely effects of their plans. It will also help in integrating disaster management into their physical plans to make our communities less risky to flood and other hazards.

Author details

Adams Osman and Benjamin Nyarko*

*Address all correspondence to: bnyarko@ucc.edu.gh

Department of Geography and Regional Planning, University of Cape Coast, Ghana

References

[1] Ashmolea Museum. The Sumerian King List. Oxford, Cuneiform Digital Library: University of Oxford; 2011.

[2] International Strategy for Disaster Reduction. Global Assessment Report on Disaster Risk Reduction. Geneva: United Nations Office for Disaster Risk Reduction. Switzerland: UNISDR Publications; 2011.

[3] Nicholls R, Wong P. Climate Change 2007: Impacts, Adaptation and Vulnerability. Cambridge: Cambridge University Press; 2007.

[4] Guha-Sapir D, Below R, Hoyois Ph. EM-DAT: The CRED/OFDA International Disaster Database—**www.emdat.be**—Université Catholique de Louvain—Brussels—Belgium

[5] McMillan HK, Brasington J. End to end flood risk assessment: A coupled model cascade with uncertainty estimation. Water Resources. 2008;**26**.

[6] Musiake K. Regional characteristics of hydrology and water resources in monsoon Asia. In: Proceeding, First International Conference on Hydrology and Water Resources in Asia Pacific Region, Kyoto; 2003, pp. 1-8

[7] Reid R. Levee leader. Civil Engineering; 2007;**77**(4):36-43

[8] Corder M. Netherlands Dike Break Fears Force Hundreds to Evacuate. 2012. Available from World Post: http://www.huffingtonpost.com/2012/01/06/netherlands-dike-break_n_1188980.html [Retrieved: March 2014]

[9] Foley J. New Problems Call for New Solutions. 2013. Available from Ensia: http://ensia.com/voices/new-problems-call-for-new-solutions [Retrieved: 15 November 2013]

[10] Yu JK. Water Is a Living Archive: Examining Myths of Where Various Urban Streams Come from- Kellog Creek. 2014. Available from Geodesign Summit 2014: http://www.lacreekfreak.wordpress.com/author/nothing301 [Retrieved: July 2014]

[11] Dangermond J. Geodesign. 2010. Available from ESRI: http://video.esri.com/watch/125/jack-dangermond-talks-about-geodesign-at-ted2010 [Retrieved: March 2013]

[12] Miller W. Introducing Geodesign: The Concept. New York: ESRI Publications; 2012.

[13] McElvaney S. Geodesign: Case studies in Regional and Urban Planning. New York: ESRI press; 2012.

[14] Dutta D, Tingsanchli T. Development of Loss Functions for Flood Urban Flood Risk Analysis in Bangkok. Tokyo: New Technologies for Urban Safety in Asia; 2003.

[15] Steinitz C. A framework for theory application to the education of landscape architects. Landscape Journal. 2010;136-143

[16] Mathur R. Geodesign: Changing geography by design. London: ESRI; 2010, pp. 4-1.

[17] Flaxman M. Fundamentals of Geodesign. California, United States of America: Esri Geodesign Summit; 2009.

[18] Fisher T. The What and Why of Geodesign. 2010. Available from Beta Noodle: http://www.noodle.org/learn/194368/geodesign-summit-2010-tom-fisher-the-what-and-why-of-geodesign-part-2-of-3 [Retrieved: 18 June 2014]

[19] Goodchild F. Toward geodesign: Repurposing cartography and geographic information systems. North American Cartographic Information Society; 2010;7-23.

[20] Abukhater A, Walker D. Making Smart Growth Smarter with Geodesign. Redlands, California: ESRI Publications; 2010.

[21] Hydro Response Limited. Flood Barriers from Geodesign Barriers. 2008. Available from Innovative Water Containment Solutions: http://www.hydroresponse.com/flood_barrier.htm [Retrieved: April 2014]

[22] Snyder K, Lally J. Communityviz Gallery. 2009. Available from CommunityVIZ: http://placeways.com/communityviz/gallery/casestudies/pdf [Retrieved: 24 June 2014]

[23] Ball M. China Embraces Geodesign for Ecological and Beautiful Urbanization. 2014. Available from Asian Surveying and Mapping: http://www.asmmag.com/2012-12-30-14-40-18/feature/6253-china-embraces-geodesign-for-ecological-and-beautiful-urbanization.html [Retrieved: 4 July 2017]

[24] Dempsey W. Making water work. Contractor; 2010;**3**:34.

[25] Ervin, S. Geodesign futures: Nearly 50 predictions. Digital Landscape Architecture Conference, Bernburg, Germany; 2012.

[26] Tanu, M. Seasonality of recession time and lag response to precipitation of the Ankobra and Pra rivers in Ghana, West Africa. 28th Conference on Hurricanes and Tropical Meteorology, Orlando, Florida; 2008.

[27] Ghana Statistical Service. Ghana Living Standard Survey: Report of the Fifth Round. Accra, Ghana: Ghana Statistical Service; 2010.

[28] Osman A, Nyarko B, Mariwah, S. Vulnerability and risks levels of communities within Ankobra estuary of Ghana. International Journal of Disaster Risk Reduction; 2016, 133-144

[29] Bollin C, Cardenas C, Hahn H, Vasta K. Natural Disaster Network: Comprehensive Risk Management by Communities and Local Governments. Washington, DC: Inter-American Development Bank Report; 2003.

Chapter 6

An Additive Statistical Modeling Approach to the Analysis of Transport Infrastructure Flood Risk-Based Resilience

Mohammad Mojtahedi, Sidney Newton and
Faham Tahmasebinia

Additional information is available at the end of the chapter

http://dx.doi.org/10.5772/intechopen.69232

Abstract

Australia is a very vulnerable region to flood events, and the frequency of flood events and damage has increased dramatically over the past decades. Although flood has impacted diverse types of buildings and built infrastructure, there has been limited research investigating flood risk management specific to transport infrastructure in Australia and the factors that might influence the resilience of the transport infrastructure to flooding. To develop an appropriate design management system for roads and bridges specific to risk assessment from flooding requires a multitude of factors to be identified and analyzed. In this study, we review the range of critical factors necessary to represent the resilience of bridges to extreme flood events and demonstrate a novel mathematical approach to evaluate the relationship between the bridge resilience and flood risk. We use additive statistical approach in arriving at a framework to evaluate the resilience of bridges. The findings confirm that metrological characteristics such as annual exceedance probability and probable maximum precipitation and structural integrity of the bridge represented by the structural age of the bridge and mechanical properties of the soils have a substantial impact on the resilience of the Australian transport infrastructure, particularly bridges located on main roads.

Keywords: resilient, risk, flood, transport infrastructure

1. Introduction

Like most countries, Australia is hugely susceptible to flood damage. The frequency of extreme flood events in Australia has increased dramatically over the recent years, and the economic loss associated with individual flood events has also increased [1]. For example, in 2010/2011, floods in Queensland, Australia, had a devastating impact on almost 10,000 km of road network and 5000 km of rail network, and severely damaged around 100 significant bridges and culverts, 400 schools, and 150 national parks [2]. In 2013, a localized flood near Brisbane, Australia, damaged 43 of the total 46 bridges in that location. On average, Australia will experience a roughly 300-fold increase in flooding events by 2100, meaning that infrastructure that is presently flooded once in 100 years will be flooded several times per year with a modest prediction of sea level rise (50 cm) [3]. There has been limited research investigating flood damage and flood risk management specific to infrastructure in Australia [4, 5]. However, previous research has tended to focus on building damage, and there has been no previous consideration of flood impact on Australian transport infrastructure and/or the factors that might influence the resilience of the transport infrastructure to flooding [6]. Transport and associated infrastructure such as roads, railways, bridges, warehouses, airports, ports, and tunnels are often exposed to the risk of direct damage from climate events. Transport infrastructure also has a vital role before, during, and after extreme weather events to reduce the vulnerability of the community more generally. Transport infrastructure vulnerability will vary by region, location, elevation, and the underlying condition of the infrastructure [7].

Elevated ocean levels can increase flood levels in the lower reaches of rivers, either by preventing floodwaters from discharging into the ocean or by filling up low-lying land and estuarine flood storage areas before the river flooding arrives. River flooding in the vast flat areas may last for one or more weeks, or even months on some occasions in some regions in Australia and can lead to extensive damage to rural towns, rural and urban roads, and rail networks [8]. In Australia, low-lying coastal areas can be inundated by storm surges, usually caused by tropical cyclones. Overflow of drainage systems in populated urban areas can also be a major problem.

The performance of transportation networks is critical for the economy and society. Maintaining structural reliability and functionality of the network under extreme hazard effects is therefore a key consideration. In any transport network, bridges and culverts are often the most vulnerable elements because of their propensity for catastrophic collapse [3, 7]. The potential economic and social disruptions due to the loss of or damage to transport infrastructure bridges are hugely significant. The possible adverse effects of increased flooding on bridges range from potential catastrophic structural collapse to increased frequency of maintenance required to ensure a minimum level of service [9]. Given that the most common form of flooding in Australia is river flooding [10] and most bridges in Australia are constructed over the rivers, it is essential to identify bridges that may be vulnerable to flood events in order to mitigate the risks associated with extreme weather events most effectively.

Effective flood risk management is aimed at increasing the resilience of transport infrastructure, but needs to balance the costs of mitigation against the benefits offered by increased resilience. Each situation and the specific community values will vary. In a general case, regular account might be taken of the following flood risk management measures for bridges:

1. The potential for and impact of ground liquefaction around bridges.
2. The slope and stability of associated riverbanks.

3. The structural performance of the bridge itself in standard risk evaluation terms.
4. A benchmark classification of failure probability across high, medium, or low probabilities.
5. A common rating measure applied to each bridge that combines several key factors to assess the vulnerability of the particular bridge structure.

Having an efficient and effective measure for the resilience of individual bridges is important to transport authorities, local councils, and emergency services. Efficiency in terms of data collection and calculation is important because there are a large number of bridges, in continuously varying circumstances, which require constant evaluation and reevaluation in normal service, the aftermath of an event, a near miss, and even following an event elsewhere that impacted a similar bridge. Nowhere is the need for an efficient and effective measure of resilience more pronounced than in the design and design management guidelines for bridges.

To develop or inform an appropriate design management system for bridges specific to risk assessment from natural hazards such as flooding requires a multitude of factors to be identified and analyzed. A recent analysis of the current design standards for bridges highlighted shortcomings specific to the resilience of bridges [11]. However, the more factors considered in design models, the greater the complexity of the design models are and the less practical the guidelines are in interpreting the resilience [12]. A bridge subject to an extreme flood event can be damaged in many different ways. For example, where the structure of a bridge is completely inundated during a flood, the damage to the bridge depends on the length of time it is submerged.

In large part, due to the potential complexities associated with evaluating the resilience and flood risk of a bridge, mathematical evaluation of bridge designs is missing from previous research. In this study, we review the range of critical factors necessary to represent the resilience of bridges to extreme flood events and demonstrate a novel mathematical approach to evaluate the relationship between the bridge resilience and flood risk.

2. Critical factors for bridge resilience and flood risk

Bridges are designed to provide an extended period of service and be resilient to particular extreme weather events. There are many and varied factors taken into account in the design of a bridge, including structural system, loading, resistance, exposure, natural environment, soil characteristics, planned maintenance, inspection regime, economic considerations, and various deficiency functions (load capacity, vertical clearance, deck width, etc.) [13]. However, the condition and performance of a bridge will vary over time in response to various environmental changes, actual maintenance practice, and changes in the quantity and magnitude of actual loads applied [12].

According to the Intergovernmental Panel on Climate Change (IPCC), a flood is "the overflowing of the normal confines of the stream or other body of water, or the accumulation of water over areas that are not normally submerged" [3]. Floods occur more often than many other types of natural disasters [14] and are generally regarded as the most catastrophic and lethal of all natural disaster types [15]. During the past century, floods have killed something in the order of 8 million people, globally [16]. Approximately 800 million people currently live in flood-prone areas across the world, and almost 10% of those are exposed to floods on average each year [17, 18].

Not all floods are of equal consequence. Leroy [19] states that time, area, and socioeconomic characteristics tend to amplify the impact of natural disasters. Ho et al. [20] show that the type of natural disaster is a good predictor of the scale of damage likely to be incurred as a direct consequence. Merz et al. [21] propose that flood type, the flood-generating process, the region or zone, and frequency are the most important characteristics to consider in predicting the impact of a flood disaster. A comprehensive literature review on flood characteristics and their relationships with flood damage undertaken by Middelmann-Fernandes [22] found that various parameters contribute to bridge damage: the depth of water, flow velocity, duration of inundation, contamination, sediment or debris load, and the age and materials of the bridge itself. A number of studies have also found that flood type, severity, and frequency are the three most useful predictors of the damage caused by flooding [3, 23, 24].

The management of flood risk reduction and flood disaster management bring further factors into consideration [25]:

- The behavior of the flood itself, including the immediate dangers and potential damage caused.
- The broad cost of flooding borne by the community.
- The projected use of the land in the future and possible lost opportunities due to potential for flooding.
- The environmental needs and impact of the source river and related floodplain areas.
- The environmental and cultural impact of mitigation measures.

This expanding scope of factors to consider around flood risk management renders the problem as a massively multidisciplinary issue. It involves high level of skills in planning, engineering, the social and environmental sciences, economics, and emergency management.

In addition to the breadth of skills required, flood risks vary depending on the temporal perspective taken. Perspectives change from a focus on existing risks, to ongoing risks, and future risks. For example, existing risks deal more specifically with the management of flood damage to maintain or return the community, properties, and infrastructure as it currently exists. Future risks deal with the potential for damage to areas not yet, but in the future possibly, developed to become of economic, social, or environmental significance. Ongoing risks target the potential local impact of generally external flood risks, where floods elsewhere might overwhelm the local infrastructure.

Flood types themselves can also vary. Risks are different where the cause of the flooding is a river blockage (fluvial), excess rainwater (pluvial), coastal inundation, lake blockage, sewer overflow, sudden downpour (flash flood), and/or storm-water drainage (urban floods). In Australia, the most common form of flooding is along rivers after heavy and/or prolonged rainfall [8]. Jonkman and Kelman [26] showed that flood damage in the United States is highest in areas near rivers, particularly in areas with deep water rivers or rivers at lower elevations. Nicholls and Tol [27] noted that the annual mean and probable maximum rainfall is other important flood characteristic to predict the potential flood damage to roads and bridge structures. However, there is strong evidence that the greater the resilience of an individual bridge or road, the lower the likely economic damage caused by floods, and that the benefits of resilience can be dramatic [5].

Godschalk [28] defined resilience as "the ultimate objective of hazard mitigation, that is, action taken to reduce or eliminate long-term risk to people and property from hazards and their effects." Other scholars have adopted and adapted this definition differently across different disciplines and currently there is no generally accepted definition. For example, Levina and Tirpak [29] proposed two key aspects for resilience: the ability of a system to withstand a disturbance without changing, implying that no damage is done; and the ability of the system to recover when damage has occurred. Further, Maguire and Cartwright [30] articulated three views of resilience: resilience as stability and resistance to change; resilience as recovery following change and return to the previous stable state; and resilience as transformation to a new and different stable state following change. Recently, Manyena et al. [31] defined resilience as the "intrinsic capacity of a system, community or society predisposed to a shock or stress to bounce forward and adapt in order to survive by changing its nonessential attributes and rebuilding itself," and again, Mojtahedi et al. [5] defined resilience as the risk reduction practices undertaken before potential disasters both to minimize the probability of the disaster occurring in the first place, as well as to enable repair and replacement during and after disasters to result in improved facilities over time. Finally, it is important to note that the concept of resilience is broader than natural disasters. Resilience also encompasses the capacity of public, private, and civic sectors to resist disruption, absorb disturbance, act effectively in a crisis, adapt to changing conditions and grow over time [32].

Park et al. [33] summarized the key challenges facing a more coherent and robust approach to resilience and risk management in the design of infrastructure which are as follows:

- A lack of comprehensive data and forecasting uncertainty both promote a focus on those aspects of risk where data does exist and the hazards are better known. As a result, very significant risk factors are ignored or discounted and the resulting risk assessment can be severely misleading.
- Current infrastructure design is based on the prevailing building codes and regulations. These codes develop and evolve slowly over time. The codes can fail to incorporate significant emergent hazards or respond effectively to new lessons from disaster events. This lag has resulted in the incremental evolution of design, rather than supporting more innovative design responses.
- The many and varied design challenges that do arise are treated as individual problems to be resolved, rather than conditions to be managed more holistically and proactively. The broad-ranging and multifaceted nature of infrastructure design warrants a comprehensive approach, but this is in direct conflict with the needs of a pragmatic decision-making process that is flexible and responsive.

The framework set out in this study will not establish parameters for detailed engineering design, but it will address some of the challenges outlined above. Most particularly, the framework will develop an approach to the evaluation of resilience that is flexible and responsive to the different situations and circumstances in which risk-assessment decisions must be made. The additive statistical method it demonstrates is a novel approach to risk assessment more generally, but is especially suited to the complexity of resilience and risk management in the design of infrastructure. The study will focus on the resilience of bridges as pivotal components in any transport network, but given the increasing frequency and severity of extreme weather events, as critical concerns for communities from major cities through to remote rural settlements.

3. An additive statistical model

When seeking to analyze a complex problem comprising a range of variables, the common approach is to apply a parametric regression method. Parametric regression defines a function in which the terms comprise a finite number of unknown parameters derived from numerical data on each of the variables of interest. In the context of transport resilience and risk management, which is highly dependent on the vagaries of individual situations, the variables of interest can vary significantly between situations. In such circumstances, regression is better defined in nonparametric terms across a set of functions. Originally proposed by Friedman and Stuetzle [34], the additive model method offers a robust and simple to interpret approach to the effect analysis of multiple variables. The additive model takes the form of a familiar regression model, but builds each model from a restricted class of nonparametric regression models. Each nonparametric model uses a one-dimensional smoother to generate linear combinations of the predictor variables in an iterative fashion.

The additive modeling approach provides distinct advantages over alternative nonparametric approaches and is entirely more general than standard stepwise regression procedures [34]. The additive model approach results in a regression model, but the relationship between each variable and the response is allowed to be flexible and/or linear in nature, as indicated in the following formula:

$$\sum_{j=1}^{n} f_j(x_j) = \sum_{j=1}^{n} \beta_j \times x_j^{\ n} \tag{1}$$

where $f(x)$ represents a linear or nonlinear relationships between phenomena.

Based on Eq. (1), we develop an additive statistical equation for analyzing transport infrastructure flood specific to bridge risk-based resilience as follows:

$$r = \alpha + \sum_{j=1}^{p} f_j(x_j) + \epsilon \tag{2}$$

r is the resilience of a bridge to a flood event, α is intercept, $f(x)$ is linear or nonlinear function between the response and the relevant indicator, and ϵ is the overall error of the model.

Additive models have the strong properties of linear or nonlinear models in so far as they are in a familiar regression form and easy to interpret, but are superior in that they relax the assumption of a linear (or known nonlinear) relationship in the data. Thus, the additive model approach is not a purely nonparametric method (which is one of the potential limitations of the proposed framework), but does represent an effective compromise between flexibility and simplicity.

To begin to build an additive model the first challenge is to identify the principal components. Choice of the principal component is dependent on the range of candidate components and the availability of data. Nonparametric regression does require larger sample sizes than parametric models because the data must support the development of a model structure as well as supply the model estimates. For the purposes of this study, a representative set of four principal components is used, but the same framework and approach can be applied to an unlimited number of principal components where relevant data are available. The principal components used in this study of resilient bridges within the context of flood risks are as follows:

x_1: likelihood of a flood event represented by the annual exceedance probability (AEP).

x_2: scale of a flood event represented by the probable maximum precipitation (PMP).

x_3: structural integrity of the bridge represented by the structural age of the bridge.

x_4: mechanical properties of the soils.

r: resilience of bridge.

3.1. Annual exceedance probability (AEP)

There are two ways of expressing the likelihood of occurrence of a flood event, annual exceedance probability (AEP) and average recurrence interval (ARI). AEP is the probability of a flood of a given or larger size occurring in any one year, usually expressed as a percentage. For example, if a peak flood discharge (PFD) of 500 m³/s has an AEP of 5%, it means that there is a 5% probability (that is one-in-20 chance) of a 500 m³/s or larger event occurring in any one year. ARI is the long-term average number of years between the occurrences of a flood as big as or greater than the selected event. For example, floods with a discharge equal to or greater than the 20-year ARI flood event will occur on average once every 20 years [25].

AEP is selected in this instance, as the data is readily available in this form. The relationship between bridge resilience and AEP is then formulated as follows:

$$y_1 = a_1 \times Ln(x_1) + b_1 \tag{3}$$

where a_1 and b_1 are the constant factors. Since there is a linear relationship between y_1 and (x_1), the possible linear regression can be determined by the following equations.

$$a_1 = \left[\frac{\sum_{i=1}^{n} (y_i - \bar{Y}) \times \left((Ln(x_i)) - \bar{X} \right)}{\sum_{i=1}^{n} \left((Ln(x_i)) - \bar{X} \right)^2} \right] \tag{4}$$

Intercept $b_1 = \overline{Y} - a_1 \times \overline{X}$, where $\overline{Y}$ is the mean value of flood risk-based resilience in different cases, and $\overline{X}$ is the mean value of the AEP in different cases. Hence,

$$\overline{Y} = \left[\frac{\sum_{i=1}^{n} y_i}{n}\right] \text{ and } \overline{X} = \left[\frac{\sum_{i=1}^{n} Ln(x_i)}{n}\right] \quad (5)$$

3.2. Probable maximum precipitation (PMP)

World Meteorological Organization defined probable maximum precipitation (PMP) as "…the theoretical maximum precipitation for a given duration under modern meteorological conditions." PMP is specific to a given storm area at a particular location at a particular time of the year, with no allowance made for long-term climatic trends. The relationship between the flood risk-based resilience and PMP can be established by the following equation:

$$y_2 = a_2 \times Ln(x_2) + b_2 \quad (6)$$

where a_2 and b_2 are the constant factors. Since there is a linear relationship between y_2 and (x_2), the possible linear regression can be determined by the following equations:

$$a_2 = \left[\frac{\sum_{i=1}^{n}(y_i - \overline{Y})\ \left((Ln(x_i)) - \overline{X}\right)}{\sum_{i=1}^{n}\left((Ln(x_i)) - \overline{X}\right)^2}\right] \quad (7)$$

Intercept $b_2 = \overline{Y} - a_2 \times \overline{X}$, where $\overline{Y}$ is the mean value of flood risk-based resilience in different cases, and $\overline{X}$ is the mean value of the PMP in different cases. Hence,

$$\overline{Y} = \left[\frac{\sum_{i=1}^{n} y_i}{n}\right] \text{and } \overline{X} = \left[\frac{\sum_{i=1}^{n} Ln(x_i)}{n}\right] \quad (8)$$

3.3. Structural age of the bridge

The structural integrity of a bridge is an obvious and critical consideration in any risk assessment. Structural integrity is directly influenced by the age of the structure, but the influence is different for different construction materials. Given the rich mix of materials used in bridge construction in Australia, and especially the preponderance of heritage bridges, a mathematical relationship between bridge structure age and the resilience of the bridge against flooding is developed. The structural age of a bridge depends heavily on the structural materials used. A range of materials and combinations of material are possible. For the purposes of this framework, the materials to be considered include brick, mortar, and cast iron.

3.3.1. Brick

One of the first tasks undertaken by the first European settlement in Australia was to search for suitable clay for brick manufacture. This was quickly found in an area that has become the southern sector of Sydney's central business district. This was in production for about

50 years until urban growth, and diminishing deposits meant that alternative sources had to be discovered. There is some irony in that these areas, which provided such large quantities of building material, were also those where seasonal swelling and shrinkage of foundation soils were to cause damage to low-rise buildings. One of the early problems was in attaining a sufficiently high temperature in the kilns in these early stages of the industry. It is clear that a range of quality of brick was produced and a careful inspection process would then select the best ones for higher-level work.

3.3.2. *Mortar*

Engineers' customarily first think of mortar as the component of structural concrete other than the coarse aggregate. This uses Portland cement, but the use of this type of cement mortar for the repair of heritage structures has frequently proved to be disastrous. One reason is that the low permeability of this type of mortar does not allow extrusion of fluid build-up within the main structural components, whether they are brick or stone. This means that there is a deterioration of these components even before the mortar, which is intended to be regarded as expendable and replaceable in a much shorter life span. In order to eliminate this problem, it is necessary to revert to the once-popular use of lime-based mortar—or partly so—which has a greater porosity, especially if the proportion of sand is much higher than is usual for concrete. One dictum is to aim for a permeability of mortar that matches that of the bricks. Another attribute of lime mortar is that is more elastic than Portland cement mortar which means that, under various types of stress, the mortar will yield rather than the greater strength of Portland cement mortar causing the main material to fracture, if weaker.

3.3.3. *Cast iron*

The use of blast furnaces in the reduction of iron oxides to some form of usable iron results in pig iron. This metal has a carbon content range of from 2.7 to 4.0%, but includes a large proportion of impurities. Such a material would have been of practical use in those early societies capable of its production, but the arrival of the industrial age demanded a more refined and consistent product. A standard method of bringing this about was by a more cautious process of reheating in a "cupola" furnace. This resulted in iron with carbon content in the middle of the above range, but still retaining useful quantities of silicon and manganese. Tests have been carried out on cast iron used in the joint connection units in New South Wales (NSW) timber truss bridges of 100 years ago. The carbon content varied from 3.0 to 3.3% with the former material having the slightly greater strength. This traditional material is referred to as "gray" cast iron in order to distinguish it from later types with improved properties. It should be pointed out that variants of gray iron are still widely used in industry although much less in civil engineering as primary components. Gray cast iron has a compressive strength in the range of 600–700 MPa. Its other main characteristic, however, is that its tensile strength is only about one-fifth of the compressive strength. What this means is that manipulation of the second moment of area of an "I" section in the interests of the economy produces a grotesquely large tension flange both in width and in thickness. As well as being a quick identifier of the material used, this section characteristic can be put to good use in that the extra-wide bottom flanges, as in the case of floor joists, can be used as the springing for shallow brick arches between adjacent joists, thereby providing a floor support structure ("jack arches"). While gray cast iron has an excellent resistance to corrosion, it has a disadvantage in its brittleness. This made it difficult to cope with the high impact nature of railway loadings as early railway engineers discovered: often the hard way. As well as its attempted use in bridges, cast iron was even used for short sections of railway track rails, and the high resulting fracture rate at least had the result of encouraging the larger scale production of wrought iron and the development of rolling mills. The main success of cast iron was in its use of columns, where the stress is primarily in compression throughout. There is a consensus against the repair of cast iron by orthodox welding technique, but there are proprietary methods available, developed from technologies used for machinery castings failures. A hazard sometimes found with cast iron is graphitization, often occurring where there is immersion in or contact with salt water and where there is a reaction with the carbon streaks within the iron. An example of this took place in the 1874 cast iron piers of the Windsor Bridge near Sydney. Although the bridge is a long way upriver, the tidal cycle of the Pacific Ocean with reduced outflow below dams has gradually increased the saline content.

The following equation would be suggested to evaluate the relationship between the resilience of bridge against flood and bridge structure age. We assume that bridge structure age is highly dependent on bridge materials such as brick, mortar, and cast iron. The relationship between the flood risk-based resilient and bridge structure age can be established by the following equations:

$$y_3 = a_3 \times x_3{}^2 + b_3 \times x_3 + c_3 \tag{9}$$

where a_3, b_3, and c_3 are the constant factors. The matrix equation for quadratic regression is:

$$\begin{bmatrix} \sum_{i=1}^{n} x_i^4 & \sum_{i=1}^{n} x_i^3 & \sum_{i=1}^{n} x_i^2 \\ \sum_{i=1}^{n} x_i^3 & \sum_{i=1}^{n} x_i^2 & \sum_{i=1}^{n} x_i \\ \sum_{i=1}^{n} x_i^2 & \sum_{i=1}^{n} x_i & n \end{bmatrix} \times \begin{bmatrix} a_7 \\ b_7 \\ c_7 \end{bmatrix} = \begin{bmatrix} \sum_{i=1}^{n} x_i^2 y_i \\ \sum_{i=1}^{n} x_i y_i \\ \sum_{i=1}^{n} y_i \end{bmatrix} \tag{10}$$

where n is the number of data points (x_i, y_i).

3.4. Mechanical properties of the soils

Determination of the mechanical properties of the soil as well as the level of the pore pressure is major aspects in the current statistical simulations. Provided that there are n layers of different soils have been reported in the different suggested sites, the simulated values for both the horizontal as well as bending stresses in different layers can be calculated by the following procedure. It should be noted that a linear strain distribution was assumed across the soil section.

The suggested procedure can be used to find a relationship between the collected data and the vulnerability of the bridges. It should be interesting to investigate that there is an inverse relationship between the mechanical properties of the soils and the damage level [35]. The reasons for this could be due to different construction methods adopted in the past or they had been rehabilitated after the previous disaster event. However, these reasons should be further scrutinized for confirmation. The resilience of the community depends on the resilience of the bridges on rural roads, but most of the bridges are vulnerable to extreme flood events. Therefore, when classifying roads for design, it is necessary to consider the impact on the community during and after an extreme event [36]. Thus, the relationship between the flood risk-based resilient and bridge structure age can be established by the following equation:

$$y_3 = a_4 \times e^{b_4 \times x_4} + c_4 \times x_4 + d_4 \tag{11}$$

where a_4, b_4, c_4, and d_4 are the constant factors.

3.5. Resilience of transport infrastructure

Globally, infrastructure is recognized as a critical element for strong economies and stable communities and it facilitates society's daily activities. Infrastructure can be considered to include everything from the physical infrastructure of roads, bridges, airports, rail, water supply, telecommunications and energy services, to the social infrastructure of health care, education, banking and finance services, emergency services, and the justice system [32]. Croope [37] states "Critical infrastructure not only responds to the needs of society for the smooth daily continuation of activities, but also provides the basis on which society exists and relies." Godschalk [28] lists two reasons behind the importance of resilience which are as follows:

1. Because the vulnerability of technological, natural, and social systems cannot be anticipated, the capability to accommodate change without catastrophic failure in times of disaster is vital.

2. People and property fare better in resilient cities when struck by disasters. Fewer buildings collapse, fewer power outages occur, fewer businesses are put at risk, and fewer deaths and injuries occur.

Societies have an increasing reliance on transportation networks for their day-to-day activities. The ability of the transport system to function during adverse conditions and quickly recover to acceptable levels of service after an event is fundamental to the wellbeing of people within society. Resilience is considered the ultimate objective of hazard mitigation, that is," action taken to reduce or eliminate long-term risk to people and property from hazards and their effects" [28].

Resilience has two main dimensions such as (i) technical and (ii) organizational. Technical dimension refers to the ability of the physical system(s) to perform to an acceptable level when subject to a hazard event [38]. In this research, we have focused on technical dimension of the resilient bridge, and therefore, we have considered four technical factors for resilient bridges such as (i) AEP, (ii) bridge structure age, (iii) mechanical properties of the soils, and (iv) PMP.

Some scholars have applied to measure the resilience of transport infrastructure. For example, Mojtahedi et al. [5] used Cox's proportional hazards regression model to determine the rate of recovery and cumulative probability that recovery occurs for transport infrastructure across regional areas in New South Wales, Australia.

4. Numerical example

4.1. Resilience of a bridge and AEP

We have assumed that the resilience of bridge is dependent on the relationship between peak flood discharge (PFD) and annual exceedance probability (AEP). A numerical example can be introduced by the following assumed data, provided that there is a possible collected data between the PFD (m^3/s) versus AEP, which has been presented in **Table 1**.

Thus, the following equation $y_1 = a_1 \times Ln(x_1) + b_1$ can be represented by $y_1 = 0.045 \times Ln(x_1) + 0.0233$. Furthermore, based on **Figure 1**, it can be observed that after reaching to PFD around 500 (m^3/s), the recorded AEP would be converged to 0.3 which can be taken into account as a potential critical value.

A same trend to examine the mutual interaction between the PFD versus AEP has been investigated by using *t*-test. After normalizing, the respond data, which is the AEP, the *t*-test values, and distribution can be presented (**Figure 2**). The same conclusion can be found that after reaching to the PFD 500 (m^3/s), the value for the AEP would be a constant value. This can show that the resilient respond after the PFD 500 (m^3/s) would be slightly different from the further increments.

4.2. Resilience of a bridge and PMP

We have assumed that the resilience of bridge is dependent on the relationship between the size of storm area and probable maximum perception (PMP). The PMP can be a function of a given size storm area at the particular location at the particular time of the year. Provided that **Table 2** presents the assumed size storm area (km^2) versus PMP, thus, the possible mathematical relationship can be $y_2 = a_2 \times e^{-b_2 \times x_2} + c_2 \times x_2 + d_2 +$.

Thus, we can have $y_2 = -0.00766 \times e^{-0.00288 \times x_2} + 0.000544 \times x_2 + 0.00022$. Based on **Figure 3**, there is a semi-linear relationship between the PMP and the size storm area.

Normalized PMP values versus *t*-test distribution are also illustrated in **Figure 4**.

PFD (m^3/s)	100	200	300	400	500	600	700	800	900	1000	1100	1200
AEP	0.23	0.26	0.27	0.29	0.30	0.311	0.32	0.324	0.329	0.334	0.338	0.342

Table 1. Peak flood discharge (PFD) versus annual exceedance probability (AEP).

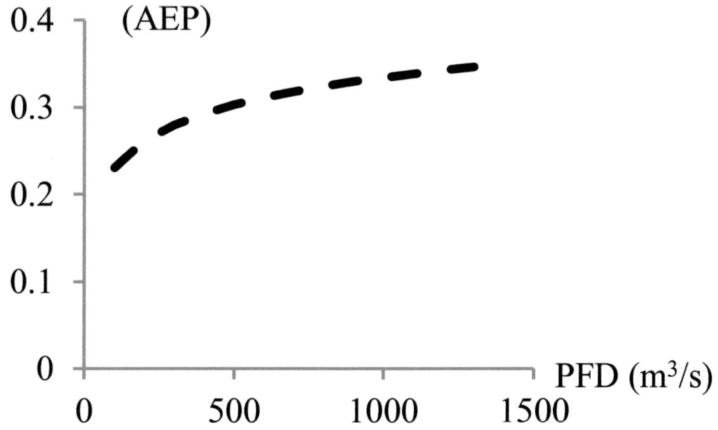

Figure 1. Peak flood discharge versus AEP.

As it can be realized in both **Figures 3** and **4**, the critical value of the PMP can be observed where there is size storm area over 60 (km^2). This is a condition where the PMP rate is more inclined with a constant rate rather than increasing sharply. Based on the numerical assumption, after size storm area over 60 (km^2) might be classified as a possible significant region/condition.

4.3. Resilience of a bridge and structure age

We have assumed that the resilience of bridge is dependent to the relationship between the bridge structure age and bridge collapsing risk. At this stage, some of the critical materials that can perform as a key role on constructing heritage roads/bridges were discussed in Section 3.3. This topic is significantly important where some of the heritage link constructions might be jeopardized by the induced flood around the NSW.

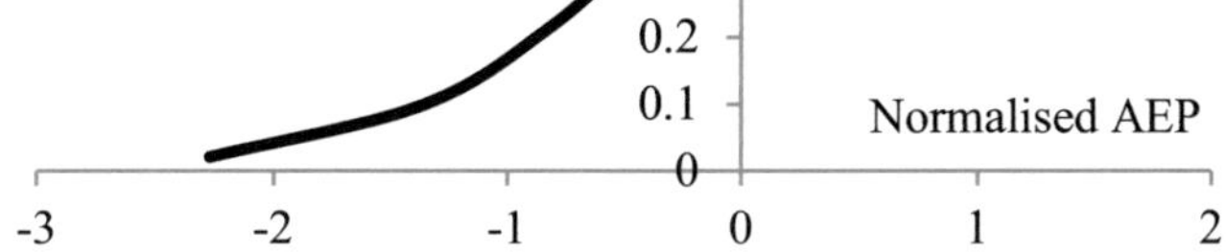

Figure 2. Normalized AEP versus *t*-test distribution values.

The size storm area	10	20	30	40	50	60	70	80	90	100	110	120
PMP	0.029	0.058	0.086	0.11	0.144	0.173	0.2	0.231	0.26	0.288	0.34	0.432

Table 2. The probable maximum perception (PMP) versus the size storm area.

Table 3 presents a numerical example of the relationship between the collapsing risk response and bridge structure age where the presented age is based on the number of decades.

As the time goes by, the risk of the collapsing of the structure increases. This somehow shows the risk assessment as a function of time. **Figure 5** demonstrates the collapsing risk response versus bridge structure age.

The same approach was taken into account to calculate the *t*-test distribution against the normalized collapsing risk response. **Figure 6** illustrates the *t*-test distribution, where it can be found that the collapse mechanism in the simulated bridges can significantly affected after 40 years where there is a possibility of starting fatigue behavior. The relationship between the collapsing risk response and bridge structure age is a quadratic function. The possible function can be presented as $0.000788 \times x^2 + 0.00011 \times x + 0.000655$.

4.4. Resilience of a bridge and mechanical properties

We have assumed that the resilience of bridge is dependent to the relationship between the strength of the soil (MPa) and bridge safety. **Table 4** presents a numerical example of the relationship between the bridge safety factor and strength of the soil (MPa).

Figure 7 presents the same trend where it can demonstrate strength of the soil versus the risk safety factor. As it can be found, the factor of safety is converged by the mean value of the strength of the soil and reached to 2 MPa.

Besides, **Figure 8** illustrates the normalized factor of safety versus *t*-test distribution.

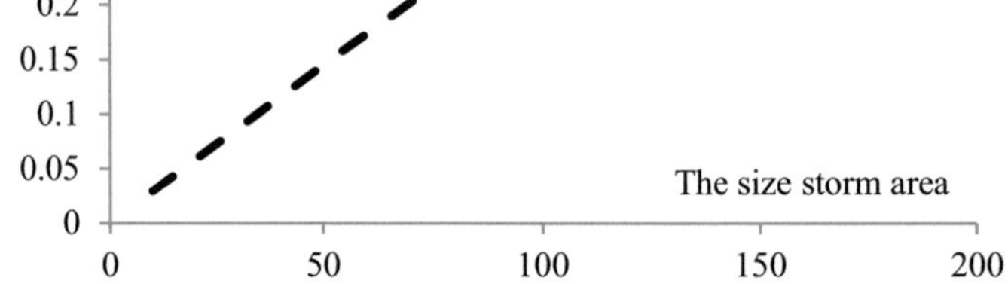

Figure 3. A semilinear relationship between the probable maximum perception (PMP) and the size storm area.

http://dx.doi.org/10.5772/intechopen.69232

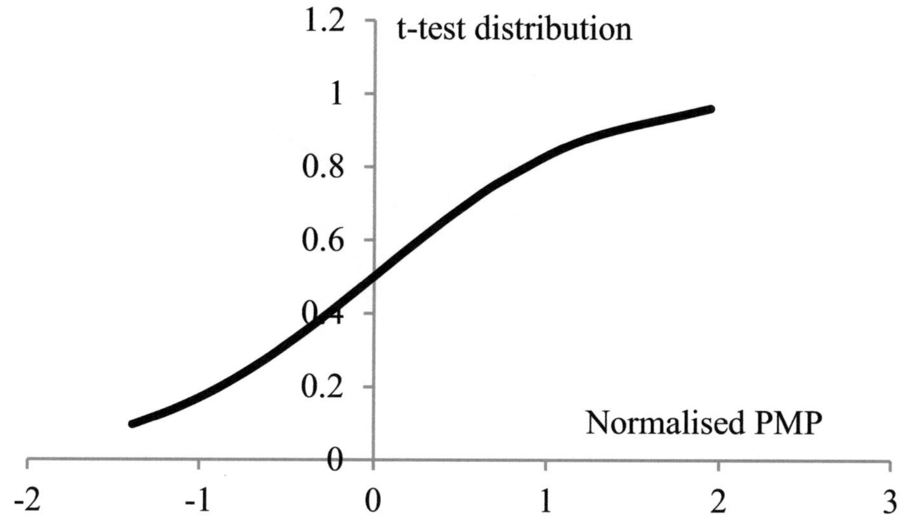

Figure 4. A normalized PMP values versus *t*-test distribution.

Bridge structure age (decades)	1	2	3	4	5	6	7	8	9	10
Collapsing risk respond	0.0001	0.003	0.007	0.013	0.02	0.03	0.04	0.05	0.06	0.08

Table 3. The Collapsing risk response versus bridge structure age.

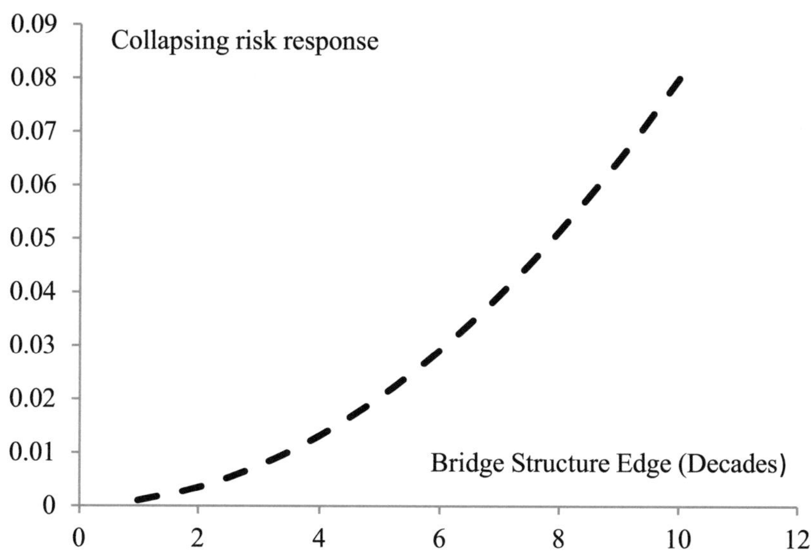

Figure 5. Collapsing risk response versus bridge structure age.

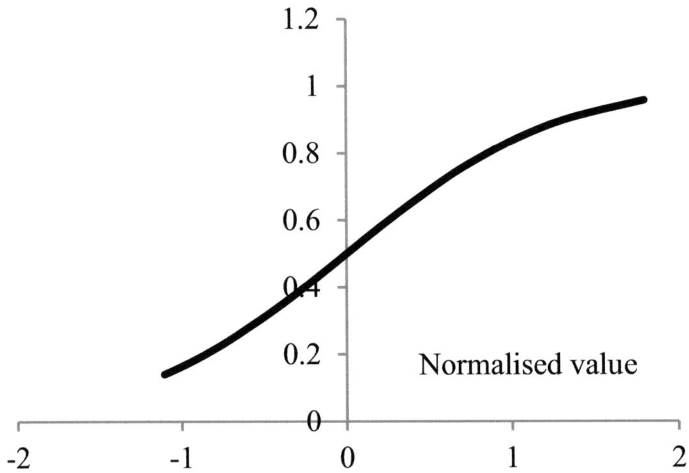

Figure 6. The *t*-test distribution against the normalized collapsing risk response.

Strength of the soil (MPa)	Factor of bridge safety
0.1	1.44
0.2	1.56
0.3	1.72
0.4	1.78
0.5	1.80
0.6	1.78
0.7	1.76
0.8	1.73
0.9	1.74
1	1.72

Table 4. Strength of the soil versus the risk safety factor.

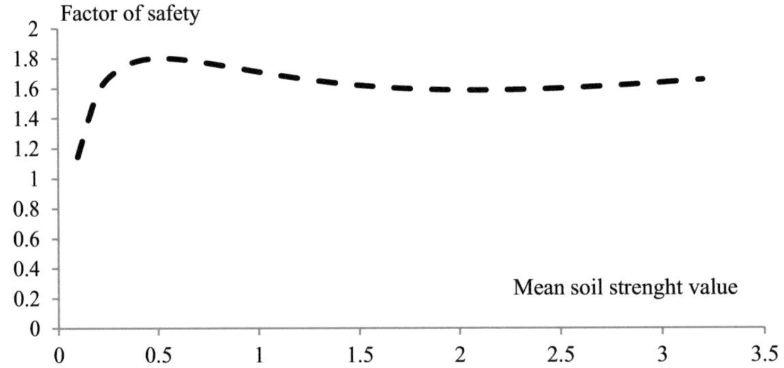

Figure 7. Strength of the soil versus the risk safety factor.

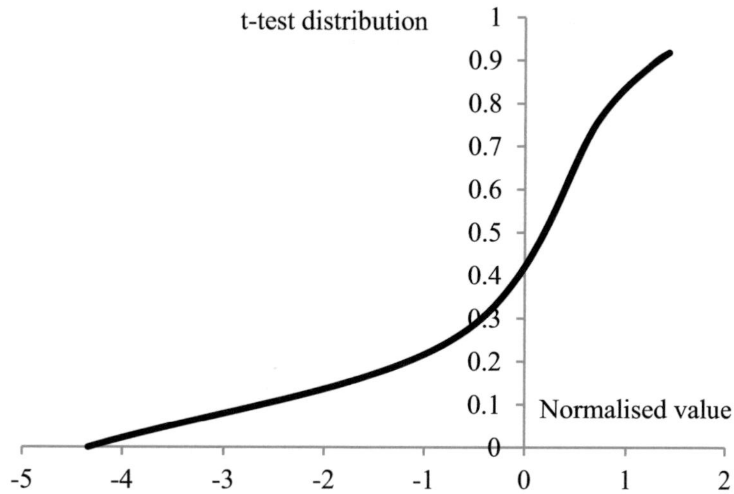

Figure 8. The normalized factor of safety versus *t*-test distribution.

4.5. Additive combination of factors for flood risk-based resilient

Provided that there is a mutual interaction relationship between the two particular indicators, for example, between x_1: AEP and x_2: PMP as well as the major response where it would be the resilient of bridge, the following equation can be considered to evaluate and determine the critical zones, $\left(\frac{x_1}{R}\right)^m + \left(\frac{x_2}{R}\right)^n = k_1$, where $m, n \geq 2$ and k_1 is a constant value.

The same trend can be extended to the interaction between the x_3: bridge structure age and x_4: mechanical properties of the soils as the major indicators and resilient of bridge as the response. Thus, $\left(\frac{x_3}{R}\right)^p + \left(\frac{x_4}{R}\right)^q = k_2$, where $p, q \geq 2$ and k_2 is a constant value. **Figures 9** and **10** present the relevant interactions of the different arrangements explained.

Provided that we assume x_2 is more critical than x_1, then the nonresilient region would be trended to the horizontal access in **Figure 9**. This condition can be shifted if the assumed condition between x_1 and x_2 is switched. The same assumption can be extended in **Figure 10** where it was assumed that x_4 is more critical than x_3. The major limitation of the suggested model is how to define the critical/noncritical condition in the unknown region where it is situated between the resilient and nonresilient regions.

An inclination can be considered to have a combination of all of the indicators ($x_{1,} x_{2,} x_{3,} x_4$) where they are interacted with the resilience of the transport infrastructure R as a major response. Thus, we can have, $\left(\frac{x_1}{R}\right)^m + \left(\frac{x_2}{R}\right)^n + \left(\frac{x_3}{R}\right)^p + \left(\frac{x_4}{R}\right)^q = k_1 + k_2$, where $m, n, p, q \geq 2$ and k_1 and k_2 are constants.

The suggested interaction diagrams as well as the overall combination of the available indicators can graphically help to determine the resilient and nonresilient (as a critical condition) regions in the presented graphs. It can also provide a reliable tool for urban/regional planner as well as the major decision makers to make a more reliable outcome schedule/scope for the communities. The suggested methods can be significantly improved by having access to the different regional/community data. Also, the suggested method should be unbiasedly tested

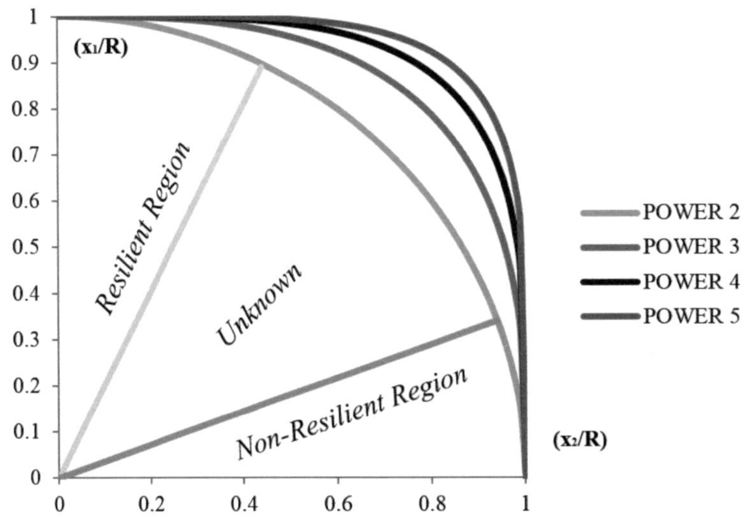

Figure 9. The interaction between x_1 and x_2 and R with the different applied powers in m and n.

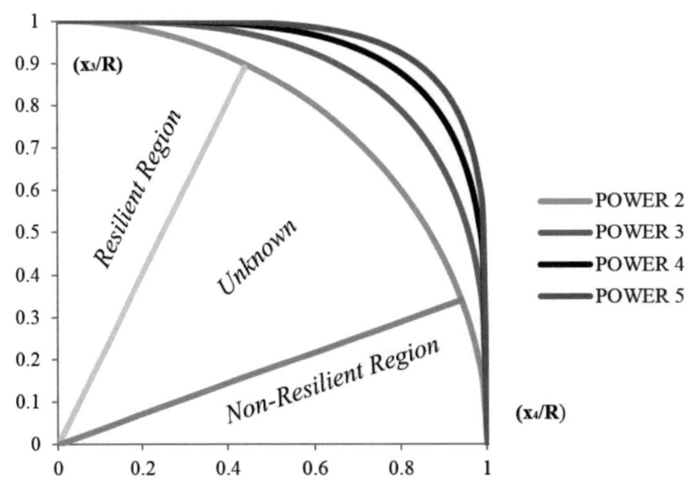

Figure 10. The interaction between x_3 and x_4 and R with the different applied powers in p and q.

in different councils in order to validate the suggested hypothesis. It can provide a reliable tool for urban/regional planers.

5. Conclusions and recommendations

Australia's transport infrastructure has long been exposed to floods. Some infrastructure is no longer fit for purpose, with design standards failing to keep pace with the resilience needs of a changing world. In addition, the meteorological factors of the specific region might exacerbate the vulnerability of transport infrastructure to flood events. Relationships between meteorological factors and transport infrastructure technical factors have never been conceptualized into integrated mathematical models for analyzing risks and measuring the resilience of transport infrastructure to flood events. Here, we used a systems approach to identify critical resilience factors for resilience measurement in transport infrastructure in Australia. The research methodology in this study covered a systematic literature review, mathematical model development, and numerical example conducted by simulation. We used additive statistical model for developing risk-based resilient decision-making system to predict the resilience of bridges. The additive model allows identifying significant predicting variables and their impacts on infrastructure planning and flood risk management. The main novelty of this research is to develop risk-based resilient equations, which would be included in both linear and nonlinear multiple regression. The equations adopt a design methods framework and reference the probability and known impact of previous floods along with the probability and aggregated costs of bridges repair work to estimate an optimum resource expenditure balance between flood risk reduction and flood disaster management. The results addressed transformations toward future resilience requires an integrated approach that allows expression of multiple perspectives on the problem, and supports an adaptive planning approach that is open for experimentation and learning. Infrastructure planners, operators, and regulators need a simple and robust method to evaluate flood risks to better manage the flood risk mitigation, preparedness, response, and recovery of our roads and bridges to extreme flood events.

This research focused on flooding as well as the effect of the natural disaster, but results can be extended to other forms of hazard communication and perception. Quantitative and qualitative analyses supported the adoption of a broad conceptualization of risk communication in both theory and practice. Additionally, the possible relationships of regulated space to communication, perception, and behavior identified in this research are relevant in other contexts where specific areas are politically delineated as hazardous. These results also contribute to more general debates over public understanding of probability and uncertainty as well as those regarding the connections between understanding, attitude, and behavior. This study confirmed that metrological characteristics such as AEP and PMP and bridge conditions such as age and mechanical properties of materials have a substantial impact on the resilience of the Australian transport infrastructure, particularly bridges located on main roads.

A number of more general recommendations for effective flood risk management relevant to transport infrastructure in Australia emerge from this study:

- Greater control of the zoning and use of land in areas prone to high AEP and PMP are required to reduce the need to construct roads and bridges in those areas.
- Upgrading bridges with brick and mortar to cast iron bridges and more resilient bridges will make them less vulnerable to flood damage and significantly reduce the costs of flooding.
- Paying significant attention to the design phase of a bridge and more comprehensive flood risk management approach to incorporate more mitigation strategies will better address the full range of causal factors.
- The fact that Australian aged bridges are not as vulnerable to floods as new bridges calls for a changed funding model for bridge maintenance and upgrades.

The suggested methods can be significantly improved by having access to the different regional/community data. Also, the suggested method should be unbiasedly tested in different councils in order to validate the suggested hypothesis. It can provide a reliable tool for urban/regional planers. Further research is needed to assess the broader flood risk variables and flood risk management capabilities particular to more vulnerable regions and relevant jurisdictions.

Acknowledgements

This research was supported by an Early Career Research (ECR) grant at the Faculty of Built Environment at the University of New South Wales (UNSW), Australia (SIR30/PS41970).

Author details

Mohammad Mojtahedi[1*], Sidney Newton[1] and Faham Tahmasebinia[2]

*Address all correspondence to: m.mojtahedi@unsw.edu.au

1 Faculty of Built Environment, University of New South Wales, Sydney, Australia

2 School of Mining Engineering, University of New South Wales Sydney, Australia

References

[1] Guha-Sapir D, Vos F, Below R, Ponserre S. Annual disaster statistical review 2011: the numbers and trends. Centre for Research on the Epidemiology of Disasters (CRED); 2012.

[2] Setunge S, Lokuge W, Mohseni H, Karunasena W. Vulnerability of road bridge infrastructure under extreme flood events. In Proceedings of AFAC & Bushfire & Natural Hazards CRC Conference; 5-7 September 2014; University of Southern Queensland.

[3] Field CB, editor. Managing the risks of extreme events and disasters to advance climate change adaptation: special report of the intergovernmental panel on climate change. Cambridge University Press; 2012. P. 151-187.

[4] Blong R. Residential building damage and natural perils: Australian examples and issues. Building Research & Information. 2004;**32**:37-41

[5] Mojtahedi M, Newton S, Von Meding J. Predicting the resilience of transport infrastructure to a natural disaster using Cox's proportional hazards regression model. Natural Hazards. 2017;**85**(2):1119-1133

[6] Pritchard R. 2011 to 2012 Queensland floods and cyclone events: Lessons learnt for bridge transport infrastructure. Australian Journal of Structural Engineering. 2013;**14**(2): 167-176

[7] Meyer MD. Design Standards for US Transportation Infrastructure: The Implications of Climate Change. Washington, DC: Transportation Research Board

[8] BOM. Bureau of Meteorology, Annual Report 2009-10. 2010

[9] Wright L, et al. Estimated effects of climate change on flood vulnerability of US bridges. Mitigation and Adaptation Strategies for Global Change. 2012;**17**(8):939-955

[10] AGO. Climate Change Impacts and Risk Management. Canberra, Australia: Australian Greenhouse Office, Department of Environment and Herritage; 2006

[11] Mohseni H, Tran H, Setunge S. In: Bittencourt T, Frangopol D, Beck A, editors. Optimisation of inspection and maintenance activities for bridge components: In: Maintenance, Monitoring, Safety, Risk and Resilience of Bridges and Bridge Networks. CRC Press; 2016. pp. 199-199

[12] Aflatooni M, et al. Synthetic rating system for railway bridge management. Journal of Civil Structural Health Monitoring. 2013;**3**(2):81-91

[13] Laman JA, Guyer RC. Condition assessment of short-line railroad bridges in Pennsylvania. Pennsylvania: The Thomas D. Larson Pennsylvania Transportation Institute. The Pennsylvania State University; 2010.

[14] Sohn J. Evaluating the significance of highway network links under the flood damage: An accessibility approach. Transportation Research Part A: Policy and Practice. 2006;**40**:491-506

[15] Alexander D. The study of natural disasters, 1977-97: Some reflections on a changing field of knowledge. Disasters. 1997;**21**(4):284-304

[16] EM-DAT. The OFDA/CRED International Disaster Database. Brussels: Université Catholique de Louvain; 2004

[17] Mojtahedi SMH, Oo B-L. Development of an index to measure stakeholder approaches toward disasters in the built environment. Procedia Economics and Finance. 2014;**18**: 95-102

[18] UNISDR. Global Assessment Report on Disaster Risk Reduction: Revealing Risk, Redefining Development. Geneva, Switzerland: United Nations International Strategy for Disaster Reduction; 2011

[19] Leroy SA. From natural hazard to environmental catastrophe: Past and present. Quaternary International. 2006;**158**(1):4-12

[20] Ho MC, et al. How do disaster characteristics influence risk perception? Risk Analysis. 2008;**28**(3):635-643

[21] Merz B, et al. Estimation uncertainty of direct monetary flood damage to buildings. Natural Hazards and Earth System Sciences. 2004;**4**:153-163

[22] Middelmann-Fernandes M. Flood damage estimation beyond stage-damage functions: An Australian example. Journal of Flood Risk Management. 2010;**3**(1):88-96

[23] Balica SF, Wright NG, Meulen F. A flood vulnerability index for coastal cities and its use in assessing climate change impacts. Natural Hazards. 2012;**64**:73-105

[24] Lamond J, et al. Flood Hazards: Impacts and Responses for the Built Environment. Boca Raton: CRC Press; 2012

[25] OEH. OEH 2005. Floodplain Development Manual: The Management of Flood Liable Land. NSW, Australia: Department of Natural Resources; 2005

[26] Jonkman SN, Kelman I. An analysis of the causes and circumstances of flood disaster deaths. Disasters. 2005;**29**:75-97

[27] Nicholls RJ, Tol RSJ. Impacts and responses to sea-level rise: A global analysis of the SRES scenarios over the twenty-first century. Philosophical transactions. Series A, Mathematical, Physical, and Engineering Sciences. 2006;**364**:1073-1095

[28] Godschalk DR. Urban hazard mitigation: Creating resilient cities. Natural Hazards Review. 2003;**4**(3):136-143

[29] Levina E, Tirpak D. Adaptation to Climate Change: Key Terms, Organization for Economic Co-Operation and Development, International Energy Agency (OECD); 2006. p. 2011-2015

[30] Maguire B, Cartwright S. Assessing a Community's Capacity to Manage Change: A Resilience Approach to Social Assessment. Canberra: Bureau of Rural Sciences; 2008

[31] Manyena B, et al. Disaster resilience: A bounce back or bounce forward ability? Local Environment: The International Journal of Justice and Sustainability. 2011;**16**(5):417-424

[32] Hughes J, Healy K. Measuring the resilience of transport infrastructure. Technical Report Research Report 546. Washington, DC: Transportation Research Board, the National Academies, 2014.

[33] Park J, et al. Integrating risk and resilience approaches to catastrophe management in engineering systems. Risk Analysis. 2013;**33**(3):356-367

[34] Friedman JH, Stuetzle W. Projection pursuit regression. Journal of the American statistical Association. 1981;**76**(376):817-823

[35] Lebbe MFK, et al. Failure mechanisms of bridge infrastructure in an extreme flood event. In: Proceedings of the First International Conference on Infrastructure Failures and Consequences; 16-20 Jul 2014; Melbourne, Australia. 2014; RMIT Publishing; p. 124-132.

[36] Lokuge W, Setunge S. Evaluating disaster resilience of bridge infrastructure when exposed to extreme natural events. Proceedings of the 3rd International Conference on Building Resilience; 17-19 Sep 2013; Heritance Ahungalla, Sri Lanka. 2013; University of Salford, Centre for Disaster Resilience; pp. 1-12

[37] Croope SV. Managing Critical Civil Infrastructure Systems: Improving Resilience to Disasters. University of Delaware; ProQuest Dissertations Publishing; 2010. 150 p.

[38] Bruneau M, et al. A framework to quantitatively assess and enhance the seismic resilience of communities. Earthquake Spectra. 2003;**19**(4):733-752

Chapter 7

Flood Management in China: The Huaihe River Basin as a Case Study

Qian Mingkai and Wang Kai

Additional information is available at the end of the chapter

http://dx.doi.org/10.5772/intechopen.69047

Abstract

The Huaihe River Basin (HRB) is a transitional river located in the transitional climate zone in China, and it has been frequently hit by big floods and suffered from flood disasters. Flood control and management of the areas are of vital importance of the Huaihe River Basin in its social and economic development. In this chapter, pioneer works of summarizing the flood management have been done for the Huaihe River in China. It first introduces flood and flood disasters of the River basin. In addition, this chapter summarizes achievements in flood control and management. Furthermore, it discusses experiences and enlightenment in flood control and management and draws conclusions for the research.

Keywords: flood management, flood control, the Huaihe River, transitional river, China

1. Introduction

China is located in eastern Asia. It is impacted by monsoonal climate and the temporal, and therefore, spatial distribution of precipitation is uneven. The terrain in China is high in the west and low in the east with mountains, plateaus, and humps, which account 67% of China's areas and basins and 33% plains. Special geographical and climatic conditions result in very serious flood and drought disasters in China. Since 1949, great efforts have been made for flood control in China. A series of policies and regulations has been formulated for flood control and management. Furthermore, numerous structural measures such as construction of reservoirs, flood detention areas and lakes, building of dykes, water gates and hydro junctions have been constructed. In addition, non-structural measures of hydrological monitoring system and flood forecasting system have been made for flood control in China.

The Huaihe River Basin (HRB) is located in the transitional climate zone in China (**Figure 1**), which is known as the transitional river of China [1]. It has been frequently hit by big floods and suffered from flood disasters, and the frequency of disaster is one time in 10 years on average [2–5]. The critical issue is that about two-third of the middle and downstream of the major rivers are prone to floods, where is inhabited 13% of people of China, 12% of cultivated land area of China, one-sixth of food product of China, and one-fourth of the food as commodity of China. Flood control and management of the areas are of vital importance for the Huaihe River Basin in its social and economic development. Strenuous efforts have been made in fighting against floods; however, there is still a long way to go.

This chapter is organized as follows: Section 1 introduces general situation of flood management in China and the Huaihe River Basin, Sections 2 and 3 summarize the study area of the Huaihe River Basin. Section 4 presents floods and flood disasters of the River basin and concerns achievements in flood control. Section 5 discusses experiences and enlightenment in flood control and management. Section 6 draws conclusions for the research.

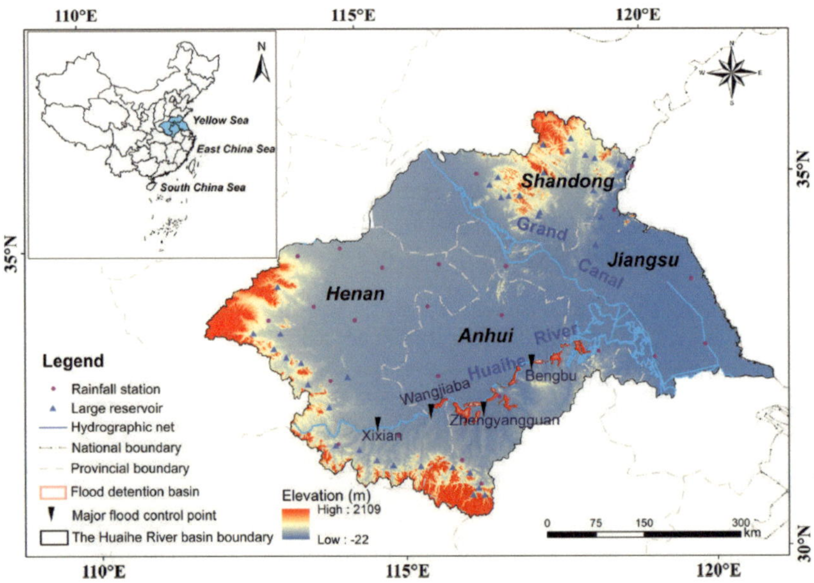

Figure 1. Location map of the study area.

2. Geographical features

The Huaihe River Basin is located in the east China, with the Yellow River in the north and the Yangtze River in the south, and its catchment area is 270,000 km^2. Starting in the Tongbai Mountains of Henan province, it flows from west to east. The upper reaches of the river are located in Henan and Hubei Provinces, the middle of the river is located in Anhui Province, and the downstream of the river is located in the Jiangsu Province. The length of the trunk stream is about 1000 km. The upper reaches of Huai river flow from the river head to the mouth of Honghe River between Henan and Anhui provinces, and the total length of the upper reaches is 360 km; the middle reaches, from Honghe River to Hongze Lake, are 490 km in length; and the lower reaches have a total length of 150 km. In the lower part of the Huaihe River Basin, there are four major outlets for floods, i.e., the Floodway to the Yangtze River, the Floodway to the Yellow Sea, the Northern Jiangsu Irrigation Canal to the Yellow Sea, and diversion waterway from the Huaihe River to new Yi River, then to the Yellow Sea.

The average annual precipitation of basin is about 875 mm, of which 50–80% precipitation is concentrated in the rainy season (June–September). Located in the north-south climate transition zone, with uneven spatial-temporal distribution of precipitation and the capture of the headwaters of Yellow river into the Huaihe River, Huaihe River flood disasters occurred very frequently. The basin-wide floods hit the Huaihe River in 2003 and 2007 of this century.

Due to the frequent flood disasters in the Huaihe River Basin, the State attaches great importance to harness of the Huaihe River. For nearly half a century, many flood control buildings have been built, including 38 large reservoirs, 21 flood storage areas, 1716 km embankment of grade I, diversion rivers such as Huaihongxinhe canal, Ruhaishuidao canal (floodway to the sea), as well as large lakes such as Hongze Lake. In addition, non-structural measures such as communication systems, hydrological forecasting system, the remote monitoring system of flood control works, remote consultation system, and flood control system have also been built. All these measures have played a positive role in the defence of floods.

3. Outstanding characteristics of the HRB

Compared with other rivers all over the world, the Huaihe River has its own outstanding characteristics:

1. HRB is located in the transitional zone from southern to northern climate of China, the weather system changes dramatically, and precipitation varies greatly in both space and time. Therefore, it is prone to flood and drought disaster. The average annual precipitation of HRB is 875 mm, the average annual precipitation in the northern area is 600–700 mm, and in the southern area reaches to 1400–1600 mm. The precipitation is concentrated in flood season (from June to September) and accounts for 70% of the annual amount.

2. The Huaihe River Basin is short of water resources, and water pollution problems have not been effectively solved yet. The total amount of average annual water resources in the Huaihe River Basin is 79.4 billion m^3, and the average amount of water resources per capita and per mu only accounts for one-fourth and one-fifth of that of China (1 mu =1/15 ha). Water is unevenly distributed in time and space, unmatched with the production and population pattern. River regulation capacity is weak, and development and utilization of water resources are difficult.

3. The Huaihe River Basin is with flat terrain, where mountainous area is less with poor flood incept and storage condition. It has a broad plain area accounting for two-third of the total area, which has flat terrain and small gradient ratio in the middle-low stream. The total

fall-head of the Huaihe River is 200 m (gradient ratio is 0.2‰). In the upstream, the gradient ratio is greater, with a fall-head of 178 m (gradient ratio is 0.5‰), and flood concentrates rapidly. In the middle reach, the gradient ratio and the fall-head of the midstream are 0.03‰ and 16 m, respectively. The river channel is curved and flat, and several parts of the river even have inverse slopes. The floods cannot flow smoothly and freely, which can easily cause flood disaster. In the middle reach, the gradient ratio and the fall-head of the midstream are 0.04‰ and 6 m, respectively.

4. Historically, with the diversion (capture) of the headwaters of Yellow river into the Huaihe River dramatically changes the natural water system, aggravates flood burden on the one hand, and increases the difficulties in harnessing the Huaihe River on the other hand. Originally, the Huaihe River is a separate river flowing into the East Sea; however, from the end of twelfth century to the middle of nineteenth century, in the 700 years of the Yellow river capture into the Huaihe River, a large number of sediments silted up the middle-low channel of the Huaihe River and made the Huaihe River lose its outlet to the Sea. It changes the natural water system to a great extent and has profound influences. It aggravated flood burden on the one hand and increased the difficulties in harnessing the Huaihe River on the other hand.

5. The Huaihe River Basin is densely populated, contradictions between human and water are sharp, and coordinated development is difficult. The Huaihe River Basin has a population of 0.178 billion, which accounts for 13% of that of China. Moreover, the Huaihe River Basin is also a main grain production area in China, with a cultivated land area of 0.19 billion mu accounting for 10.5% of the total amount of China; grain yield of the Huaihe River Basin takes up 17% of the total amount of China, and commodity grain of the Huaihe River Basin constitutes 25% of that of China. There are so many trans-boundary rivers that numerous conflicts of interests between different regions occur in terms of drainage and water resources utilization and protection. As a result, water affairs and conflicts are complex, and therefore, it increases the difficulties in river harness and management.

4. Floods and flood disasters

4.1. Floods characteristics

The Huaihe River Basin is located in the transitional zone from southern to northern climate of China, the weather system changes dramatically, and precipitation varies greatly in both space and time. Therefore, it is prone to flood and drought disasters. The average annual precipitation of HRB is 875 mm, the average annual precipitation in the northern area is 600–700 mm, and in the southern area reaches to 1400–1600 mm. The precipitation is concentrated in flood season (from June to September) and accounts for 70% of the annual amount. The maximum records of point rainfall of different durations in the Huaihe River Basin are close to or higher than the corresponding maximum records of China and the world. Extremely large flood peaks caused by such high intensity and extensive coverage rainstorms result in serious flood disasters (**Figure 2**)

The magnitude of the maximum floods varies greatly from one area to another over the whole basin, and the important events are given in **Table 1**.

http://dx.doi.org/10.5772/intechopen.69047

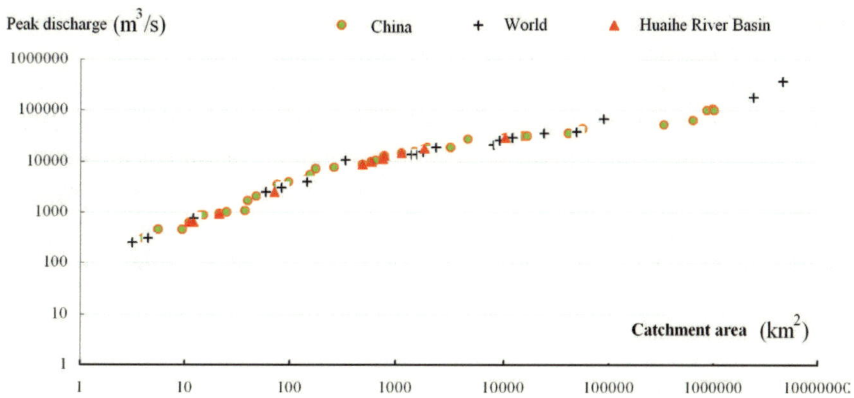

Figure 2. The relationship between catchment area and peak discharge in the Huaihe River Basin, China, and the world.

River	Gauged station	Catchment area (km²)	Observed	
			Discharge	Year
Huaihe mainstream	Wangjiaba	30,630	17,600	1968
Huaihe mainstream	Lutaizi	88,630	12,700	1954
Huaihe mainstream	Bengbu	121,330	11,600	1954
Huaihe mainstream	Sanhe	158,160	10,700	1954
Shihe southern major tributary	Jiangjiaji	5930	4550	1968
Pihe southern major tributary	Hengpaitou	4370	6420	1969
Shayinghe northern major tributary	Fuyang	36,606	3310	1965
Guohe northern major tributary	Mengcheng	15,475	2080	1963
Yihe major tributary	Linyi	10,315	15,400	1957
Shuhe major tributary	Daguanzhuang	4529	4250	1974

Table 1. The maximum flood discharge for major rivers in HRB (Unit m³/s).

4.2. Flood disasters

According to statistics, during the 2256 years from 246 B.C. to 2010, totally 1946 flood and drought disasters had occurred in the Huaihe River Basin, among which, number of flood disaster is 1008, while number of drought disaster is 938, which means disaster almost happens every year. Number of basin-wide flood and drought disasters is 340 (number of flood disaster is 268 and number of drought disaster is 72), and frequency is approximately once every 6.6 years on average. From the diversion (capture) of the Yellow river into the Huaihe River in 1194, flood disasters became more frequently. From the thirteenth century to the nineteenth century, there were 165 flood and drought disasters occurred in the HRB, one time in every 4.2 years on average.

Due to the unique natural features of the HRB, in recent years, large-area flood and drought disasters happen frequently, and local flood and drought disasters happen annually. According to statistics, from 1949 to 2008, average annual flood disaster area of the HRB is 25.29 million mu (**Figure 3**), among which, number of annual flood disaster area above 30 million mu reaches to 15 accounting for 25% of the number of statistical years; number of annual flood disaster area higher than 40 million mu reaches to 11 accounting for 18.3% of the number of statistical years; and number of annual flood disaster area above 50 million mu reaches to 8 accounting for 13.3% of the number of statistical years. **Table 2** shows statistics of flood disasters for three basin-wide floods in the HRB.

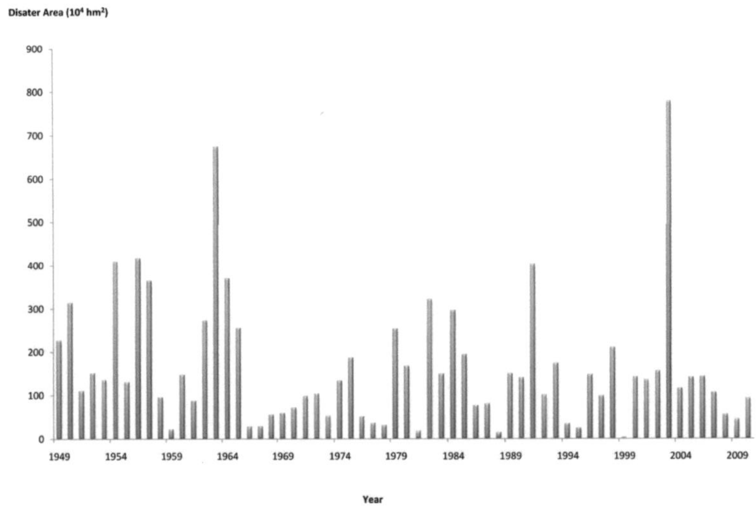

Figure 3. Disaster area in different years in the Huaihe River Basin.

Year	Disaster area (10^4 hm^2)	Affected people (10^4)	Destroyed houses (10^4)	Immigrant (10^4)	Direct economic losses (10^8 RMB)
1991	401.6	5423	196	226.1	339.6
2003	259.1	3730	77	207	286
2007	158.7	2472	11.53	80.9	155.2

Table 2. Statistics of flood disasters in 1991, 2003, and 2007 in HRB.

5. Achievements in flood control

Since the foundation of the new China, for the purpose of effectively reacting to flood and drought disasters and reducing losses, under the guidance of "jointly considering storage and discharge of floodwater" principle for harnessing the Huaihe River, with 60-year continuous river harness, tremendous achievements have been made in engineering construction of

http://dx.doi.org/10.5772/intechopen.69047

the Huaihe River harness. Laws and regulations, i.e., "Flood Control Law of The People's Republic of China," "Flood Control Regulation of The People's Republic of China," "Drought relief regulation of The People's Republic of China," and "Administrative regulation for flood storage and detention areas," have been established, and basin flood prevention and dispatching programs have been improved, and all these play a vital role in reducing flood and drought disasters.

5.1. Structural measures

Overall arrangement: In the upper reaches, carrying out water-soil conservation and constructing reservoirs are important to intercept and store floods. In the middle reaches, taking structural measures to dredge & broaden river course, construct dykes and flood diversion & storage areas to increase channel flood discharge capacity. In the lower reaches, taking structural measures to excavate river channels for enlarging flood discharge capacity. The overall layout of structural measures is shown in **Figure 4**.

Water and soil loss have accumulatively been harnessed with an area of 40,000 km^2.

About 6300 reservoirs have been constructed, with a total storage capacity of 30 billion m^3, and 40 of them are large reservoirs, with a total storage capacity of 20 billion m^3 and a flood control capacity of 6.2 billion m^3. Seventeen flood detention areas and large lakes for controlling flood have been constructed, with a total storage capacity of 3.59 billion m^3 and a flood storage capacity of 2.63 billion m^3. **Figure 3** illustrates the major structures in the Huaihe River Basin.

Artificial channels have been constructed with a length of 2100 km. Different types of dikes have been constructed with a length of 50,000 km, and the length of key dike is 11,000 km. River channel discharge capacity has been significantly promoted, channel discharge capacity of the

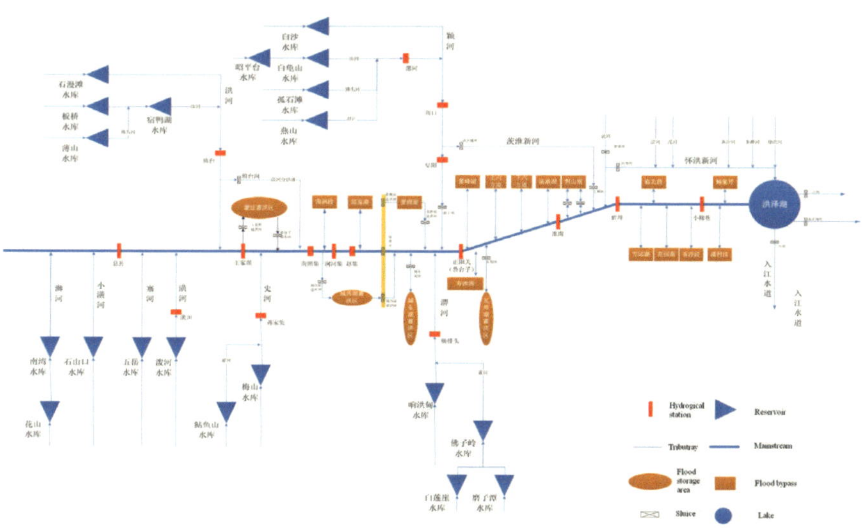

Figure 4. Simplification of major water conservancy projects of the Huaihe River.

upper mainstream has been enhanced from 2000 to 7000 m^3/s, channel discharge capacity of the middle mainstream has been enhanced from 5000 to 7000 m^3/s to 7000 to 13,000 m^3/s, and channel discharge capacity of the lower mainstream has been enhanced from 8000 to 18,270 m^3/s. In addition, 1200 sluices have been constructed.

At present, flood control standard of the mainstream in the upstream is once-10-years, and flood control standards of the key flood protection areas and important cities in the middle and lower reaches were promoted to once-100-years. Flood control standards of the important tributaries can reach to once-20-years to once-50-years. Specifically, in the upper mainstream, channel discharge capacity has been enhanced from 2000 to 7000 m^3/s. In the middle mainstream, channel discharge capacity has been enhanced from 5000–7000 to 7000–13,000 m^3/s. In the lower mainstream, channel discharge capacity has been enhanced from 8000 to 18,270 m^3/s.

5.2. Non-structural measures

So far, non-structural measures of flood prevention regulation command system jointly form system mechanism for flood prevention and disaster reduction.

5.2.1. Legal and institutional system for flood control

China has already promulgated laws and regulations concerning water, e.g., water law, flood control law, regulation on river channel, regulation on flood control, etc.

(1) Flood control law

The Law of Flood Control was adopted at the 27th Meeting of the Standing Committee of the Eighth National People's Congress on August 29, 1997, and promulgated by Order No. 88 of the President of the People's Republic of China on August 29, 1997. It is the first law on the prevention and control of natural disasters in China. It is also a very important law following the Water Law and the Law of Soil and Water Conservation in water domain in China.

Numerous administrative regulations were promulgated to clarify the responsibilities of responsible parties concerned in the flood control, such as flood Control Regulation of the People's Republic of China (2005 Revision), Regulation of the People's Republic of China on the Administration of River Courses, Guidelines for safety and construction of flood detention areas and flood control operation plan, etc., which have played an vital role in flood control and management in China. However, with the rapid socio-economic development, population growth, and acceleration of urbanization, numerous problems and new challenges have risen in terms of flood control, which can be summarized as follows: (a) Lack of legal consciousness on the importance of flood planning and at the same time the approved flood plan was not fully observed or strictly enforced; (b) Flood protection standard was relatively low; (c) There was no effective means for river channels management, e.g., sand excavation; (d) No effective management of flood plains area; and (e) Lack of funds in flood control and infrastructure construction. Therefore, there is an urgent call for specific law on flood control to ensure that necessary measures could be implemented legally.

(2) Institutional Arrangement

(a) Ministry of Water Resources (MWR)

Ministry of Water Resources (MWR), the Chinese Government Department responsible for water administration, was founded in October 1949. In order to clarify the responsibilities among different ministries/departments under the State Council, Ministry of Water Resources was reorganized on July 22, 1988. In accordance with the stipulations of the State Council of the People's Republic of China, the function and responsibility of the department were summarized as follows:

- ensure rational development and utilization of water resources in China,
- formulate water resources development strategies, plans, and policies in China,
- provide draft legislations,
- promulgate water administrative rules and regulations,
- undertake integrated water resources management and supervision,
- take charge of water resource protection and water conservation,
- organize, coordinate, and supervise the work of flood control and drought relief, and be responsible for control of soil and water losses.

(b) Flood control and drought relief commanding headquarters (FCDRH)

The flood control and drought relief commanding system has been constructed at national, river basin, and local levels (**Figure 5**). On a national level, the General Commander is a vice premier of the State Council. Its members are from administrative leaders of governmental departments and the military, who are responsible for organizing and guiding efforts in flood control and drought relief throughout the whole country [6–9]. In the six major river basins, namely, the Yangtze River, Yellow River, Huai River, Hai River, Songhua River, and Pearl River, Flood Control and Drought Relief Headquarters are constructed, and they take the

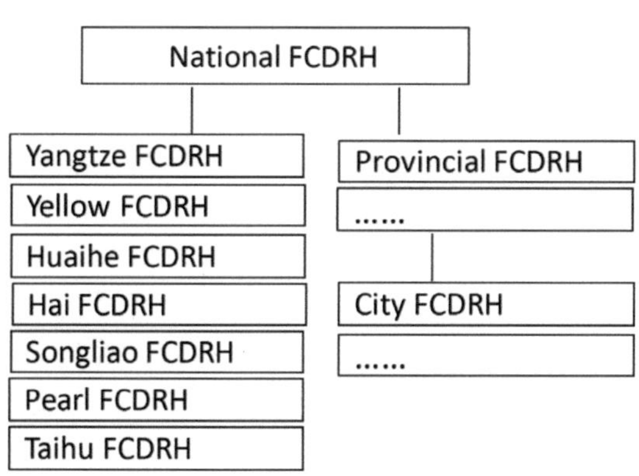

Figure 5. Institutional framework of Flood Control and Drought Relief Commanding Headquarters.

same responsibility of flood control and drought relief on a river basin level. In local governments that undertake flood and drought tasks, flood control and drought relief commanding headquarters are constructed as part of the local governments.

(c) The Huaihe River Basin Commission

The Huaihe River Commission is a river basin authority dispatched by the Ministry of Water Resources of China to exercise water administrative functions in the Huaihe River Basin and Shandong Peninsula, which is responsible for basin design and planning, flood control and drought relief, integrated water resources management, etc.

In accordance with the stipulations of the Ministry of Water Resources(MWR), the Commission is given the following major mandates in Huaihe River Basin (HRB):

1. It is in charge of making the integrated planning and related professional planning for HRB, such as water resources developing planning, annual water projects construction implementing planning, and so on.
2. It is in charge of HRB's water resources management and supervising.
3. It is in charge of checking the water body acceptance capacity for pollutant, and monitoring the water quality for the water functional areas located in the boundary of provinces.
4. It is in charge of dealing with routine work of the Huaihe River Flood Control and Drought Relief Headquarters. This includes the organization, coordination, supervision, and direction of flood control for HRB, and execution of operations of flood control and drought prevention for major river basins and key water projects.
5. It is in charge of building, managing, and operating for important water projects.
6. It is in charge of enforcing the laws and regulations relative to water administration.
7. It is in charge of organizing water and soil conservation in HRB, including development of engineering measures for water and soil conservation, and organization of the monitoring and overall prevention and control of soil and water losses.

(d) The local water resources management agency

The local water resources management comprises four levels, i.e., the provincial, prefecture, country, and the village (town). Overall, it has the most of main functions and responsibilities as that of the central government. Specifically, there are also direct legal duties of flood control and drought relief.

5.2.2. The flood forecasting and warning system

(1) The flood monitoring system

Hydrological information and flood forecasting provide basic information for flood control. By 2014, totally 329 hydrological stations (hydrologic information including information of precipitation, water level, and discharge), 220 stage gauging stations, and 2488 rain gauges

http://dx.doi.org/10.5772/intechopen.69047

had been constructed. For those stationary gauging stations, they collect rainfall information using rainfall recorder, water level information using telemetering stage recorder, and discharge information using acoustic doppler current profilers (ADCP) or current meter. In addition, 1489 water quality monitoring stations, 324 moisture stations, and 3024 ground water monitoring stations had been constructed; all of these constitute the hydrological network and also the flooding monitoring system over the Huaihe River Basin. During flood period, with the aid of water information transmission system, 1250 stations (including important hydrological stations, stage gauging stations, and rain gauges) are mandated to collect, transmit, and share hydrological information from single state level to provincial, river basin, and the national levels. It runs at a regular time interval ranging from 6 minutes to 6 hours as stipulated on the basis of the requirement of flood forecasting for the river system [2].

(2) Information transmission

The hydrological information is transmitted through the national telecommunication system. Most of the hydrological information collected are transmitted using client software of the transmission system. For those river reaches and water projects of special importance, short wave radio stations were established to ensure more effective information transmission. Furthermore, data collection and management have been computerized in the real-time flood forecasts on the state level, the river basin level, the province level, and the municipal level.

(3) Flood forecasting and warning

Flood forecasting and warning were made by hydrological bureau or flood control office, which is the member of the Flood Control and Drought Relief Headquarters (FCDRH) at various levels ranging from municipal level to the country level. Faced with emergence, the FCDRH will issue warning via government at different levels. Before the flood season in every year, annual meeting of FCDRH at different levels is held to ensure preparatory work is well organized. In addition, on a national level, there are about 1000 hydrological stations that conduct river flood forecast as requested, and there are 66 hydrological stations conducting river flood forecast in the Huaihe River Basin. Numerous models, e.g., the Antecedent precipitation index (API) model, the Xinanjiang model, and tank model, have been employed for flood forecasting (**Table 3**) [9–13]. Most of the hydrological bureau above municipal – level established flood forecasting system in the flood forecasting [14] (**Figure 6**). In addition, a new real-time flood forecasting platform FEWS_HUAIHE has also been established for probability forecasting (**Figure 7**).

1	P-Pa-R relation curve	7	Xinanjiang Model
2	P+Pa-R relation curve	8	Grid-xinanjiang Model
3	Zm~Qm, Z~Q	9	Tank Model
4	Unit hydrograph	10	TOPMODEL
5	Muskingum	11	TOPKAPI
6	Tacit knowledge	12	HBV

Table 3. Flood forecasting models applied in the Huaihe River Basin.

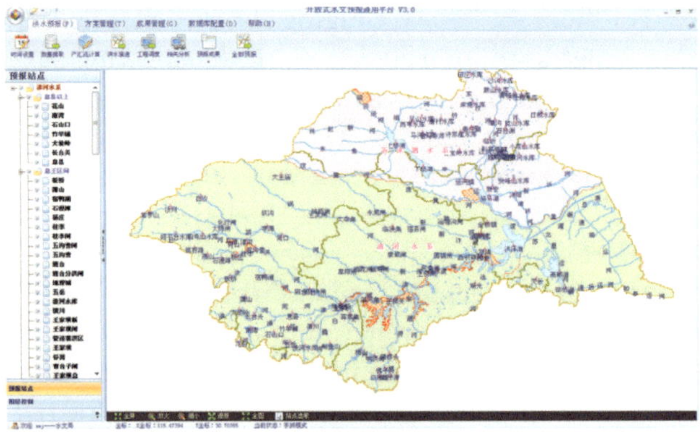

Figure 6. Real-time flood monitoring, forecasting, and warning system for the Huaihe River Basin.

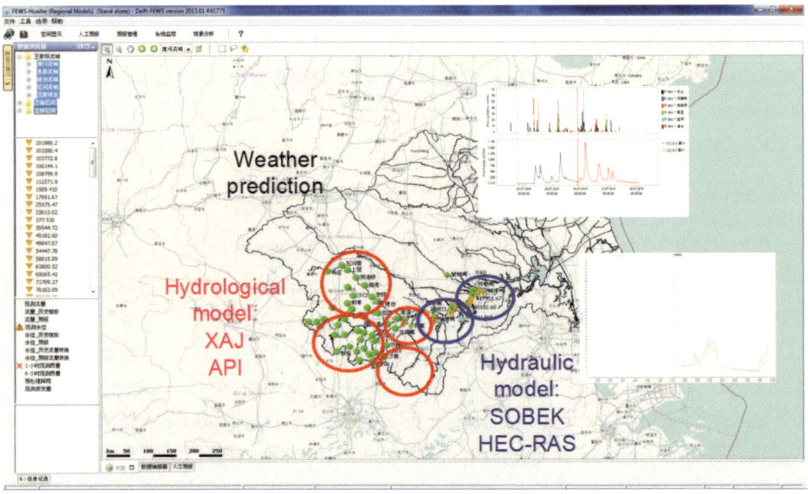

Figure 7. Probability flood forecasting system for the Huaihe River Basin.

(4) Decision support system for flood control

The major flood control measures can be summarized as "upper-stream storage, middle-stream passage, and lower-stream discharge" in accordance with the principle of flood prevention and regulation. Decision support system for flood control was designed for real-time control of structures in the Huaihe River Basin in case that big floods hit the basin (**Figure 8**). It has the following functions: reservoir regulation (**Figure 9**), flood bypass regulation (**Figure 10**), sluice regulation, hydrojunction regulation as well as joint dispatching of hydraulic structures, etc.

(5) Flood disaster assessment system

Flood disaster assessment system is designed for evaluating losses of flood disasters during the whole process in several key flood-prone areas in the Huaihe River Basin. It has the major function of pre-disaster evaluation, real-time evaluation, and post disaster assessment (**Figure 11**).

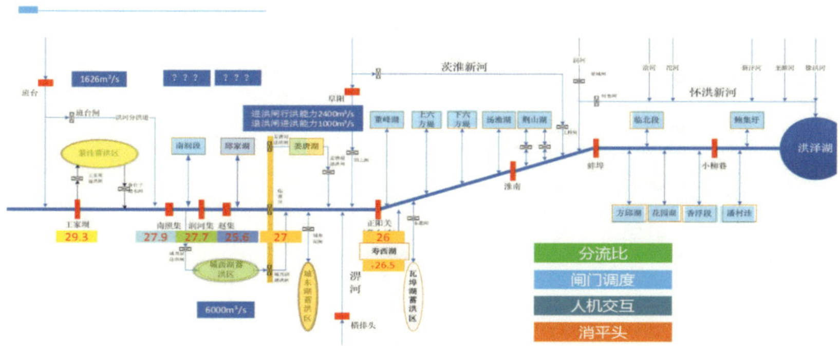

Figure 8. Topological diagram of hydraulic structures for the Huaihe River Basin.

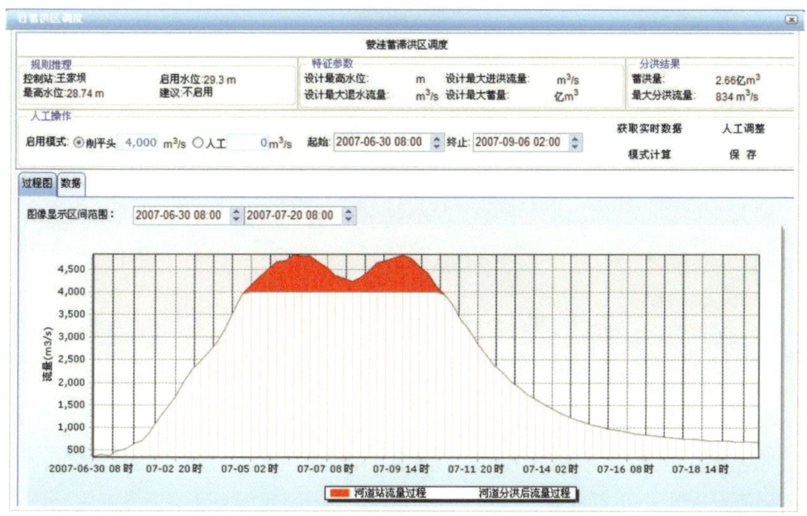

Figure 9. Interface of flood bypass regulation.

(6) Flood risk mapping

Flood risk mapping identifies flood hazards, assesses flood risks, and partners with provinces and communities to provide accurate flood hazard and risk data to guide them to mitigation actions. Flood risk mapping is an important part of flood regulations and flood insurance requirements. It maintains and updates data through Flood Insurance Rate Maps (FIRMs) and risk assessments. FIRMs include statistical information such as data for river flow, storm tides, hydrologic/hydraulic analyses, and rainfall and topographic surveys (**Figure 12**).

In summary, an integrated system of flood monitoring, forecasting, and warning system, including weather prediction, flood monitoring and forecasting, flood dispatching, and flood control discussion, has been preliminarily established in the Huaibe River Basin (**Figure 13**), which plays a vital role in basin flood control. However, system for flood risk management, e.g., probability flood forecasting system, flood risk dispatching models for reservoirs and hydro-junction, still stays at a starting stage.

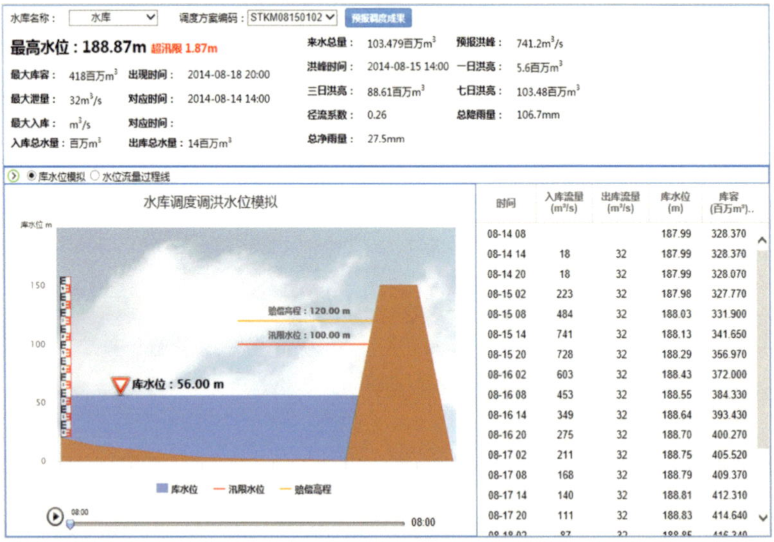

Figure 10. Interface of reservoir regulation.

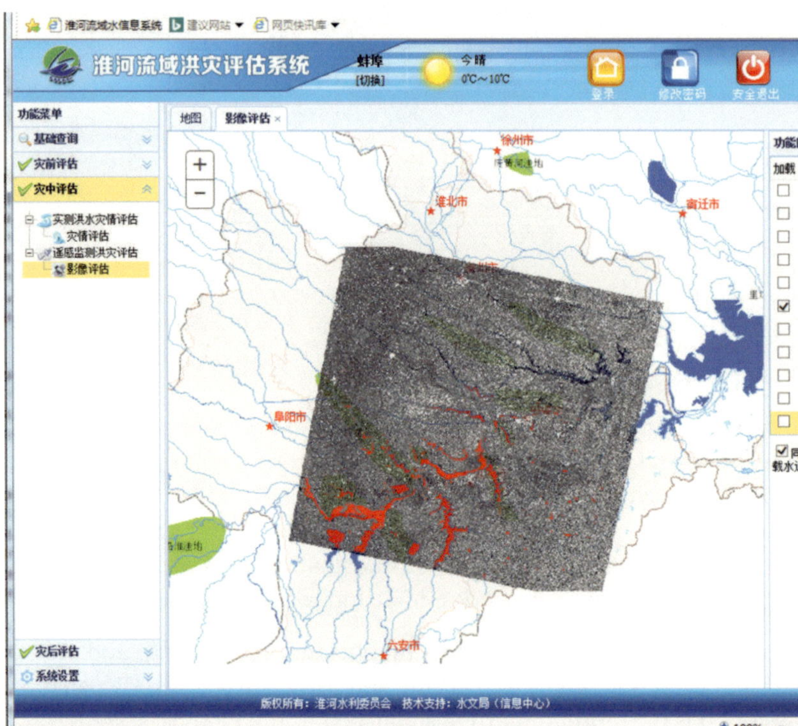

Figure 11. Flood disaster assessment decision support system.

http://dx.doi.org/10.5772/intechopen.69047

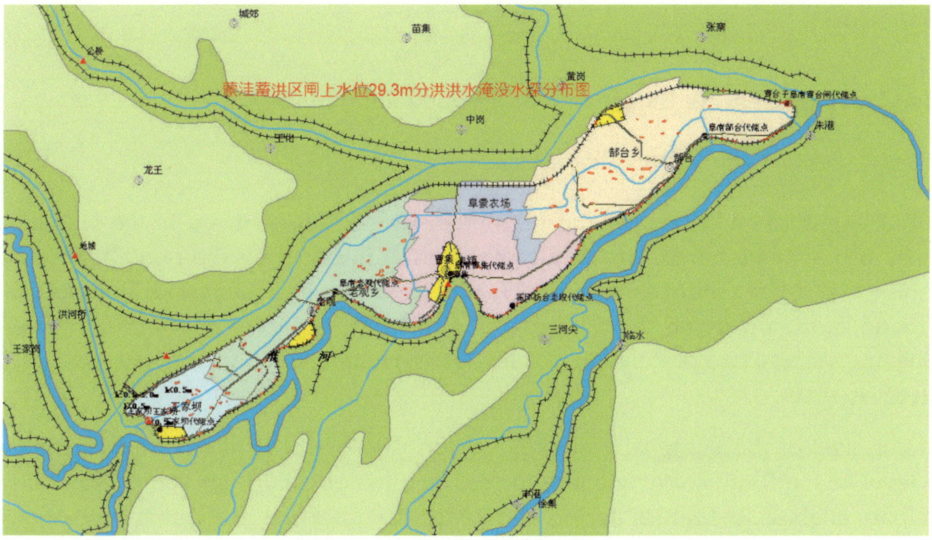

Figure 12. Flood risk mapping for a key area in the Huaihe River Basin.

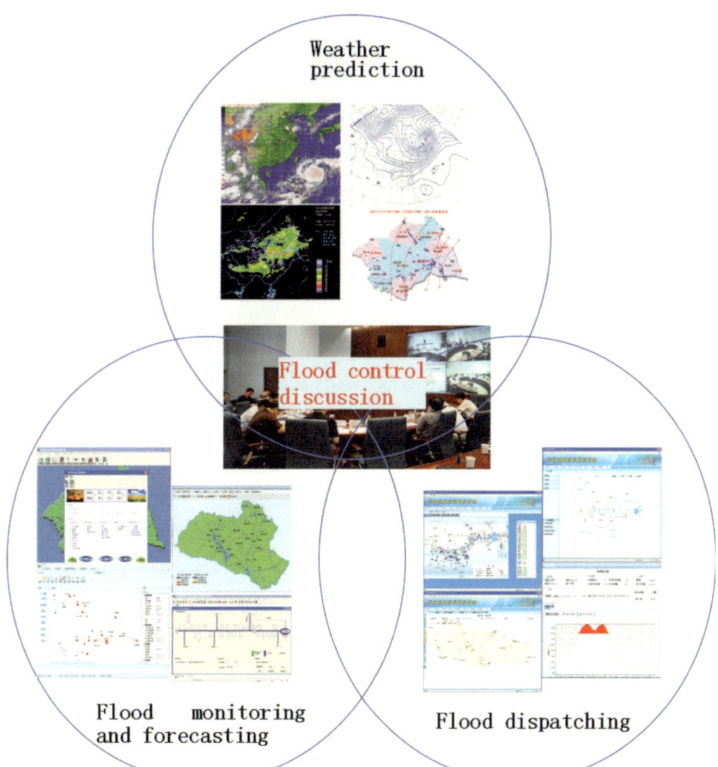

Figure 13. Integrated system of flood monitoring, forecasting, and warning system.

6. Case study

6.1. Flood management for the Huaihe Basin-wide flood in 2007

6.1.1. Floods in 2007 in the Huaihe River Basin

In 2007, a long-time, wide-range, and high-intensity rainfall occurred in the Huaihe basin, in which area average rainfall reached to 465 mm. Influenced by the rainfall, a multi-peak flood appeared in the mainstream of the Huaihe River. Among them, four flood peaks have been found in the upstream of Wangjiaba of Huaihe River, three peaks in the Wangjiaba-Linhuaigang section, two peaks in the Linhuaigang-Huainan section, and one peak in downstream of Huainan.

The water level of the Wangjiaba-Runheji section rose over the guaranteed stage, and the water level of the Runheji-Wangji section rose to a historically high value. According to a preliminary analysis, the return period of the Mid-Huaihe River is about once-in-15-years to once-in-20-years, while that of the downstream is about once-in-25-years.

The fierce flood lasted for a long time, affected 2.5 million hectares crops, and caused a direct economic loss of 15.5 billion Yuan. However, compared with floods in 1991 and 2003, the loss was reduced by 54.3 and 45.7%, respectively. Why did the same magnitude floods make such differences? It is primarily resulted from scientific regulation of the Huaihe River flood.

6.1.2. Flood management

Flood regulation is an important part of the Huaihe River flood Management work. In 2007, all levels and branches of government facing a serious flood situation took many measures comprising of "intercept, discharge, store, divert" in accordance with "the programs for the defenses of the Huaihe River floods."

(1) Intercept flood with the reservoirs on upstream

In the general flood defense process, the first step is to discharge flood with river channels, and the second step is to intercept water with reservoirs and then to store flood by detention basins. In 2007, the 18 large reservoirs in the Huaihe River upstream, such as Suyahu, Nianyushan, Meishan, and Xianghongdian, totally stored flood with an amount of nearly 2.1 billion cubic meters, which greatly released the flood defense pressure.

Although the reservoir played a significant role in the Huaihe River flood in 2007, in actual operation, it also faced many difficulties. At first, the Suyahu Reservoir and Meishan Reservoir, which are closely related to the flood control and prevention, are facing different levels of safety problems. Reservoir regulations are subject to the safety of reservoir projects, if a reservoir has safety problems, which means that the project itself is prone to the flood, and it cannot intercept flood according to design standards. The second is the difficulty of forecasting. At present, it is difficult to accurately and timely forecast sudden heavy rain in Huaihe Basin for our technology. Third, the provincial conflicts need to be coordinated. Most of the reservoirs in the upstream of the Huaihe River are located in Henan Province. How to weigh the pros and cons has always been a difficult issue.

Suyahu Reservoir regulation is a typical example. The reservoir is located in Henan Province. On July 15, the reservoir water level reached to 54.76 m, over the flood control water level l 2.26 m, and a lot of land was flooded and therefore reservoir operation was facing great pressure. At the same time, as the downstream part of the river exceeded the guaranteed stage, flood situation was very tense. The Flood Control and Drought Relief Headquarter (FCDRH) of Huaihe River, Henan Province, and Anhui Province launched flood consultations repeatedly and regulated Suyahe Reservoir carefully, and finally intercept 500 million cubic meters for downstream on the premise of guaranteeing projects safety, and thus the masses in this reservoir area have paid a heavy price.

(2) Pre-discharge capacity of rivers and lakes

Under normal circumstances, it takes 2 days for the formation of flood peak in mainstream of the Huaihe River from the beginning of rainfall to a flood peak appears. Short-term weather forecasts can basically predict the location and quantity of rainfall. It provides an opportunity to use the project ahead of time. The Bengbu Gate is the important control hydro-junction in the middle reaches of Huaihe River. When rainfall started in upstream and a large flood forecast was reported, the Bengbu Gate opened all 40-hole gates to pre-discharge the water. As a result, when the flood arrived, the water level can be reduced from 17.85 to 16.03 m. The reduction of 1.82 m of water level provided an opportunity to receive the upstream runoff and drainage.

Hongze Lake is the largest lake in the Huaihe Basin, and Sanhe Gate is the largest controlling gate of the lake. On July 4, before the floods arrived at Hongze Lake, the FCDRH of Huaihe River coordinated with the FCDRH of Jiangsu Province and opened Sanhe Gate and began to discharge water. When flood discharge and water level reach to a warning value in the upstream, open Sanhe Gate of Hongze Lake in the lower stream in advance to decrease water level for discharging flood freely. Through analysis and calculation, Sanhe Gate discharged a cumulative capacity of 16.8 billion cubic meters for flood prevention from July 4 to July 31 for totally 28 days and reduced the water level of the lake by 0.2 m, all of which win an advantageous opportunity in flood prevention.

(3) Make full use of detention basins for flood storage

The detention basins are important parts of the Huaihe River flood control system. When large floods occur, the use of the floodplain detention can reduce the flood volume, decrease flood peak, and lower the water level of rivers, which is a major feature of the flood prevention of the Huaihe River. The Mengwa Detention Basin is the first one in the Huaihe Basin. Due to the important location that involves the two provinces, it is scheduled by State Flood Control and Drought Relief Headquarters (FCDRH).

On July 8, the water level raised rapidly due to a heavy rain. The water level of Wangjiaba hydrological station, which is known as the Huaihe River Flood Control barometer and flood weathervane, reached quickly to 29.3 m, the guaranteed stage. At this point, the 18 large reservoirs in Huaihe River upstream were in full load operation, and the flood control capacity was exerted to the most, many rivers exceeded the guaranteed water level, and the dangers of embankment appeared ceaselessly.

Facing the severe situation, the FCDRH promoted the flood emergency response level to grade II on July 9 and to the highest level I on July 10. This is the first time to start the highest level of flood emergency response. In accordance with request of the grade I response, the Mengwa Detention Basin was operated timely when the water level of Wangjiaba of the Huaihe River reached 29.3 m.

However, there are several real problems in storing floodwater, namely the migrant relocation and resettlement, inundated farmland, heavy loss of property, and waste of human and material resources. To operate, or not to operate? That's the question. At that time, the SFCDRH made an analysis and found that there would be three negative results when not using detention basin or using not properly. At first, inundated farmland in the upstream would be increased to 2530 hectares, with an increased area of 1340 hectares more than the use of the Mengwa Detention Basin. The second is the raising of the Huaihe River water level, which would increase the difficulty of flood prevention and threaten safety of the mainstream embankment. Lastly, probability of the use of the detention basins in downstream would be increased. However, the use of the Mengwa Detention Basin means that more than 3000 people have to be transferred, 12,000 hectares farmland would be flooded, and the part of the infrastructure in the region to be destroyed. After the five consultations by the relevant authorities, it was determined finally that the Mongwa Detention Basin would be opened. According to analysis, the basin had diverted water for 46 hours, stored flood with a volume of 250 million cubic meters, lowered the Wangjiaba water level by 0.2 m, reduced the water level of the downstream of the Zhengyangguan pass by 0.1 m, and has played a significant role in clipping flood peak. Subsequently, the Huaihe River Basin started using nine more detention basins diverted floodwater with a total capacity of 1.5 billion cubic meters. The use of these detention basins shorten the duration of Wangjiaba stage at high level nearly 20 hours.

(4) Divert water with the channels timely

It can relieve the pressure of flood prevention in the main stream by regulating diversion rivers to discharge a part of the flood water. Both Huaihongxinhe canal and Ruhaishuidao canal (canal to the sea) played the role of accelerating flood discharge in the Huaihe River flood prevention in 2007.

Huaihongxinhe canal is an artificial diversion river that passes through Jiangsu and Anhui provinces. In late July, the Huaihe River water level was still high. After soaking in high level water, much more dangerous situation of the embankments appeared. On the one hand, the rescue personnel were tired due to the long struggle; on the other hand, the high temperatures of 38°C caused inconvenience for the rescue team and relocated people. In order to reduce the water level of Huaihe River and relieve the defensive pressure of the downstream from Bengbu city as soon as possible, the FCDRH of the Huaihe River actively coordinated Jiangsu and Anhui provinces for reasonable operation of Huaihongxinhe canal, thereby reducing the dike defense and emergency pressure, and 110,000 metastatic masses returned home 2–4 days ahead of time and returned to a normal production and living order.

http://dx.doi.org/10.5772/intechopen.69047

6.2. Room for river project in the Huaihe River Basin

6.2.1. Development of "Room for river" idea in the Huaihe River Basin

From 1950 to 1969, there were 10 flood storage areas (total area is 3788 km^2 and flood storage capacity is 10.18 billion m^3) and 21 flood bypasses (total area is 1301 km^2 and flood discharge capacity is 1000~3500 m^3/s when fully open).

The HRB has constructed five flood storage areas (total area is 3300 km^2, and flood storage capacity is 9.3 billion m^2) and totally 21 flood bypasses. These flood diversion and storage areas constitute as part of the flood control engineering system and played important roles in preventing basin-wide floods in the past.

However, there are also outstanding problems, e.g., too many flood bypasses occupy river's room, evacuation of 1.8 million people is a hard work, etc.

Since mid-1980s to the twentieth century, problems of flood diversion and storage areas have received a great deal of attention, new ideas about the Huaihe harness have progressively formed, i.e., properly reducing number and area of flood diversion and storage areas, enhancing river channel discharge capacity, reducing the probability of using flood bypasses, and reinforcing security construction of people's safety.

By 2010, number of flood bypasses along the Huaihe mainstream reduced from 21 to 17, of which four flood bypasses were abandoned and returned room to river with an area of 24.3 km^2, other flood diversion and storage areas returned room to river with an area of 96.4 km^2 by partial dike retreat, and totally returned 120.7 km^2 area to river.

By the plan completion in 2020, only six of the 17 flood bypasses will be preserved, among which two of the them will be abandoned, three of them will be changed to flood storage areas, and five of them will be adjusted to flood protection areas (**Figure 14**).

Totally, 230 km^2 will be returned to river, width of river channels will be increased 100~2500 m, and flood discharge capacity will be increased from 800 to 1200 m^3/s. The operation standard of the preserved flood bypasses will be promoted to once in 10–20 years (**Figure 15**).

From the start time of the flood diversion and storage areas construction to now, flood diversion and storage areas have experienced from "conflicts between human & water", to "co-existence of human & water" and finally to "harmony between human & water", which also reflects development of idea about "Room for river" in the HRB.

What follows take Jingshanhu flood bypass as a case study.

6.2.2. Room for river: Jingshanhu flood bypass as a case study

Jingshanhu flood bypass is about 4 km long from east to west, and 17 km wide in south-north direction, with an area of 72.0 km^2. It plays a vital role in raising the flood-diversion capacity and decreasing water levels of Bengbu and Huainan cities.

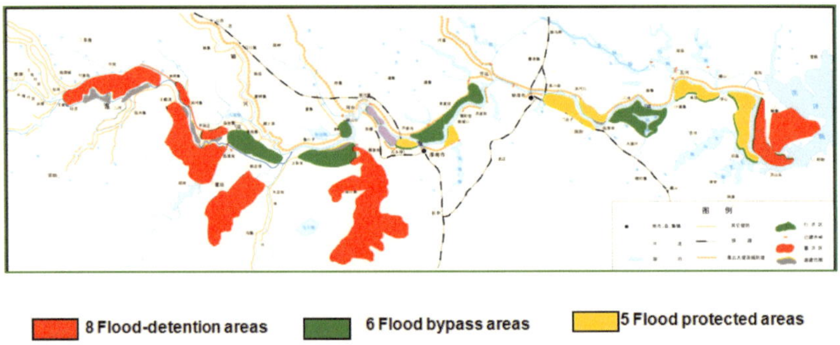

Figure 14. Adjustment of flood detention areas and flood bypass in 2020.

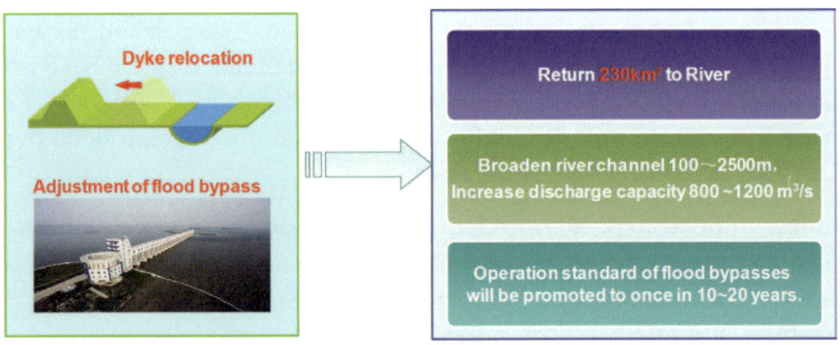

Figure 15. Achievements after completion of adjustment of flood detention areas and flood bypass.

The major problems before the harness were as follows: (1) The river channel is narrow, river channel for flood discharge is only 500 m wide, and maximum discharge capacity is only 5700 m^3/s. (2) There are 10,000 people in the Jingshanhu flood bypass and most of them live in the plain area, which involves a large amount of works when carrying out immigration. (3) The flood bypass operates so frequently that it causes serious property losses. (4) The controlling gates of the flood bypass are so inconvenient that it cannot be timely and effectively operated. (5)Reconstruction after disaster involves hard works.

By retreating, reinforcing, and newly constructing dykes with a length of 42 km, constructing one withdrawal sluice and one intake sluice, and security construction, e.g., retreat road construction, reinforcement of Zhuangtai (small villages on raised ground with higher elevation), communication and warning facilities, etc., the chance of flood diversion was promoted from once-in-7-years to once-in-15-years, and it has significant effects.

After the control measures were implemented, the following achievements have been made:

1. The problems of local people's security have been solved. There are only 855 people living in the flood-protected villages.

2. The flood dispatching measures have been strengthened. In 2007, the Huaihe River Basin was hit by a basin-wide large flood event again and the newly-built intake and withdrawal sluices firstly came into use and showed good effects.
3. The flood passage of the mainstream has been enlarged. After the implementation of dike reconstruction and reinforcement, the width of the dike inside reached to 600 m, and the release discharge of the reaches lower than Zhengyangguan pass in the mainstream can reach to 8000 m^3/s without using the flood bypass.
4. Probability of using the flood bypasses has been promoted from once-in-7-years to once-in-15-years.
5. If inundated, economic losses of peasants have been compensated by our country and it is conductive to resume production.

Table 4 summarizes control measures and achievements of Jingshanhu flood bypass.

Control measures	Achievements
Relocating, strengthening, and newly constructing 42 km dykes and dredging 3.7 km river channels	Returning 4.5 km^2 lands and enlarging river channels with a width of 600 m, and thus discharge capacity of these reaches has been increased to 800 m^3/s
Constructing one withdrawal sluice and one intake sluice	Intake and escape sluices have been constructed for timely and effective operation
Constructing inside security facilities	Inside security facilities, e.g., retreat road and Zhuangtai (village on raised ground), etc., have been constructed
Most of the people living inside have been relocated to other safe areas outside	A total of 800 people have been relocated to safe villages on raised ground and the rest 9000 people have immigrated to flood protection areas and got rid of flood threat
Constructing drainage pumping stations and improving drainage system inside	Two drainage pumping stations and improving drainage system inside have been constructed

Table 4. Control measures and achievements of Jingshanhu flood bypass.

7. Experiences and enlightenment

China is a flood-prone country, and flood disasters occur frequently. The Chinese Government attaches great importance to flood management and drought relief, and great efforts and achievements have been made with aid of the structural and non-structural measures. However, flooding is still a big issue in China. The ability to control floods needs further improvement of the non-structural measures, including the relevant laws, monitoring networks, warning and forecasting, and social management [14].

Both the structural and non-structural measures are very important for flood control and management. However, for the over-standard flood cases, the non-structural measures, such as the hydrological monitoring and flood forecasting, become much more important.

Flooding is still a big issue in China. The ability to control floods needs further improvement of the non-structural measures, including the relevant laws, monitoring networks, warning and forecasting, and social management.

7.1. Experiences in flood management

Flood disaster, which occurred suddenly and inevitably, is different from the general emergency disasters. It is determined by the natural conditions of the Huaihe Basin. By review of the Huaihe River flood regulation, it is found that the keys of success are still dependent on the following factors:

1. The complete flood control structural system is the foundation of flood regulation and management. The standard and quality of the flood control works are directly related to flood management and projects operation. Although the standard of the flood control system in Huaihe Basin is not high enough, it is compounded by a wide range of works with high correlation. Their operation is very flexible, especially the flood regulation of detention basins and diversion channels like Huaihongxinhe canal and Ruhaishuidao canal (floodway to the sea), and plays a vital role in flood control. In future, we should strengthen the construction of structural system for flood prevention and increase the flood control standard appropriately.

2. Accurate hydrologic forecasting is the prerequisite of the flood regulation. The forecast accuracy and the lead time will directly affect the correctness and timeliness of the flood regulation decision. In recent years, the extreme weather events have increased significantly from global perspective. The sudden strong rainfall is unforeseeable, and it becomes a new issue for flood prevention work. In the future, we should continue to strengthen early warning system, to optimize flood forecasting model, and to improve forecast accuracy and quality. Particularly, there is an urgent call for constructing an ensemble flood forecasting system integrating with numerical weather models, distributed hydrologic models, hydraulic models, and real-time control models.

3. Scientific analysis and judgment of the flood are the key to flood regulation. Flood regulation should consider all factors as an integrated system, including upstream and the downstream, both sides of river, global and local, flood control and drought relief, as well as economy and society. To balance different interests is a hard nut to crack in decision-making. In future, we need to continue to strengthen the flood management and risk management and to resolve the problems in laws and regulation, mechanism, technology, and other issues.

4. Advanced technology and perfect plan are the effective support for flood regulation. Huaihe River floods in 2003 and 2007 are not only a test of the flood structural system, but also a full inspection of the Huaihe River non-structural system. All of advanced technology and comprehensive plan of flood monitoring and forecasting, flood control scheduling, emergency management mechanism, and the joint regulation of the flood engineering system play an important role in joint operation of flood control structural system. In future, we should learn from domestic and international flood management experience, promote the application of high technology, constantly improve the flood control, and strive to reduce disaster losses.

7.2. Outlook on flood management in the Huaihe River Basin

In the future, we have to change from flood control to flood management and finally achieve the goal of flood risk management.

1. Room for River. In terms of river planning and training, it is necessary to make more room for flood and formulate relevant schemes and measures by comprehensive analysis of flood. Based on the structural system comprising of reservoirs, dykes, and flood detention areas, we have to enlarge passageway for small and middle magnitude floods in river harness for the purpose of releasing such kind of floods. To construct flood detention areas to receive extra floods that exceed the river channel capacity aims to solve the big flood issues.
2. Live with floods. In order to realize harmony between human and nature, first and foremost, an integrated flood control and management system should be well established on an operational level. In addition, in order to achieve a shift from flood control, flood management to flood risk management, there is urgent call for raising awareness of flood risk management both for the leader and the public.
3. Flood risk management. We do not have to control all floods of different magnitude, and we cannot bear enormous flood disaster risk. We have to avoid risks actively, take preventive measures as high priority, live far from flood disasters, and control flood disaster risk to a certain extend. We have to determine reasonable flood protection standards, and the standards should not be too high or too low. We have to classify function of flood structures in a reasonable way, because flood disaster risks could propagate if function of flood structures is not classified correctly or properly. We have to share risks. Construction of flood structures can absolutely cause risk propagation; therefore, we have to treat risk propagation in a right way and different regions have to share flood risk and construct flood risk compensation mechanism and flood insurance system.

Acknowledgements

The study was financially supported by Non-profit Industry Financial Program of MWR of China (201301066 and 201501007) and National key research and development program of China (2016YFC0402700).

Author details

Qian Mingkai and Wang Kai*

*Address all correspondence to: wangkai@hrc.gov.cn

Hydrologic Bureau of Huaihe River Commission, Bengbu, China

References

[1] Huaihe River Commission (HRC). Handbook of Water Resources in the Huai River Basin. Beijing, China: Science Press; 2003. p. 454

[2] Huaihe River Commission (HRC). Huaihe River Year Book 2014. Bengbu, China: Huaihe River Press; 2014. p. 386

[3] Huaihe River Commission (HRC). The 1991 Huaihe River Storms and Floods. Beijing, China: China Water Publication; 2010. p. 180

[4] Huaihe River Commission (HRC). The 2003 Huaihe River Storms and Floods. Beijing, China: China Water Publication; 2006. p. 275

[5] Huaihe River Commission (HRC). The 2007 Huaihe River Storms and Floods. Beijing, China: China Water Publication; 2010. p. 306

[6] Yan J, Jun X, Gang-sheng W. Simulation of environmental change in response to operation of dams in Huaihe Basin. Water Science and Engineering. 2009;**2**(3):27-36

[7] Li ZJ, Bao HJ, Xue CS, Hu YZ, Fang H. Real-time flood forecasting of Huai River with flood diversion and retarding areas. Water Science and Engineering. 2008;**1**(2):10-24

[8] Zhen-kun MA, Zi-Wu, B.S. F, Zhang M, Su Y-l. Flood risk control of dams and dykes in middle reach of Huaihe River. Water Science and Engineering. 2014;**1**(7):17-31

[9] Xiaotao C. Changes of flood control situation in China & adjustments of flood management strategies. In: Wu, B.S et al., editors. Flood Defence '2002. New York: Science Press; 2002

[10] Jingping E. Report for the National Workshop on Flood Control and Drought Relief Headquarters [Internet]. 2003. Available from: http://www.cws.net-ch/minister/index.asp/2003-01-16

[11] Zhitong Z. Report for the National Workshop on Flood Control and Drought Relief Headquarters [Internet]. 2004. Available from: http://www.cws.net-ch/minister/index.asp/2004-02-10

[12] Jingping E. Report for the National Workshop on Flood Control and Drought Relief Headquarters [Internet]. 2005. Available from: http://www.mwr.gov.cn/ztbd/2005fbzrhy/20050110/45413.asp

[13] Xiaotao C. Large-scale flood control activities in the new era calling the guidance of reasonable theory: One discussion on the flood risk management with Chinese features. Water Resources Development Research. 2001;**2001**(4):1-6

[14] Zhang J, Liu Z. Hydrological monitoring and flood management in China. In: Tchiguirinskaia I, Thein K, Hubert P, editors. Frontiers in Flood Research. IAHS Publ; 2006. pp. 93-101

Chapter 8

Two-Dimensional Hydraulic Modeling and Geotechnical Analysis of Earthen Regulation Dams Located at Arroyo Las Viboras Watershed in a Major Transboundary Mexico-USA Metroplex: How an Ordinary Rain Event Caused Major Damage Related to Extraordinary Flooding?

Oscar Sotero Dena Ornelas, Diane Irene Doser,
Oscar Fidencio Ibañez Hernández,
Griselda Janeth Obeso Cortez and
Miguel Angel Galdean Vega

Additional information is available at the end of the chapter

http://dx.doi.org/10.5772/intechopen.68853

Abstract

Despite low annual precipitation rates, major cities located in deserts are prone to flash flooding events, which incur important economic costs and in some cases loss of human life. The northern Chihuahua desert metroplex of Ciudad Juarez, Chihuahua-El Paso, Texas, is a region where seasonal summer rains result in flash flooding. One of the most important flooding zones in Ciudad Juarez occurs at the Arroyo Las Viboras. Detailed light detection and ranging (LIDAR) terrain models were incorporated during two-dimensional modeling to understand flow conditions for two specific ordinary rain events. In addition, electrical resistivity tomography and seismic geophysical studies were conducted to examine if the hydraulic infrastructure was properly emplaced from a geological and geotechnical perspective. The results showed that Las Viboras watershed is almost hydrologically unprotected since the major dam is not regulating water flow volumes. The Camino Real transmountain road is improperly acting as a retention wall and geotechnical and geophysical findings indicate that Puerto La Paz dam has to be moved to a new location since it is currently located on top of plastic clays. El Filtro dam poses a higher risk because.Finally, three more dams are required to be constructed upstream in order to ensure hydrological resilience of the watershed.

Keywords: desert, flash flooding, bi-dimensional modeling, geotechnics, resilience

1. Introduction

Despite low annual precipitation rates of less than 300 mm/year [1], the urbanized areas of the northern Chihuahua desert paradoxically experience flash flooding events as a consequence of seasonal summer rains that often release more than 50% of the yearly rain in only 48 hours [2]. The main population center located in this semiarid region is the twin city, transboundary metroplex formed by Ciudad Juarez, Chihuahua and El Paso, Texas. The denser populated area corresponds to the Mexican city of Juarez, with more than 1.5 million inhabitants distributed along the western piedmont of the Sierra de Juarez (SDJ) mountain range and the Rio Grande River (RGR) basin's main depocenter, defined by the axis of the Rio Grande River (**Figure 1**). This growing population experiences severe property damage on a yearly basis as consequence of seasonal rain events.

Ciudad Juarez's hydrology is divided into eight basins according to the municipal research and planning agency [2]. Although all the basins become problematic flooding areas during seasonal rains, flooding at Las Viboras watershed's downstream area, located in the Anapra

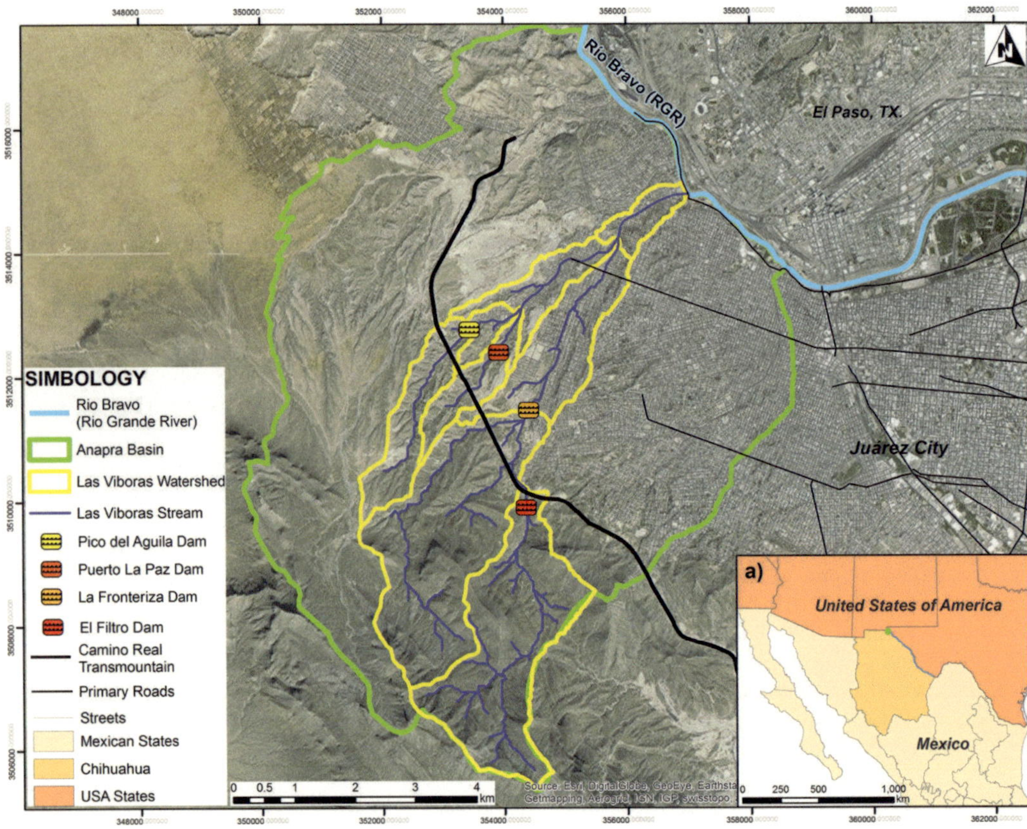

Figure 1. Location of the Cd. Juarez, Chihuahua–El Paso, Texas Metroplex and Las Viboras watershed, earthen regulation dams and hydrologic network.

http://dx.doi.org/10.5772/intechopen.68853

basin (**Figure 1**), has been historically one of the most complicated water volumes to be managed efficiently. Anapra is the only basin in Juarez that drains directly into the RGR. The hydrological protection of Las Viboras Arroyo Watershed depends on four existent earthen regulation dams: Pico del Aguila, Puerto La Paz, La Fronteriza and El Filtro (**Figure 1**). However, ordinary rain events in November 2016 recorded important flooding and high flows at the Rio Grande River delivery point, precisely indicating a lack of flow regulation. In this research, we have combined a holistic approach by integrating hydrology, bidimensional hydraulic modeling and geological-geophysical methods to investigate how the ordinary rain event of November 4, 2016 resulted in severe loss of life and property in Las Viboras Arroyo watershed. We also explored what would happen if El Filtro dam would have failed during a 500 year return period (YRP) event and how the Camino Real transmountain road culverts operate under hydraulic stress.

2. Background

2.1. Hydrology

Ciudad Juarez's hydrology is divided into eight basins according to the municipal research and planning agency [2]. Three basins are located at the Sierra de Juarez (SDJ): Anapra, Centro and Jarudo, whereas the Aeropuerto basin, adjacent to Jarudo basin, is located in a nearly flat area. The Chamizal and Rio Bravo basins are located next to the Rio Grande River (RGR), and the Barreal and Acequias basins are of endhorreic type.

As expected, the hydrological drainage network is chiefly controlled by the topography of the SDJ. This mountain range reaches elevations of 1800 m above sea level, resulting in a relative height difference of nearly 470 m with respect to the surrounding basin floors and of 490 m with the RGR.

In spite of the dramatic topographic contrast, the only basin draining directly into the RGR is the Anapra. Las Viboras Watershed is the main watershed in Anapra basin. This system, Las Viboras, collects runoff volume from a catchment area of nearly 22 km^2 [3]. Although the watershed is supposed to be protected by three regulation dams, historically this watershed has experienced some of the greatest damage associated with hydro-meteorological events in Ciudad Juarez. In 2006, during a 10 year return period (YRP) event, the hydrological resilience of this watershed was demonstrated as being poor since the most important dam, in terms of capacity, was overtopped during the rain event, requiring authorities to evacuate population located downstream due to the imminent risk of the dam's wall breaking [3]. After the rain event, the dam had to be artificially breached by constructing a 70 m^3/s channel to prevent water accumulation. As a temporary solution a new retention wall was emplaced upstream of the dam to mitigate water flow. However, and regretfully, the temporary solution became permanent. Now, Las Viboras system seems to be even more compromised, in hydrological terms, than in 2006, as consequence of the lack of infrastructure maintenance and a poor urbanization plan that allowed several new low-income neighborhoods to be developed downstream without regard to the hydrological safety.

2.2. Geologic and tectonic setting

Structurally, the major tectonic feature controlling recent basin formation and deposition style in the area is the Cenozoic Rio Grande Rift. Extension in the rift has produced an asymmetric intramontane basin system controlled by normal faulting that has overprinted the Laramide-related compressional structures. Among the Laramide-related structures, the Sierra de Juarez is the major landmark in the region. The stratigraphic sequence of the SDJ is composed of Mesozoic geologic units of a transgressional marine sequence deposited in the Chihuahua Trough [4, 5]. Shales and sandstones characterize the lower part of this sequence, whereas the middle part is formed of limestones and shales, topped with post reef limestone facies [6]. Volcanic activity occurred in the Paleocene with the emplacement of acid to intermediate composition intrusive igneous rocks. Well-consolidated deposits, such as conglomerates, silts and sands, were deposited discordantly during the Plio-Pleistocene [7].

The top of the stratigraphic column is composed of interbedded sequences of coarse clastic sediments and fine grain materials transported from the more topographically elevated areas and deposited in the basins as alluvial, lacustrine or aeolian sediments. The basin fill is mainly Pliocene with some thinner quaternary units on top.

Basement units that outcrop in the nearby SDJ include the Cretaceous Cuchillo, Benigno, Lagrima and Finlay formations [8]. Each formation has very distinctive hydraulic and mechanical properties as a consequence of the compositional transitions between massive coral limestones (Benigno and Lagrima formations) to interbedded limestone-sandstone (Cuchillo) and limestone-shale (Finlay) formations that are less permeable.

In terms of geologic elements to provide structural support for proper dam foundation, emplacement and construction, the topographic closures flanked by competent rock units are present only in the highlands of SDJ. The best geological units, the massive limestone rock units are related to the Aurora group, which are the most competent. On the other hand, flaky shales (Ojinaga and La Casita formations) associated with Turonian and Tithonian horizons, respectively, should be avoided [8, 9].

3. Methodology

First, we used a precise light detection and ranging (LIDAR) derived digital terrain model (DTM) to determine watershed boundaries and update other physiographic elements for calculations of surface hydraulic flow. We also modeled the flow conditions through bidimensional hydraulic modeling of the Naviers-Stokes equations using the IBER software package to investigate how hydraulic structures such as culverts and earthen regulation dams are actually operating and why dangerous high velocity and turbulent flows are recorded downstream even in ordinary rain events. In addition, we also investigated the subsurface geologic setting by conducting high-resolution electrical resistivity tomography [10], which combined with seismic refraction and S-wave velocity studies [11], allowed us to infer the foundational and structural integrity of Camino Real culverts and regulation dams located in Las Viboras stream drainage system.

http://dx.doi.org/10.5772/intechopen.68853

The first step in this study consisted of modeling the hydrology of the watershed to estimate the runoff volume transported by Las Viboras watershed drainage network for ordinary rains. Then, once the hydraulic flow was calculated, geological, geophysical and geotechnical studies were conducted at several structures.

3.1. Hydrologic modeling

To calculate the total runoff flow and water volume collected along the drainage network of Las Viboras watershed for ordinary events, we distributed a 2 YRP event, or 50% probability of exceedance rains in 24 hours, across the watershed. The watershed boundary was defined from a high resolution LIDAR 1 m × 1 m cell size DTM utilizing the software package Hec-GeoHMS [12], which runs on the GIS software platform ArcGIS.

The watershed geometry, drainage network and the physiographic parameters: area (km^2), slope (10–85%), length (m), concentration time (s) and lag time (s) were obtained from the Geo-HMS run and then validated by hand calculations. The runoff and/or infiltration were estimated applying the curve number (CN) method, also referred as "Soil Conservation Service (SCS) runoff Curve number" [13], which defines the rain abstraction or how much rain infiltrates into the ground in terms of the soil group, hydrological conditions and land-use of the catchment area. Both the SCS CN and the impervious percentage were derived from high-resolution aerial imagery analysis and fieldwork. The precipitation-runoff transformation, hydrogram estimation and hydraulic parameters (flow and volume) at each hydrologic element were estimated as a result of the hydrologic simulation calculated from storm data for 2, 50, 500 and 10,000 YRP and physiographic parameters as input data into the HMS software package [14].

3.2. Hydraulic modeling

Because of the complexity of Las Viboras arroyo system, we decided to apply two-dimensional modeling to simulate arroyo hydraulics. The software utilized (IBER) for the calculations solves the 2D Saint-Venant equations [15] using the finite volume method, which relies on a nonstructured finite discretization of the terrain. The mesh representation obtained in this way allows the software to be able to represent almost any surface geometry, being able then, to model subtle river features. To perform the calculation, a 1 m × 1 m DTEM derived from LIDAR data was utilized, the hydrologic parameters were input as a rain hyetograph for 2 and 500 YRP rain events, and bed roughness was modeled with manning roughness coefficients. The results are a grid representation of the output parameters such as water depth, velocity, Froude number, etc.

3.3. Geoscience: geology, geophysics and geotechnical

Geological studies are traditionally applied to determine the best wall dam emplacement location based upon topographic conditions, water-tightness of the reservoir, slope stability along the reservoir perimeter and availability of construction materials. The emplacement depends upon the dam foundation requirements, which in turn, are a function of the type of dam, dynamic moduli of the soil such as strength and deformation, depth to foundation and geohydrological properties such as permeability. In this case, since the Earthen Regulation Dams (ERD) were already constructed and information about their construction process is no longer available, geoscience approaches, mainly geophysical methods to determine geotechnical parameters, were applied not only to infer foundation characteristics but also to reveal, in an indirect way, the structural integrity of the culverts and three dams. More specifically, three geophysical methods were applied: Electrical Resistivity Tomography, Seismic Refraction and multichannel analysis of seismic waves (MASW).

Multielectrode electrical resistivity tomography (ERT) is a geophysical method that allows the user to infer the subsurface conditions in terms of the electrical contrast existing between different subsoil and geologic units. The contrast relies primarily on the grain size, water and mineralogical content of the units [16]. Therefore, this approach is appropriated to infer conditions related to earthen dams and foundation [22]. The high resolution data were collected along several profiles atop each dam's crest. We used a Terrameter LUND System with electrodes connected every 2 m to a multicore set of cables attached to a selector switch that controls the injection of current and ground resistance readings in accordance with a predetermined acquisition protocol. The ABEM instrument computer automatically selects the current injection, which ranges from 200 to 500 mA. To obtain a reliable measurement of the ground resistance, 4 stacks per reading were averaged to obtain the actual recorded resistance value. The electrode arrays applied in the field were the Wenner and dipole-dipole configurations. Details of these arrays can be found in Refs. [10, 19].

Once the data were collected in the field, the electrical resistance field for each layout was geometrically corrected using the software package ERICGRAPH© [18]. The apparent resistivity pseudosections were then inverted using a robust L1 norm algorithm, which minimizes the sum of the absolute values of the resistivity field spatial variations in order to map lateral heterogeneities [10]. The inversion was then implemented with the computational package RES2DINV to obtain the true electrical resistivity field and true depth of penetration sections that gave the least possible error between the observed and calculated data.

Editing of the noisy data and half electrode spacing cell width was applied during modeling since the data showed high surface resistivity variations [19]. As a last processing step, the inverted data were corrected for topographic effects and converted into ASCII format to develop a geographically referenced geoelectrical database. This was further post-processed in Oasis Montaj to render a resistivity voxel to better visualize the underground electrical resistivity distribution.

Seismic studies were conducted because the elastic properties of soils and subsurface can be inferred by properly mapping both compressional and shear wave velocity fields [11, 20]. The dynamic moduli: shear modulus, Young's modulus and Poisson's ratio provide a characterization of the soil-structure interaction and liquefaction potential of subsurface layers [20]. In this study, the seismic refraction lines were deployed with a geophone separation of 6 m, and the MASW lines with the same length, except for the culvert analysis, where a 3 m spacing between geophones were applied. The processing of the compressional velocity (Vp) field consisted in the picking of first arrivals observed at each shot gather [21]. Then a tomographic inversion iterative process was applied to minimize the error misfit, since abrupt lateral variations were expected due to topographic and subsurface heterogeneities. For the S-wave dispersion studies (MASW), the dispersion curve was extracted by analyzing all traces in a shot gather and applying the Tau-P transform to obtain a coherent energy section relating *Vs* with frequency. The dispersion curve picked from the *Vs*-frequency spectra was then inverted for a 1D depth-velocity (*Vs*) model. By then applying this procedure from one shot gather to another and using a common mid point (CMP) mapping of the seismic energy, we were able to generate a 2D section representing the shear wave velocity field.

4. Results

4.1. Hydrology

One of the main results of the hydrologic analysis of Las Viboras watershed shows how the paving process, linked to the urban development, has negatively affected the hydraulic condition of this watershed, since almost all the rain volume is conveyed downstream as runoff as consequence of the increase in impervious surfaces. According to the hydrologic modeling parameters, this watershed has a catchment area of 22 km^2. It is intended to be hydrologically protected by four regulation structures: The Pico del Aguila dam, regulating the north branch runoff flowing down from the SDJ, the Puerto La Paz dam, regulating volumes from the central branch of Las Viboras drainage system, La Fronteriza dam, regulating flow from the south-central and southern branches of Las Viboras and El Filtro, regulating upstream flow on the southern branch of Las Viboras. As consequence of the 10 YRP event of 2006, La Fronteriza dam failed by overtopping. It is now breached to avoid water accumulation since the dam is structurally compromised. The El Filtro dam, located upstream, was then emplaced to temporarily regulate the southern branch runoff, thus decreasing the hydrologic pressure downstream. We modeled each hydraulic structure, including Earth Regulation Dams and culverts, to determine their hydraulic operation, including storage, pool elevation, inflow and outflow. The dams were analyzed for ordinary rains (2 YRP) up to extreme events (10,000 YRP), whereas culverts were analyzed for 500 YRP rain events. The hydraulic operation for each dam for 500 YRP and 10,000 YRP rains is shown in **Figure 2**.

4.1.1. Culverts

The natural drainage system of the Sierra de Juarez was modified as consequence of the construction of the transmountain road, El Camino Real, in 2008. In the area covered by the Las Viboras system watershed, there are 10 culverts to allow runoff to freely flow downstream. The slopes of those structures are greater than 3%, and depending on the flow, they are either pipelines or concrete rectangular structures with a cross section of 4.5 m × 4.5 m. The hydrologic analysis of the culverts showed that they operate satisfactorily up to rain events of 500 YRP. However, the culvert located at the discharge of the El Filtro dam shows that downstream of El Camino Real, the discharge channel makes a very sharp turn, almost 80° to the north. Two-dimensional hydraulic modeling is required to determine if there is a hydraulic jump as a consequence of the sharp channel geometry.

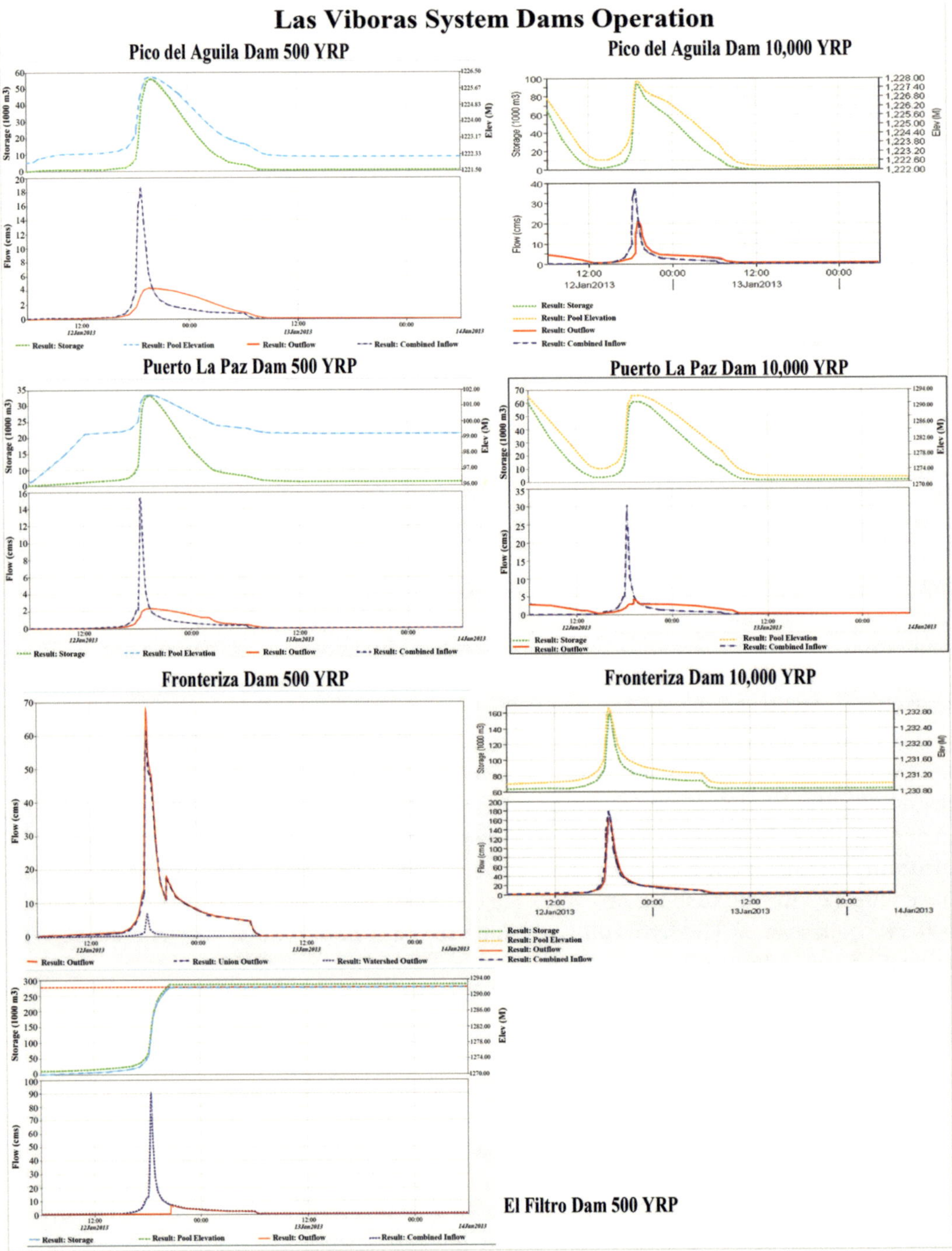

Figure 2. Models of hydraulic operation of dams for 500–10,000 YRP. The results were obtained by 1D hydrologic modeling.

4.1.2. ERD Pico del Aguila

The Pico del Aguila dam has both a riser spillway and emergency spillway. The outlet pipeline has a diameter of 75 cm. The dam crest reaches an elevation of 1221.50 m above sea level and the maximum impoundment capacity is 61,226 m^3. The GEOHMS model shows that the total catchment area for this hydraulic structure is 2.11 km^2, and the model results indicate that this structure operates properly for ordinary rains and extraordinary rain events of up to 500 YRP with a freeboard of 1.55 m. However, for extraordinary or extreme events, such as the 10,000 YRP, the hydraulic operation is seriously compromised since the water flow would overtop the dam's wall, eventually breaching the structure.

4.1.3. ERD Puerto La Paz

This regulation dam is designed to regulate the runoff volumes moving downstream from the southern central part of the highlands of Las Viboras system, which are located at the scarped flanks of the Sierra de Juarez. The results of the hydrologic run show that the dimensions of the reservoir and wall of this regulation dam are able to operate satisfactory up to a 500 YRP rain, with a freeboard of 1.55 m. However, the analysis for a 10,000 YRP rain with full reservoir indicates that the structure will be overtopped by the water volume. Thus, the dam is not hydraulically operating properly according to Mexican law, which requires that any dam should be able to transit a hydrogram corresponding to a 10,000 YRP runoff, assuming that the reservoir is full. In spite of this failure in hydrological safety of the Puerto La Paz dam for extreme events, the structure can very efficiently handle ordinary rains.

4.1.4. ERD El Filtro

This structure is not even a dam since it lacks a riser spillway, emergency spillway and outlet. This structure is just a pile of dirt and pieces of rock obtained from the materials excavated during construction of El Camino Real that were piled up at the pilot channel and berms of the stream. Although this structure has retained water from ordinary rains not exceeding 5 YRP, it eventually drains out through the interstitial spaces of the rocks that were piled up to form the embankment. The elevation of this dam reaches a height of almost 13 m above the main stream channel. The hydrologic analysis of this structure shows that is capable of retaining water volumes for rains up to 100 YRP, but is overtopped in a 500 YRP event. If this occurs the structure will dramatically be breached causing major flooding downstream. This structure was intended to be a temporary solution due to the lack of water regulation downstream as a consequence of the La Fronteriza dam failure in 2006, but its poor construction and the lack of an emergency spillway, makes this structure a manmade hazard that jeopardizes life and property downstream.

4.1.5. ERD La Fronteriza

La Fronteriza dam was designed to regulate runoff volumes from the SDJ and avoid flooding-related problems downstream. This dam has a wall with a design height of 12 m from the stream's pilot channel to the dam crest. It has a riser spillway, outlet and an emergency spillway. The emergency spillway is no longer in use since several years ago poorly planned urban development allowed the construction of houses on top of the emergency channel. The original design storage volume of this dam was 300,000 m^3, but sediment transport and accumulation has severely reduced its storage capacity by up to 90,000 m^3 [3]. This structure failed by overtopping in 2006, when a 10 YRP rain occurred in the metroplex area. The spill over the crest was the consequence of the diversion of one tributary stream of the southernmost branch of the Arroyo Colorado. The water flow was diverted and connected downstream through a structure known as "La Gasera," which is a diversion wall and channel. The extra volume received by the La Fronteriza dam during the 2006 rain, coupled with the reduction of storage capacity due to sedimentation, resulted in the overtopping failure of this structure. Following this event, the dam's wall was intentionally breached with a channel with a hydraulic section and slope capable of transporting water flowing at rate of up to 70 m^3/s, as revealed by the hydrologic analysis.

4.2. Hydraulics

The bidimensional hydraulic modeling we carried out consisted of modeling two distinct scenarios. First, the November 4, 2016 rain, associated with a 2 YRP, was modeled. Once several pitfalls were identified in the ordinary rain modeling and hydrological results, a second rain scenario was modeled with a design storm associated with a 500 YRP rain event. The 500 YRP event would generate, according to the 1D hydrologic modeling results, a breaching scenario for El Filtro dam. This second scenario was designed to simulate the breaching or failure of El Filtro to evaluate both El Camino Real culvert operation and downstream effects.

The main result for the 2 YRP modeling showed the lack of water regulation at La Fronteriza dam is a relative concept since the width of the reservoir prevents all water from flowing freely downstream through the channel at 70 m^3/s, recording reservoir depths up to 2 m even for ordinary rains (**Figure 3**(a)). Nevertheless, this regulation is not sufficient since this drainage network branch is still the major contributing factor to the nearly 1 m water depths reached downstream of this location. In the same way, the bi-dimensional model at Pico del Aguila dam shows that this structure is properly regulating, reaching water depths of up to 1.60 m behind the dam. This is also the case for the Puerto La Paz dam, where runoff volume is properly regulated reaching depths of 1.80 m (**Figure 3**) behind the dam. Although this combined runoff flow regulation effectively diminishes the rate of water flowing downstream along this tributary stream, water depths of 0.40 m are still reached before the intersection of Las Viboras drainage with the RGR (**Figure 3**). This analysis also indicates that El Camino Real's culverts are retaining water, even for a 50% PE rain; a result not anticipated by the HMS modeling (**Figure 3**). Then, El Filtro dam is retaining, rather than regulating, a water volume flowing from the southern tributary watershed with a catchment area of nearly 6 km^2. For heavier rains it is anticipated that this poorly constructed wall, lacking riser spillway and emergency spillway or outlets may fail, perhaps just increasing the hydrologic risk rather than reducing it as forecasted by the HMS modeling results.

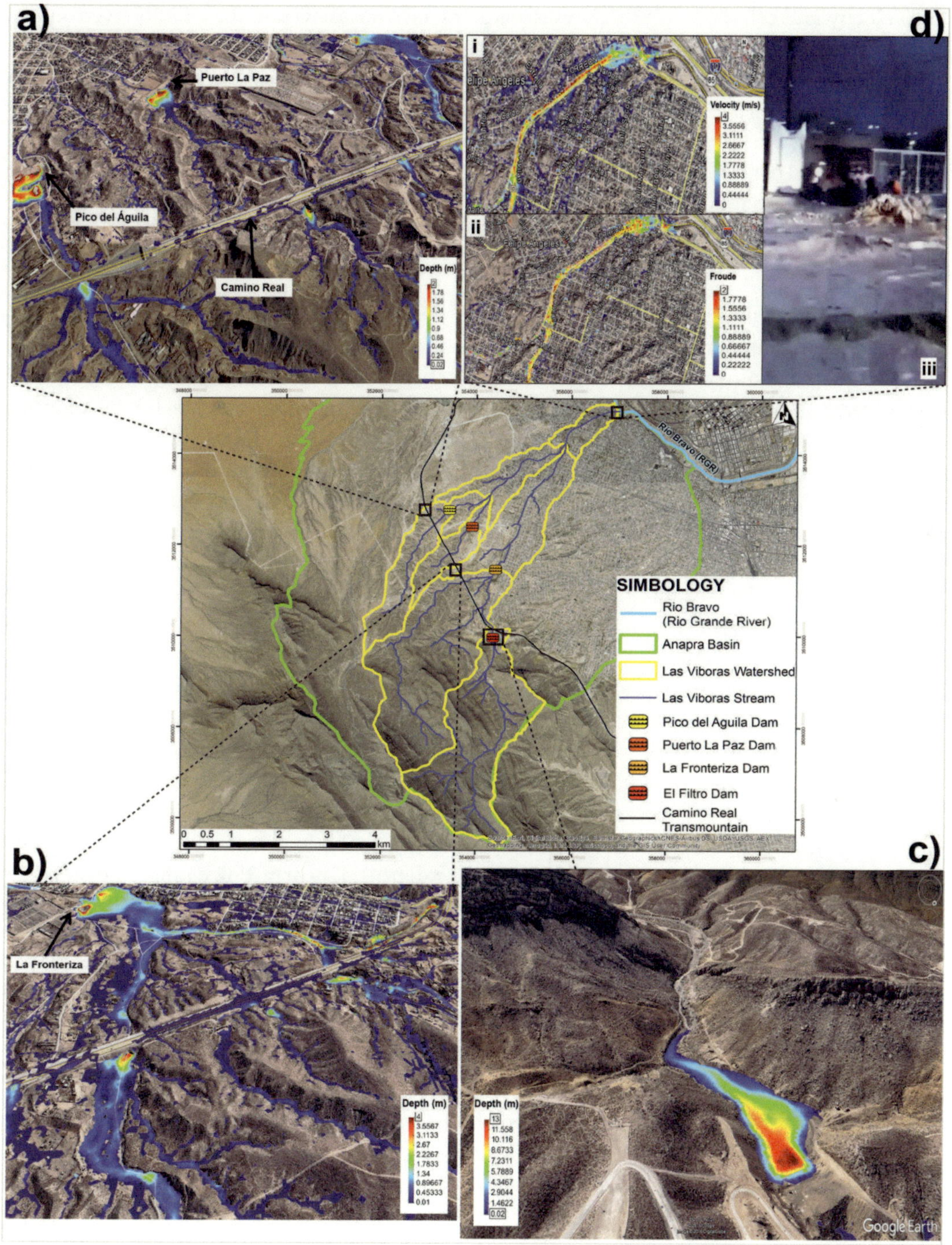

Figure 3. The results of 2D hydraulic modeling. (a) Water levels at Pico del Aguila and Puerto La Paz dams. (b) Water levels at culverts along El Camino and at La Fronteriza Dam. (c) Water levels at El Filtro Dam. (d) Velocity (i) and celerity (ii) results, (iii) turbulent flow at Las Viboras delivery point into Rio Grande River, photo taken by Azteca Noticias news.

The velocity and celerity analysis downstream of Las Viboras system for a 2 YRP even shows that after the tributary and main channels have joined the flow velocity is high, recording values of 4 m/s (**Figure 3**(i)). The celerity analysis resulted in Froude number is greater than 1 (**Figure 3**(ii)), thus a high velocity turbulent flow is present at Las Viboras' delivery point to the RGR. This is a consequence of the lack of proper regulation at the main channel due to the non-operational condition of La Fronteriza dam. The turbulent flow was the hydraulic feature that caused major damage in November 2016, causing severe economic and loss of life, not as consequence of an extreme rain event linked or related to climate change, but for an ordinary rain event with a 50% likelihood to occur yearly.

The second hydraulic scenario corresponds to El Filtro dam breaching for a 500 YRP event. The results are consistent with the hydrology, resulting in a runoff volume of 260 m^3/s flowing through the simulated breach. This volume shows that the 4.5 × 4.5 m culvert located downstream is not able to efficiently conduct the flow (**Figure 4**), yet the 1D calculations showed it as capable of transporting 360 m^3/s. The hydraulic jump generated at the upstream culvert opening reaches depths of 12 m, so that El Camino Real's embankments act as a regulation dam, albeit they are not designed to do so. The detailed bi-dimensional modeling of the breached El Filtro dam-culvert system, also shows that the closed turn of the channel at the downstream culvert outlet produces turbulent and erosive velocities which are scouring the diversion wall of "La Gasera" located only 35 m to the north (**Figure 4**). As consequence of these results, the system was modeled again for a 2 YRP, but with the hypothetical scenario of removing El Filtro dam. The results show dramatic downstream effects; the downstream culvert is completely overtopped with water depths of 4 m both sides of the culvert, whereas water depths increase up to 2 m for the last branch of the drainage network. To conclude the hydraulic analysis, field work was conducted to identify erosional features such as gullies at the slope toe of "La Gasera" wall. The result was even more complex than predicted by the model; the wall scouring process related to several ordinary rain events show that the culvert is partially occluded with debris scoured from the wall. The hydraulic jump caused by the occlusion at the culvert outlet is recorded by water depths of nearly 3 m, as shown by a water line traced upstream across the culvert passage (**Figure 4**).

4.3. Geotechnical assessment

Hydrological analysis is not the sole criteria to determine if a regulation dam is operating safely. Structural analysis is a key factor to provide an assessment of the structural integrity of the wall and foundations. Several methods were applied in this study to do this. First, a visual inspection of the wall showed very evident fractures and vertical displacements along the wall. This was followed up with other geological and geophysical analyses outlined below.

4.3.1. Pico del Aguila

A geological reconnaissance was carried out along the wall and reservoir perimeter. This showed no presence of solid bedrock, but only a tertiary polymictic conglomerate composed

http://dx.doi.org/10.5772/intechopen.68853

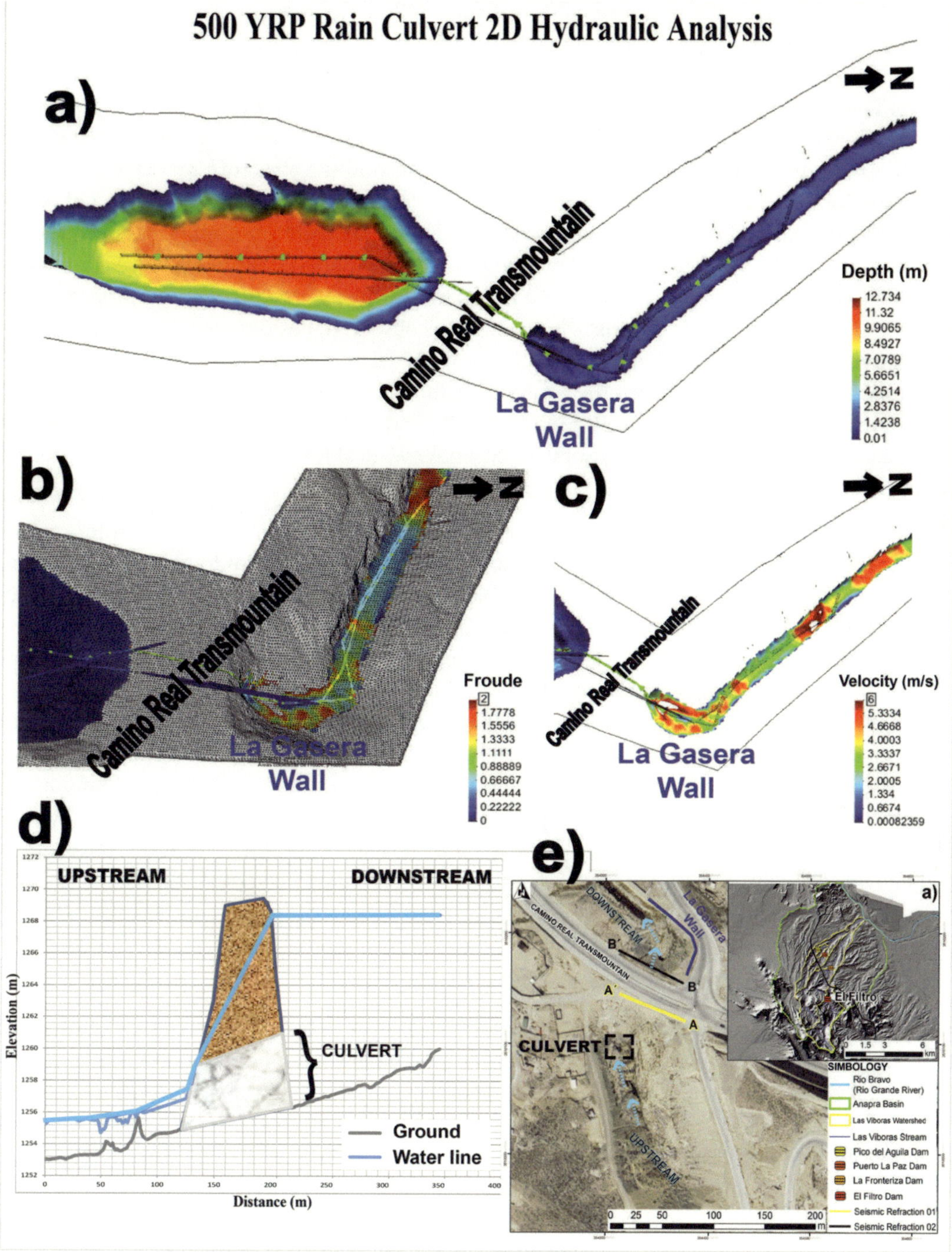

Figure 4. Hydraulic modeling results for the breaching of El Filtro dam. Inset (a) Water depth. Inset (b) Froude number. Inset (c) Velocity. Inset (d) Cross section showing how El Camino Real acts as retaining wall.

of interlayered beds of sand, gravel, silts and clays [17]. Also, a visual inspection of the wall showed the presence of a displaced block in the vicinity of the spillway. This block is readily identifiable, even in recent satellite imagery. No fractures and/or cracks were observed elsewhere. The geoelectrical analysis (**Figure 5**) of the wall and foundations revealed that, in general, inside the wall body, an interbedded sequence of high, low and high resistivity values are observed. At depths greater than 10 m beneath the wall the structure seems to be resting on a foundation with high electrical resistivity. The electrical resistivity layers do not show a quasi-horizontal and homogeneous pattern. Layers are actually updipping between model coordinates (MC) 70 and 120 m. At 130 m along the profile, a low-resistivity vertical anomaly is identified at the base of the wall. From 140 to 200 m MC the layers are down dipping. Beyond 200 m MC another vertical low-resistivity anomaly is observed. Besides showing a lower range in resistivity values, this anomaly is vertically propagated across the whole wall at MC 230 m. This resistivity anomaly matches the location of the vertical down dropped block observed at the surface, atop of the wall crest. Finally, by the end of the section, the layers beneath the emergency spillway are quasi-horizontal, until a new vertical low-resistivity anomaly is found again at MC 265 m.

The seismic data collected along this structure is a set of two refraction lines that were concatenated. The final *Vp* seismic section is displaced respect to the electrical tomography by 50 m, meaning that the seismic starts at MC 50 m on the geoelectrical section. The compressional velocity field (**Figure 5**, center) shows the presence of a very low velocity layer of 400 m/s along the whole seismic line, except at MC 140 m where this layer pinches out. The velocity field shows layers with velocities ranging between 400 and 600 m/s from the surface to 8 m in depth. Below this depth a nearly quasi-horizontal layer of 650 m/s bears the foundation of the wall. The seismo-stratigraphic pattern observed within the wall body (depths from 0 to 8 m) is characterized by the presence of quasi-horizontal layers pinching out into a high velocity structure found at the middle of the section with values ranging between 600 and 700 m/s where the isovelocity contour of 600 m/s reaches depths of only 3 m. No direct probing has been done at this structure to validate these nondirect methods and to assign these geophysical layers specific lithological and geotechnical parameters .

4.3.2. ERD Puerto La Paz

The geological reconnaissance of the topographic closure revealed no presence of bedrock flanks to anchor the structure. Actually, the only geologic unit identified both in the wall flanks and along the reservoir perimeter is a tertiary-age polymictic conglomerate composed of interlayered beds of sand, gravel, silts and clay [17]. Geophysical analysis of the wall and foundations through ERT (**Figure 6**, top) revealed that inside the wall body the strata are vertically displaced, as indicated by the undulating contact horizon geometry. The dam appears to rest on top of a very low resistivity (<8 ohm-m) non-homogeneous body located at 10 m depth. The seismic refraction analysis (**Figure 6**, middle) shows a vertical gradient in compressional velocities ranging from 350 m/s at the top of the wall and 700 m/s at the bottom of the originally designed wall that had a height 10 m above the river channel [1]. At depths greater than 10 m from the wall top, the velocity field reveals a ca. 800 m/s structure. This

http://dx.doi.org/10.5772/intechopen.68853

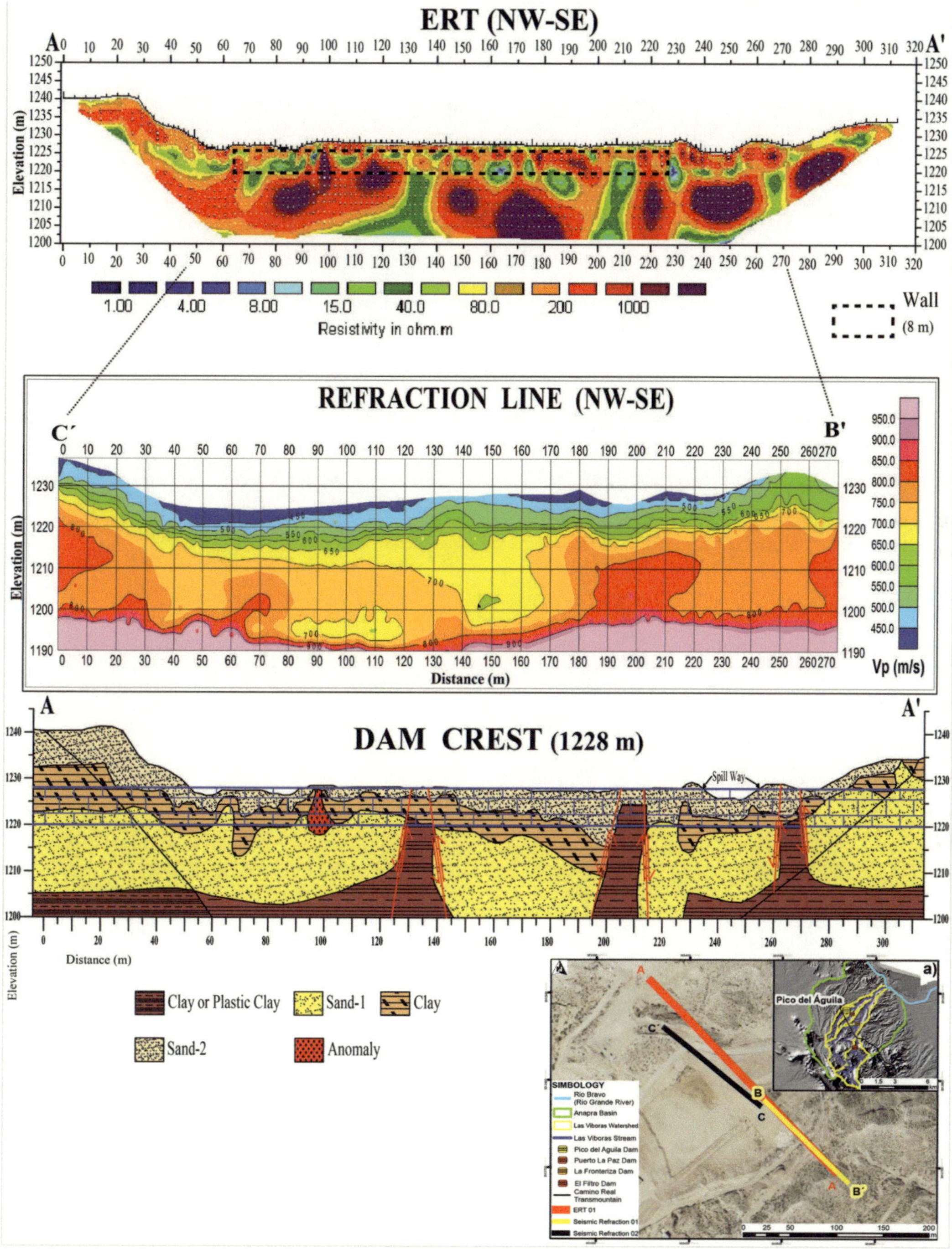

Figure 5. Geotechnical and geophysical assessment results at Pico de Aguila EDR. Top image: geoelectric profile. Middle image: *Vp* Refraction profile. Bottom image: interpreted results. Inset map at bottom shows location of geophysical surveys relative to extent of dam.

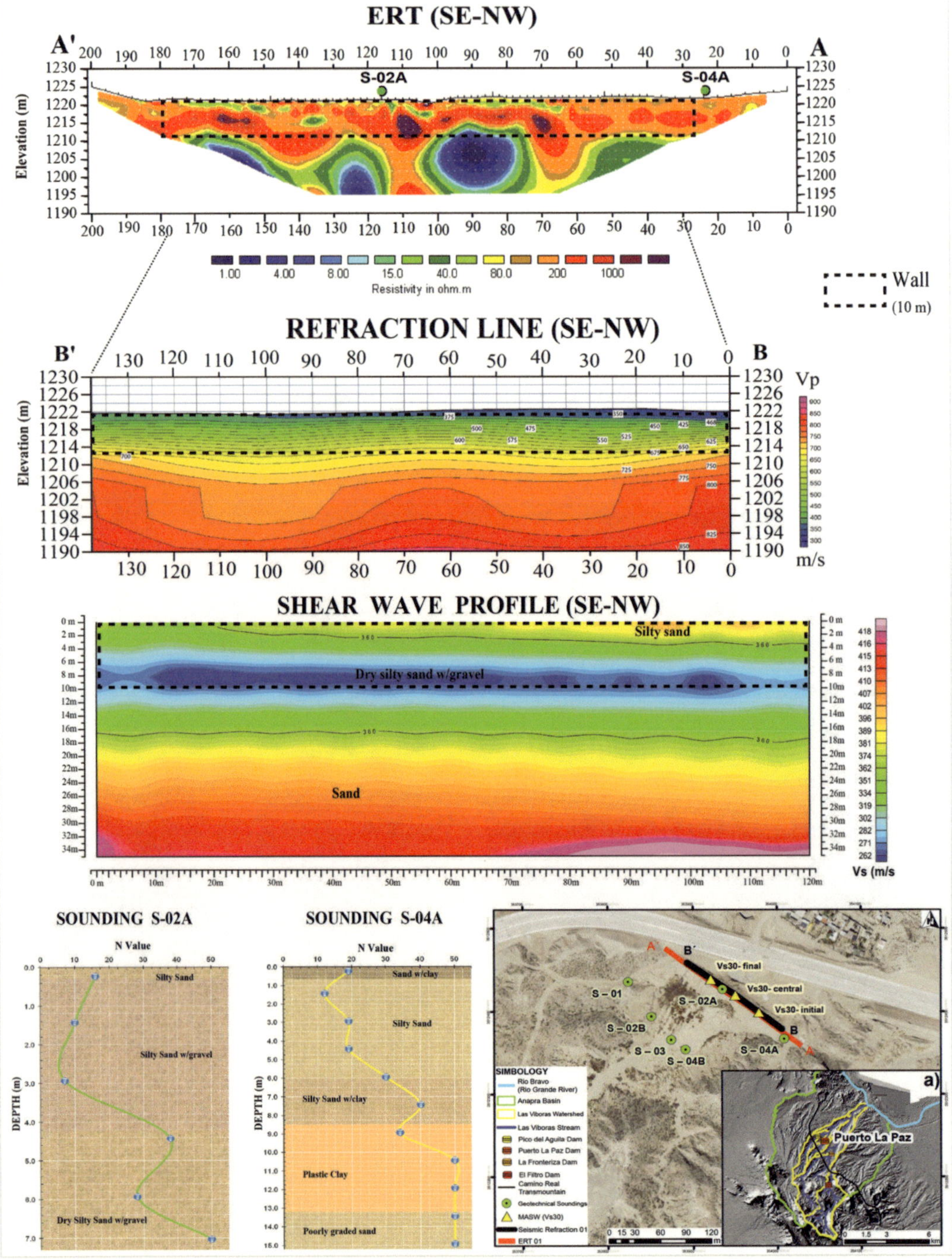

Figure 6. Geotechnical and geophysical assessment results for the Puerto de Paz dam. Top image: geoelectric profile. Position of soundings are indicated relative to dam are indicated. Middle Images top: *Vp* Refraction profile. Bottom image: *Vs* profile. Bottom left shows direct probing soundings. Bottom right shows geophysical and geotechnical survey locations with respect to the dam.

feature matches the depth and location of the low resistivity (<8 ohm-m) body observed in the ERT profile. The *Vs* section (**Figure 6**) does not show the same behavior observed in the *Vp* field, i.e., a positive velocity gradient as function of depth. The S-wave velocity (*Vs*) field shows a 270 m/s low shear velocity zone (*LVsZ*) located between depths of 6 and 10 m. The velocities above the *LVsZ* range from 360 to 380 m/s. The velocity field increases from 360 m/s at a depth of 10 m to 420 m/s at 20 m in depth. The *Vp*/*Vs* and Poisson's ratios are respectively: 1.66 and 0.22 at depths of 4 m, 2.2 and 0.37 at the *LVsZ* and 2.34 and 0.39 at 10 m in depth.

Although direct soundings are available to validate the geophysics, there is only one sounding located atop the wall, and the maximum depth reached is 7 m. This sounding reveals that the materials that form the wall are mainly silty sands. The load capacity parameter N from the standard penetration test (SPT) starts at 16 at the top of the wall and decreases to 7 at 3 m in depth, increases up to 38 blows at 4.5 m and decreases again to 28 at a depth of 6 m. Another sounding is also available at the right shoulder of the wall. This sounding shows the presence of silty sands forming the body of the wall, but since this sounding goes deeper, it records plastic clays at the bottom of the dam wall, which corresponds to the resistivity values of less than 8 ohm-m observed at these depths.

5. Discussion

5.1. ERD Puerto La Paz

The regulation dam Puerto La Paz is hydrologically capable of handling rains up to 500 YRP, but is hydraulically compromised for a 10,000 YRP volume since the emergency spillway flow exceeds the required freeboard of at least 0.91 m, as required by the Bureau of Reclamation and Mexican law. Although this dam seems to be able to handle runoff volumes even for non-ordinary rains, its structural condition, as revealed by the geotechnical assessment through geoelectrical, seismic studies and direct soundings, is seriously compromised. The wall's foundation rests on top of a thick plastic clay layer which seems to have experienced differential loading effects resulting in the wall's very poor condition observed at surface. Furthermore, the tomographic geoelectrical profile (**Figure 6**) shows quasi-vertical strata displacements matching the displacements observed at surface. The most evident displaced layer is a high resistivity unit of 200 ohm-m atop a very low resistivity unit (<8 ohm-m) interpreted as plastic clay as a result of the correlation of the direct soundings with geophysics. Therefore, this structure poses a serious hydrological hazard, which combined with the overpopulated, low-income neighborhoods located downstream, significantly increases the risk as a consequence of population vulnerability.

5.2. ERD Pico del Aguila

Although no direct geotechnical data are available for this dam, the geological scenario, which is similar to the Puerto La Paz (PLP) dam since it is located only 1 km from Pico del Aguila, allows us to make a correlation of electro-stratigraphic and seismo-stratigrafic units with specific lithologies found by direct exploration at the Puerto La Paz (PLP) dam. The interpreted lithological section (**Figure 5**, bottom image) shows that the body of the wall would be then composed of an interlayered sequence of silty sands associated with resistivities greater than 200 ohm-m. The interbedded stratum with resistivity values between 15 and 60 ohm-m is associated with a sandy clay unit. A nearly vertical high resistivity anomaly, located at M.C. 90 m, is perhaps associated with the presence of an abandoned concrete sewer line emplaced across the dam's body. Foundation of the dam is resting on a very high electrical resistive package (>1000 ohm-m) associated with clean sands and or gravel. Seismically, the interbedded sequence is not observed, but this might be a limitation of the method, unable to model velocity inversions [21]. But in contrast, a positive velocity gradient, starting at 350 m/s at the surface and increasing to over 700 m/s at the wall bottom, is observed in the velocity field. The *Vp* values at the foundation horizon are then defined by the 700 m/s isovelocity contour, interpreted as associated with Tertiary sediments composed by sand and gravel, which makes it a competent strata. Although the foundation layer is competent, the presence of the highly warped region of 600–700 m/s *Vp* velocities in the middle of the seismic section, reveals that the material forming the body of the wall was severely perturbed by deformation and movement. These materials appear to have been re-emplaced and compacted "a-posteriori." This process definitely does not lead to proper linkage or anchoring of the original layering to the recently emplaced material. This feature must be considered a serious flaw, which seriously compromises the structural integrity of the wall. Furthermore, the quasi-vertical low resistivity anomalies indicate clay bodies are intruding the structure in as similar fashion to that observed within the Puerto La Paz dam wall. Although no direct probing of the material is available, these inferred clay structures compromise the structural integrity of the wall, as evidenced by cracks and openings propagated along the wall slope at the same location as the clay bodies that are revealed by the geoelectrical data.

6. Conclusion

The geological, geophysical and geotechnical assessment of El Camino Real road culverts and dams revealed that, as expected, El Camino Real structure is not designed to prevent internal erosion; that the Puerto La Paz and La Fronteriza dams were severely compromised and it was definitively necessary to breach them. The El Filtro dam is also structurally compromised and poses a major danger if a 500 YRP event occurs again. This analysis also showing that the Pico del Aguila dam is structurally compromised. Finally, the 500 YRP modeling of Las Viboras watershed under actual conditions reveals that El Filtro dam would be breached due to hydrostatic pressure and overtopping. Hence, the risk for a 500 YRP event is severely increased downstream as consequence of failure of the dam.

In terms of risk, if we consider the risk function as defined as the concatenation of three elements: conditional elements, triggering elements and vulnerability [23], then the structurally compromised dams, geology and abrupt topography are the conditional factors, whereas the extreme and ordinary hydro-meteorological events are the triggering elements. On the other hand, the overpopulated neighborhoods represent the vulnerability element, which all together define Las Viboras specific risk function. On this point, this study has detailed proven that from a structural and hydrological point of view, the hydraulic operation of earthen regulation dams is seriously compromised for any rain event (not necessarily an extreme rain event). Then, the hazard function resulting from joining conditional and triggering elements is high. This hazardous condition, linked to the overpopulated neighborhoods, with more than 30,000 people, located downstream of Las Viboras dams, constitutes and even higher risk function. While Las Viboras Dam is perhaps the most complicated watershed in terms of flooding, the vulnerability element (people), should be analyzed for the whole Juarez-El Paso Metroplex system, with nearly 2 million people. Then, although each watershed is modeled independent, the final results should be portrayed including the system as a whole, considering three key elements:

(1) Modeling and design of hydraulic infrastructure with 1000 YRP rains as consequence of climate change effects, which have shortened the occurrence interval for heavy rain events.

(2) Anapra basin drainage outlets directly into the Rio Grande, posing a risk not only for Las Viboras population but also for El Paso downtown area if a heavy rain event occurs, since the RGR is not going to be able to handle this runoff coupled with runoff from other regions upstream.

(3) The hydrologic effect of a planned ~10 m high border wall that will be placed on the United States river bank to decrease the flow of undocumented immigrants from Mexico also needs to be examined. It is likely to prevent flood water from reaching downtown El Paso, but will increase the flooding risk to Mexico as water will be deflected into downtown Juarez. In addition, if the border wall is not designed as a water retention structure capable of bearing hydrostatic load, then failure of the wall in an extreme rain event could be catastrophic for El Paso.

In terms of resilience, we adopted the socio-hydrologic resilience framework [24], since it incorporates the interaction between social and hydrologic elements as a system, considering not only the effect of human activity on the hydrologic ecosystem but also the impacts of the hydrology on the society. The socio-hydrologic resilience is then more precisely defined as a function of three system´s capacities: absorptive or tolerance capacity, adaptation or response capacity and transformation or ability to change capacity. Once a more detailed ad hoc resilience definition was available we compared the tree capacities for two specific tempo-rain scenarios for Las Viboras system: 2006 with a 10 YRP and 2016 with a 2 YRP.

In 2006, for a 10 YRP, the resilience function, shows how the manmade diversion of the Colorado River at La Gasera wall, stimulated a undesired hydrological adaptation of the system that caused severe erosion paths due to the nearly 90° deflexion channel curvature at La Gasera wall, and an excessive storage volume that resulted in an overtopping failure of La Fronteriza Dam in 2006. The other hydraulic structures at Las Viboras system operated satisfactorily. Therefore, we may conclude that the socio-hydrologic resilience function in 2006 was nearly enough for a 10 YRP rain event. This means that even though the system did not fully absorb the rain event, it certainly buffered it with a satisfactory rain-runoff conversion. In terms of the nonanticipated system adaptation, widening channels and erosion were properly absorbed too, but not the excessive transported volume of water that overtopped a dam. Thus, the transformation capacity (i.e., ability to change with infrastructure) operated nearly properly by compensating, maybe on the safety edge, but still properly, since no problems were reported downstream, although La Fronteriza dam's breaching risk was imminent once it started to overtop.

In 2016, the presence of the Camino Real with sub-dimensioned culverts, the non-operational condition of La Fronteriza Dam and the structurally compromised Puerto La Paz, Pico del Aguila and El Filtro dams constitute a decrease in the transformation capacity. The absorption or tolerance capacity has also decreased since paving process has increased impervious surfaces. Then, if transformation and absorption capacities have been diminished, the hydrologic adaptation of the system has resulted in 2016 in very negative effects that were not present on the susceptibility inventory in 2006. First, erosion of materials of the down slope of La Gasera wall has nearly blocked the Camino Real culvert downstream of El Filtro dam and culverts along the Camino Real are retaining water as shown by the 2D modeling even for ordinary rains. La Fronteriza dam´s lack of regulation conveys high water volumes of water and sediments from a nearly 4 km^2 watershed. This resulted in water depths reaching 1.4 m at Las Viboras highly urbanized discharge area. High velocity and turbulent flows were present downstream along the whole main stream. The turbulent flow has practically mechanically destroyed the once claimed "water resistant" hydraulic concrete road surface near the watershed outlet, because this concrete can bear laminar flows but not the mechanical stress superimposed by turbulent flows. Also, the non-laminar flows and high transport velocities of heavy sediment loads resulted in the mechanical destruction of the underground sewer line with an immediate cost of nearly 25 M USD. And finally, not only property loss has been reported as consequence of high velocity and turbulent flows but also life loss, since two people were killed by or as consequence of the November ordinary rain event.

In summary, Las Viboras System was considerably less sociohydrologic resilient in 2016 than in 2006. This is a direct consequence of a multifactorial function where the poor local governance is the common denominator. In other words, the almost non-existent maintenance hydraulic structures maintenance, poor urban planning (i.e., developing even more house complexes downstream) and the lack of political will and compromise to foresee the need of hydraulic infrastructure in major city that is prone to flashflooding, have resulted in zero lobbying to access available federal resources. Thus, as consequence, this watershed is not even able to properly manage 2 YRP runoff volumes. This means that socio-hydrologic resilience function is practically null, since even ordinary rain events claim property and life loss.

This diagnosis reveals that the solution for this watershed is to increase the socio-hydrologic resilience by focusing on the transformation capacity, rather than in the absorption and the long-term adaptation capacities. Three dams have to be constructed upstream of the current dams' locations. These new dams would be located between highly competent rock flanks and rock basement to ensure structural stability, in order to warrant hydrological regulation and retention of fine grained, water suspended sediments. These three dams would replace

El Filtro dam, which is more a risk than a protection, and the non-operational La Fronteriza dam, which should not be rebuilt at the same location due to the high volumes of sediment that have severely diminished its storage capacity. The Pico del Aguila dam, although working properly for ordinary rains, should be rebuilt with an emergency spillway to transit the hydrogram peak associated with a 10000 YRP rain; this will ensure its hydraulic and structural safety. The Puerto La Paz dam is completely ruined, with evident gullies exposed at the surface and internal erosional features and a poor foundation as revealed by geophysical methods. However, based on hydrological considerations, a new dam is required in the same area. The geophysics carried out at the containment area has revealed that competent Tertiary conglomerates lacking plastic clays are located 100 meters upstream. 1D modelling of the hydrology with the wall displaced 100 m upstream shows that the structure would still be able to store and regulate volumes associated up to a 1000 YRP event and safe transport of water from a 10000 YRP event if a proper emergency spillway is emplaced. This technical solution is the key component for the system to change to an equilibrium or resilient condition. Finally, absorption capacity should be increased or at least preserved with the design of a resilient master plan for urbanization for the whole watershed. In addition, the adaptation capacity should be viewed not only as the response of the system under change or stress conditions, but also as the ability to learn from previous flood events and understand that the socio-hydrological system is a two-way road. If we impact the hydrology, the system response will have a later impact on society, so a great effort focused on education and outreach is definitely necessary to insure increasing resiliency to flash flooding events.

Author details

Oscar Sotero Dena Ornelas[1]*, Diane Irene Doser[2], Oscar Fidencio Ibañez Hernández[1], Griselda Janeth Obeso Cortez[1] and Miguel Angel Galdean Vega[1]

*Address all correspondence to: oornelas@uacj.mx

1 Universidad Autónoma de Ciudad Juárez, Ciudad Juárez, México

2 University of Texas at El Paso, El Paso, USA

References

[1] Comisión Nacional del Agua. Estudio de Factibilidad para el control Integral de avenidas en las cuencas Zona I Anapra y Zona II Centro en Ciudad Juárez. 1st ed. Ciudad Juárez: CONAGUA; 2009. 255 p

[2] Instituto Municipal de Investigación y Planeación. Plan Sectorial de Manejo de Agua Pluvial. 1st ed. Ciudad Juárez: Instituto Municipal de Investigación y Planeación; 2004. 384 p

[3] Instituto Municipal de Investigación y Planeación. Atlas de Riesgo de Ciudad Juárez. 2nd ed. Ciudad Juárez: Instituto Municipal de Investigación y Planeación; 2016. 420 p.

[4] Haenggi W. Tectonic history of the Chihuahua trough, Mexico and adjacent USA, Part II: Mesozoic and Cenozoic. Mexican Geological Society Bulletin. 2002;**1**(L):38-94

[5] PEMEX. Provincia Petrolera de Chihuahua. PEMEX Exploración y Producción, Subdirección de Exploración. 2013;**1**(1)

[6] Keaton J, and Barnes J. Paleoseismic Evaluation of the East Franklin Mountains Fault, El Paso, Final Report,. 3rd ed. Texas: U.S. Geological Survey National Earthquake Hazards Reduction Program; 1996. 134 p

[7] Collins E, and Raney J. Tertiary and quaternary structure and paleotectonics of the Hueco Basin, trans-Pecos Texas and Chihuahua, Mexico. Bureau of Economic Geology. 1991;**2**(91):44

[8] Drewes H, and Dyer R. Geological map and structure sections of the Sierra Juárez, Chihuahua, México. U. S. Geological Survey Miscellaneous Investigations. 1993;**2287**(1):2

[9] Dena O, Obeso G, Leyva J, Domínguez M, Hernández V, Granados A, and De la Cruz S.Estudio geológico y geofísico de la ladera sur del corte del mirador Hidalgo en Ciudad Juárez, Chihuahua. Ingeniería, Investigación y Tecnología. 2011;**12**(4):439-459. DOI: 14057743

[10] Loke M. Electrical Imaging Surveys for Environmental and Engineering Studies. 1st ed. Swedich: LUND University; 2000. 31 p

[11] Yilmaz O. Engineering seismology with applications to geotechnical engineering. Investigations in Geophysics: Society of Exploration Geophysicist. 2015;**71**(3):371-512

[12] Ackerman C. HEC-GeoRAS GIS Tools for Support of HEC-RAS using ArcGIS User's Manual. 1st ed. USA: U.S. Army Corps of Engineering Hydrologic, Hydrologic Engineering Center; 2011. 97 p

[13] Scharffenberg W, Flemming M. Hydrologic Modeling System HEC-HMS User's Manual. 2nd ed. USA: U. S. Army Corps of Engineers Hydrologic Center; 2010. 305 p

[14] Hydrologic Modeling System HEC-HMS Application Guide. 11th ed. USA: U.S. Army Corps of Engineers Hydrologic, Hydrologic Engineering Center; 2008. 97 p

[15] Myrland J. Two-dimensional hydraulic modeling for flood assessment of the Rio Rocha, Cochabamba, Bolivia [thesis]. Uppsala University: Department of Erath Sciences Program of Air, Water and Landscape Science; 2014. 200 p

[16] Palacky G. Resistivity characteristics of geological targets. In: Nabighian M, editor, Electromagnetic Methods in Applied Geophysics-Theory. Society of Exploration Geophysiscist. 1987;**5**(3):53-129

[17] Servicio Geologico Mexicano. Carta Geologica y Magnética de Campo Total Escala 1:50,000 de Ciudad Juárez, Chihuahua. 1st ed. Chihuahua: SGM; 2008. 2 p

[18] Dahlin T, and Linderman J. ERIGRAPH Software Developed in Cooperation between ABEM. 1st ed. Lund University and Terraohm: ABEM; 2007. 60 p

[19] Loke M. TUTORIAL: 2-D and 3-D Electrical Imaging Surveys. 2nd ed. LUND University: ABEM; 2004. 31 p

[20] Essien U, Akankpo A, Igboekwe M. Poisson's ratio of surface soils and shallow sediments determined from seismic compressional and shear wave velocities. International Journal of Geoscience. 2014;**5**(5):1540-1546. DOI: 10.4236/ijg.214.512125

[21] Prem V. Sharma. Environmental and Engineering geophysics. 1st ed. University of Copenhagen: Cambridge University Press; 1997. 477 p. DOI: 10.1017/CBO9781139171168

[22] Dahlin T, Sjödahl P, Friborg J, and Johansson S. Resistivity and SP surveying and monitoring at the Sädva Embakment Dam, Sweden. In: Procs. 5th European ICOLD Symposium; 25-27 June; Geiranger, Norway: Lund University of Technology; 2001. pp. 107-113

[23] González V, Ferrer M, Ortuño L, and Oteo C. Ingenieria Geologica. 1st ed. Madrid, España: Prentice Hall; 2002. 744 p

[24] Mao F, Clark J, Karpouzoglou T, Dewulf A, Buytaert W, and Hannah D. A Conceptual Framework for Assessing Socio-Hydrological Resilience under Change. Hidrology and Earth System Sciences, European Geosciences Union. 2016;DOI: 10.5194/hess-2016-499

Chapter 9

Understanding Flood Risk Management in Asia: Concepts and Challenges

Saleem Ashraf, Muhammad Luqman,
Muhammad Iftikhar, Ijaz Ashraf and
Zakaria Yousaf Hassan

Additional information is available at the end of the chapter

http://dx.doi.org/10.5772/intechopen.69139

Abstract

In this chapter, an attempt is made to review the behavior of flood in Asian region and mechanism of flood risk management adopted among Asian nations. Flood is the most frequent natural disaster at present and vulnerability is widespread across the globe. Though, Asian region is on a knife-edge. Distribution of natural disasters in Asia followed by economic damage and human killing is illustrated in this chapter. In addition, discourse of China, Pakistan, India, Bangladesh, Indonesia, Nepal, Vietnam, and Sri Lanka on flood risk management is examined. Flood risk management policies framed by these nations over the period of time are synthesized. Research and investment on forecasting, planning, preparedness, assessment, evaluation, and mitigation of flood risk are explained. This synthesis can present a pathway for better response and flood management for debated Asian countries through filling the identified policy gaps. This chapter also urges a need of holistic and inter-countries research and cross country analysis followed by increased funding for sustainable management of risk.

Keywords: floods, vulnerability, risk management, policy, research

1. Introduction

Overflowing of water in dry land is generally known as flood. Intergovernmental Panel for Climate Change explained that flood is the situation of water overflow or the excessive accumulation of water on specific areas which are not submerged usually. Flash floods, river floods, urban floods, coastal floods, and sewer floods are different forms of floods usually prevail subject to varying mechanism. Fundamentally, risk of flood may prevail anywhere.

Flood can befall when riotous amount of water arches across a piece of land. The source of water may be river, lake, or the sea. If there is a hole in a defense built to hold water, it could also bring flood. Furthermore, flood may prevail pertinent to excessive rains and profligate melting of glaciers. Erratic weather patterns have become a prime reason behind flood occurrences as well. Researchers endorse precipitation as significant determinants of intermittent climate. Threshold for flood occurrence upsurges with annual rise in average rainfall.

Flood is an outcome of climatic changes (**Table 1** reflects climate change analysis of different countries) and influenced by climatic determinants vis-à-vis temperature patterns (fast melting of ice in case of rise in temperature, snow, soil freezing when temperature goes low) and precipitation (duration, intensity, frequency, timing, form-snow or rain) [1]. Brakenridge et al. [2] were of the view that chances of flooding increases if river is close to sea level. Floods

Country	Impacts
Vietnam	Copsey et al. [4] presented that humans are in notice of climatic variations. Temperature is on continuous rise, and rainfall patterns are less predictable now. People residing around Central Coast and Red River Delta have discerned high intensity of extreme weather events.
Bangladesh	Cyclones and numerous floods hit Bangladesh since past due to ever increasing sea level. People are experiencing high temperatures, less rainfall and erraticness of weather as compared to last decade. Unpredictable weather is posing serious threats to loss of resources and livelihoods. Indian dams, urban growth, industry development and farming practices are surpassing impacts on climate [5].
India	High temperature, less rainfall and erratic behavior of weather patterns are prominent in Indian states. Water shortage has become major concern for India as compared to other allied Asian nations. This acute shortage is directly or indirectly associated with climate change. Majority of Indians is feeling extreme weather patterns and on verge of climate-related risks [6].
Sri Lanka	Sri Lanka is seriously vulnerable to enhanced intensity of floods, droughts, landslides, erratic rains, increased temperature and abrupt sea level rise. The vulnerability is in direct connection with climate change [7].
Pakistan	About half (44%) of people in Pakistan have experienced climate change impacts. The percentage is much higher than any other Asian nation. Increased temperature, unpredictable rains, elevated extreme weather events and augmented infestation of pests and mosquitos are some outcomes from climatic severities. Inconsistent rain is affecting crops and access to water for drinking and irrigation is downsizing. Upwelling of pests and mosquitoes is affecting agriculture and health of humans and animals as well [8].
Nepal	Nepal is fronting rise in temperature and less predictable rains. Likelihoods of floods and droughts have been increased in last decade. People are in agreement that climate is changing. These changes are bringing significant decline in farm productions. Human health is also vulnerable to adverse impacts of climate change [9].
China	In China, people discerned upturn in temperature, erratic rains, and extreme weather events. One-fifth (20%) people perceived that impact of climate change is high, though the percentage is lower among all Asian nations. People from Nanchong and Sichuan provinces reported a wide increase in rainfall and an enormous fall in agricultural productivity [10].
Indonesia	Plantation of trees and population of animals has decreased. People are of the view that climate has been changed. They believe that weather is hotter and less predictable now. Frequency, intensity, and patterns of rains have been transformed with sudden increase somewhere while abrupt decline in other areas [11].

Table 1. Climate change impact analysis in different Asian countries.

sway by basin drainage conditions like preexisting level of water in river followed by ice or snow cover, status and characteristics of soil (moisture contents in soil and permeability), existence of dams and reservoirs, and urbanization rate in the region. Hasty failures of impeding structures evolve floods, landslides, and compact conveyance channel of water [3]. Abiding rains, reckless melting of glaciers, and mounting temperature are widely debated flood risk sources. Consequently, each year, floods of varied magnitude and scale prevail in different parts of the world. Floods prevailed in India, China, and Pakistan in mid-2010 memorizes a damage of unprecedented scale. Colombia and Australia were reported victims to devastating flooding in the same year. Many other countries including Uganda, Namibia, Mozambique, South Africa, Mexico, Columbia, Brazil, United States, China, Korea, Pakistan, India, Thailand, Philippines, and Cambodia witnesses high material damage and causalities associated to floods [3]. In 2012, "killer floods," inducing more than 50 fatalities each, occurred in Madagascar, Niger, and Nigeria in Africa; Bangladesh, China, India, North and South Korea, the Philippines, and Russia in Asia; and Argentina, the United States, and Haiti in America.

2. Floods in Asia

Asia is on the verge of natural disasters and has witnessed numerous disasters with devastating impacts. During 2015, about 62.7% of total global disasters were recorded in Asian region. Approximately 167 natural disasters were recorded in Asia during 2015 which is far greater than the natural disasters (155) occurred during the period of 2005–2014. Among disasters, meteorological disasters appeared raised up to 30% as compared to period of 2005–2014. China, Indian, Philippines, Indonesia, and Pakistan were the countries major hit by natural disasters with average disasters of 45, 36, 21, 15, 10, and 10, respectively. On an average, these countries faced 55% of total natural disasters prevailed during 2015. More or less 69 million people were affected from reported disasters across the Asia. However, annual average of victims appeared lesser than average (158 million victims) during 2005–2014. It may be concluded that with the passage of time, frequency of natural disasters is on the rise, which may be more elevated in the forthcoming years.

A number of researchers [12–15] are in agreement that Bangladesh, China, and India are more vulnerable to natural disasters due to their geographical locations and extensive human and material losses could be anticipated. Bangladesh is the most flood prone country among allied Asian nations. Bhutan, Indonesia, Pakistan, Papua New Guinea, Sri Lanka, Thailand, Timor-Leste, Uzbekistan, and Vietnam were declared most vulnerable countries in Asian region by Asian Development Bank [16]. Ministry of Environment and Forest [17] reported that flooding is a usual and recurrent phenomenon in Bangladesh because of river overflows, poor drainage, and erratic monsoonal rains. Floods of 1987, 1988, and 1998 inundated more than 60% area of Bangladesh. Flood of 1998 in isolate caused 100 deaths and left 30 million people homeless.

India is collateral to Bangladesh in case of vulnerability to floods. Forecasting is in favor that by 2080, Mumbai city will be three time more vulnerable to floods [18]. During July, 2005, Mumbai city faced 944 mm of rain around 24 hours [19]. This flooding exhibited significant

Years	Top 10 countries by number of natural disasters occurred (2006–2015)									
	1	2	3	4	5	6	7	8	9	10
2006	China	US	India	Philippines	Indonesia	Afghanistan	Vietnam	Pakistan	Bangladesh	Romania
2007	US	China	India	Indonesia	Philippines	Pakistan	Japan	Mexico	Algeria	Haiti
2008	China	US	Philippines	Indonesia	India	Vietnam	Colombia	Kenya	Thailand	Australia
2009	Philippines	China	US	India	Indonesia	Brazil	Mexico	Vietnam	Bangladesh	Australia
2010	China	India	Philippines	US	Indonesia	Mexico	Russia	Australia	Vietnam	Pakistan
2011	Philippines	US	China	India	Indonesia	Mexico	Brazil	Nepal	Japan	Guatemala
2012	China	US	Philippines	Indonesia	Afghanistan	India	Russia	Japan	Bangladesh	Haiti
2013	China	US	Indonesia	Philippines	India	Vietnam	Japan	Brazil	Afghanistan	Bolivia
2014	China	US	India	Japan	Philippine	Indonesia	France	Mexico	Thailand	Nepal
2015	China	US	India	Philippine	Indonesia	Pakistan	Australia	Japan	Chile	Bangladesh

Source: *Author compiled the data from "Annual disaster statistical review reports 2006–2015. Centre for Research on the Epidemiology of Disasters, Institute of Health and Society (IRSS), Université catholique de Louvain – Brussels, Belgium".*

Table 2. Top 10 countries by number of natural disasters occurred during 2006–2015 (10 years).

loss on household level. China is on verge of acute and severe flooding in future as well. Monsoonal rains during 2011 cemented a destructive flood in Thailand and a huge loss of 45 billion USD was recorded. Economic growth sliced down to 1.5% from estimated growth of 3.5–4% in 2011. Similarly, 9% land area of Japan comprising 41% population and 65% economy support is vulnerable to floods [20]. Pakistan faced back to back floods in 2010 and 2011 authenticating wide losses. Brakenridge et al. [2] recorded 2000 instant fatalities in result of monsoonal flooding. Unpredictable nature of monsoonal rains is likely to increase in Bangladesh and China [21]. Peduzzi et al. [22] unveiled that Vietnam, Cambodia, and Bangladesh will remain on edge of flooding.

Centre for Research on the Epidemiology of Disasters (CRED), Institute of Health and Society (IRSS), Université catholique de Louvain—Brussels, Belgium published report entitled "Annual Disaster Statistical Review 2014; The numbers and trends" each year indicating natural disasters statistics. Data mentioned in **Table 2** are accumulated summary edited by author by consulting statistical reports from 2006–2015. Data reflect that from 2006–2015, each year, majority of countries are from Asia among top ten countries hit by natural disasters. China, India, Philippines, Bangladesh, Pakistan, and Nepal are Asian countries widely vulnerable to natural disasters.

Figure 1 is illustrates the prevalence of natural disasters during 2000–2016 in Asia. Flood is the most prominent disaster with share of 41.27 in Asian region. During the period of 2000–2016, Asia faced more or less 2692 disasters (floods, earthquakes, landslides, extreme temperatures, storm, drought, wildfire, and epidemic). Storms posed one fourth (26.82%) share among disasters followed by 11.44% occurrence of earthquake. However, there is a significant gap between occurrence of flood and other disasters. Generally, it can be said that Asian region is more vulnerable to floods at present and in future as well. Distribution of floods in selective countries of Asian region is represented in **Table 3**.

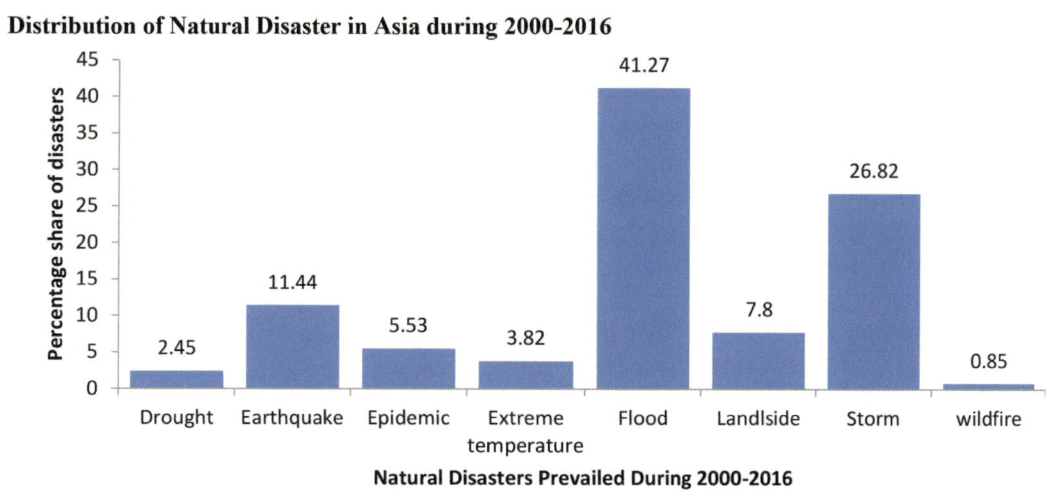

Figure 1. Distribution of natural disasters in Asia.

Country	Number of floods	Total deaths	Total affected	Damage
Pakistan	57	6584	46,001,293	18,707,148
India	148	23,592	293,642,535	42,294,347
Bangladesh	32	2764	72,706,135	3,014,000
Sri Lanka	31	211	309,071	2,000,000
Indonesia	109	3553	4,769,651	5,853,633
China	173	11,078	829,598,958	125,938,474
Japan	16	192	565,408	12,197,000
Nepal	23	2067	2,323,988	69,729
Vietnam	57	2612	17,474,738	3,123,502

Table 3. Total number of floods in some selective countries of Asia (2000–2016).

3. Asian discourse on flood risk reduction

Human security is of best entrust for the state governments across the world. Government bodies stay responsible for the planning and execution of public safety and vulnerability mitigation programs. Each country documents plans and strategies for sharing risk knowledge, prevention and mitigation, and preparedness. A brief overview of Asian countries' policies and intervention on flood risk management is discussed below. For a brief overview, extensive literature indicating policies, planning, programs, and outcomes was kept under debate. Cross-country comparison is documented, which could harness felt needs of future.

3.1. Pakistan

Pakistan lies in south Asian part and shares border with China, Afghanistan, India, and Iran. Geographically, Pakistan is on the verge of natural disasters. Pakistan witnesses occurrence of earthquakes in 2005 and 2013, floods in 2007, 2010, 2011, and 2012 costing damage of millions. However, floods are reported more devastated. Inadequate storage of water has been one of the prime reasons behind occurrence of floods in country. Efforts regarding water safety and security are scanty, regardless of constructions of dams for the storage, and water security is inevitable to mitigate floods. However, efforts regarding flood risks management are underway in form of different action since inception [23]. For substantial reduction of disasters losses and future security, country embarked National Disaster Management Commission (NDMC) and National Disaster Management Authority (NDMA) in 2006 and 2007 respectively with a vision to boost flood risk awareness, prevention and mitigation, and preparedness among masses. Authority was further devolved to province level with nomenclature of Provincial Disaster Management Authority (PDMA). Purpose of devolution was to implement action plan on grass root level. National Disaster Risk Reduction Policy (DRR) was formulated in 2013 to highlight and deliver comprehensive guiding agenda for addressing flood risk; which is increasing at pace. Policy acclaimed building positive outcomes across the communities, but a number of experts negated the success and outcomes of the policy. DRR reflected partial success and usually narrated as a tale of contradictory perspectives. DRR revealed decisive weakness with the prevalence of back to back floods in 2010, 2011, and 2014. The occurrence of floods summarized poor performance of policy department and failure in achieving goals due to nonimplementation of proactive strategies. Usually, flood hit rural areas and rural poor communities paid a high cost for these disasters. This damage not only represented a poor implementation of policies but also cemented that "disasters are for the poor." Criticism made by Zeeshan and Khan [24] on sluggish early warning system authenticated the failure of policy. Early warning system has been practiced using radio and emergency flood risk announcements through loud speakers of Mosque. System in practice is so traditional and meager in effectiveness. Public was not trained about evacuation procedures and traditional information sharing system was not capable for this objective. Traditional approach was also not enough to make people aware about safe locations and places to dislocate. Inclusion of "Mass Media" in flood risk policy could harness noteworthy potential in disseminating flood risk and management awareness among community. On contrary, aspect is missing in policy [25]. Financial security of flood victims and migrants is another missing link in policy. The policy holds gaps concerning institutional vulnerability, resources gaps, and institutional coordination and collaboration to accomplish policy goals. Local knowledge and management activities on domestic level are not recognized by disaster reduction experts neither included in policy, which could be another productive step in flood risk management. It can be stated that Government of Pakistan is still struggling to lay plans in wake of catastrophe management [26]. Disasters management and implementation plans are insufficient to meet requirements of risk reduction. There has been no accountability framework for disaster management bodies in instance of poor performance in risk management [27].

Country is allowing natural disasters to occur with increased frequency despite of existence of disaster bodies is clear message of departmental failures. Prediction and forecasting about weather and raining patterns made by meteorological department are less reliable. Occurrence of disastrous drought in "Tharparker (a district in Baluchistan province)" followed by routine human deaths is a dent on performance of disaster management bodies.

Research in Pakistan on research on flood risk and their management is scanty. Research is not sufficient to disclose that how science and technology could be utilized to tackle societal decisions making in context of risk management. Incorporating research in policy formulation on flood risks could address the management of risks in better fashion. For the purpose, social sciences may be included in research agenda for better social overview and community perceptions on risk and their management. There is no database to reflect disaster statistics in Pakistan. Past experiences and research findings are not gathered in Pakistan. Social research in assessment and perception of community on different aspects (for example, about construction of dams) could unveil the felt needs.

3.2. India

India has faced a number of floods in the past and persists vulnerable to more disasters in future. Carnage exhibited by disasters urged Indian Government to design and implement disaster management programs. "State disaster management plans" was the foremost policy documented by Indian Government. These plans inferred that all Indian states are obligatory to compile disaster management papers reflecting preparations outline, actions to minimize risk, and responses to cope with associated threats. State government in association with district governments execute rescue, disaster, relief, and rehabilitation functions with support of national government [28]. Regarding disaster management plans and strategies, Bahadur et al. [29] endorsed implemented programs condensed sound foundation for coordinating disaster management at state level and construct significant reduction in vulnerability of disasters. ICIMOD [30] revealed that coordination among departments (i.e. flood forecasting centers and central water commission, meteorological department and central relief commissioner) helped in saving numerous lives. National programs put more emphasize on human resource development. Preparation of flood risk maps on districts level has been in practice to enrich mitigation goals since the 1997 [30]. Government of India [31] has reported economic loss of approximately 100 million Indian rupees in 2009 due to floods.

Government of India [31] highlighted more focus of disaster reduction programs on safety and protection of vulnerable communities and strengthen coordination and partnerships with local communities to enhance flood risk mitigation. This aspect of policy posed reflection of true participatory approach. However, despite of sound planning, program showed partial success. Poor coordination with local communities, which was subject to extensive workload, and poorly transparent structure were the leading weaknesses. Feeble organizational cooperation between Brahmaputra Board, CWC, and Ganga Flood Control Commission put the working of these departments under criticism. Meager interdepartmental cooperation displayed deprived services followed by below-average effectiveness [32]. Failure in forming floodplain zoning in India raised quandaries, which appeared a barrier of flood management on national level. Floodplain zoning was the mandate of National Water Policy executed in 2012 [33]. Unfortunately, the objective was not met. Absence of enforcement mechanism and demarcation of zones were additional barriers in wake of flood risk management [32]. Findings of Government of India [34] report agreed poor governance behind sluggish implementation of integrated water resource management which was anticipated as leading flood management strategy.

Government of India launched National Disaster Management Authority in 2005 with responsibilities for constructing risk reduction policies and yearly plans. Das [35] agreed with success of NDMA in enhancing capacities and skills development of managerial and field staff. NDMA staff served the society through skill provision which was effective in managing disasters. Separate local as well as national level funds for disaster relief and mitigation activities assured the minimal pressure on public exchequer for embryonic disagreements [36]. Flood risk management in India is least discussed in research and literature. Studies and literature are mainly focused on risk assessment only, while flood insurance, structural control, and integrated management of flooding are least discussed.

3.3. Bangladesh

Bangladesh is also vulnerable to floods like collateral Asian nations. Flood prevailed during 1988 inundated 89,000 square kilometer area nationwide and 1517 human deaths were recorded. After a decade, during 1988, another devastating flood hit Bangladesh and caused inundation on 100,000 square kilometers followed by 918 deaths. Numerous noteworthy floods have prevailed in different years like 1922, 1954, 1955, 1974, 1984, and 1987 with considerable losses [37]. About 70% land of Bangladesh stays exposed to flooding with likelihood of potential impacts on national economy [38, 39]. National Water Plan was implemented in 1980 with aim of balanced water use. In mid-nineties, Flood Action Plan was executed in Bangladesh. Later on, the plan gave origin to National Flood and Water Management Strategy in 1996 with recommendations of general public participation in environmental impact assessment at implementation phase of flood action plan. This action was kept mandatory for all water-related projects in future. National Water Policy was introduced in 1999 followed by subsequent emergence of National Water Management Plan in 2001 with vision to implement different courses of action for water conservation and management. Water-associated disaster management initiative further aided the foundation of Comprehensive Disaster Management Plan (CDMP) and Disaster Management Guidelines with involvement of different stakeholders for execution of different activities to strengthen disaster management capacity at local and national level. Planning of postdisaster rehabilitation and relief was major constituent of disaster management plans. Government of Bangladesh enjoyed smooth control of floods through water management groups, which were helpful in controlling water and allowing specific amount of flooding flow according to hydrology and land topography [40, 41]. Samuels et al. [42] endorsed the effective management water and land and trainings on natural resources management which resulted reduction in of river loads to keep embankments protected and easy carriage to the Bay of Bengal via amended capacity of rivers.

Bangladesh witnesses adoption of structural as well as nonstructural methods of flood prevention and management [39, 43]. Paula and Hossain [44] illustrated more preference of structural flood management techniques as compared to nonstructural technique of flood management. Despite structural methods of flood prevention not being economically conducive, they require huge investment and are not environmentally friendly [33, 45–48]. Over dependency on structural management techniques lead to internal drainage issues followed by siltation and waterlogging [49].

Government of Bangladesh integrated adaptation to climate change into legislation [50] for effective mitigation of floods. Intergovernmental panel for climate change (IPCC) has praised the effective policies of Bangladesh Government which resulted vital improvements in disaster management, reduction in flood fatalities, and enriched adaptive capacities at households' level [51]. This improvement was subject to increased coordination between concerned sectorial departments and concrete planning followed by practical implementation [52, 53]. Early warning system ensured local participation and produced promising outcomes regarding flood risk management [54].

Though efforts of Bangladesh government in managing floods appeared with some remarkable results, a number of policy gaps still persist. Huq and Bracken [55] opined that current flood relief regulatory framework of Bangladesh is only operative for response and relief during times of flooding which shows substantial gaps and sluggish necessities to uplift community adaptive capacities and tendencies to combat future vulnerabilities. Relief, rehabilitation, and rescue operations effected by National Government are deficient in considering needs of communities [44]. National top-down approach of planning repeatedly failed to convey effective and in-time flood mitigation services [56–58]. Most of the efforts were framed for local participation, but community participation in practical was just at early state in flood-related projects rather than at implementation and execution phase [55]. Unnecessary extraction of upstream water particularly in dry season (November–May) usually paved the way to freshwater reduction and increased salinity level which affected agricultural production and other requirements [59]. Early warning system is not effective according to expectations [44]. Government of Bangladesh also evolves lack of coordination among departments, budgetary scrapes, and rigorous institutional frameworks which are suppressing the results of flood management measures.

Research institutions emphasizing on flood management possess serious gaps. Research in Bangladesh is mainly focused on concept of vulnerability and vulnerability to natural disasters [60–62]. Very few researches are so far carried out on geographic location, community cultural, socioeconomic, and political attributes to vulnerability [48, 63–65]. Assessment of community indigenous knowledge on coping of flooding is poorly addressed [39, 43, 66]. Moreover, fewer researches are conducted on assessment and evaluation of structural flood management techniques and embankments projects [67–69]. Shajahan and Reja [70] orated research gaps in flood mitigation activities and evaluation of flood management strategies and institutional management [38, 49]. General outlook entails that with the passage of time research on flood risk management is partially focused.

3.4. China

China is most vulnerable country of Asian region being exposed to disasters more often. According to the estimate, more than 200 million people every year are affected by disasters. Floods utterly distresses lives of people in China [71]. Economic loss per year is more than 100 billion yuan about a quarter of the entire world [72]. Flood occurrence frequency in China is greater than the average of world. This frequency of floods may predict more economic loss in future as well. For future security, Chinese Government is utilizing finest options to combat flood risks. Government of China introduced national integrated system to contest flood risks with integration of geographic information systems, remote sensing, global positioning system, and other techniques for assessment, monitoring, and evaluation of flood exposures. Integrated system reflected pivotal role in flood management and executed as key part of flood management system at National Flood Control Headquarters, China [73]. Forecasting, monitoring, evaluation, simulation, and analysis techniques hold potential to reduce floods [74–76], and China kept all the techniques under consideration. Ministry of Science and Technology China provided financial assistance and laid foundation of National Professional and Operational Integrated System in 1995 for evaluation and monitoring of stern natural disasters.

Forestation was most preferred initiative undertaken by China to mitigate floods. In last three decades, coverage of forests is elevated on consisted basis across the China, and reduction in Floods has been documented in this period. In 1998, China faced a flooding with precipitation rate of equivalent to floods prevailed in 1980, 1983, and 2002; in these years, flooding was less serious. However, now situation is reversed, and floods have become severe, which is directly associated with destruction of forests and reduction in forest area. For instance, forest coverage in Yangtze River at current is 10% which was 30–40% in 1950s [77]. Wang et al. [78] reported that forest coverage reflected decline from 1970s to the end of 20th century. Deforestation is result of fulfillment of human needs and exploitation of natural resources in nonjudicious way. This deforestation leads contribution in poor management of floods and enhanced vulnerability of flood risks [79, 80].

China has elongated history regarding management of floods beginning with DaYu and Gun's flood management policy [81]. "Dujiangyan" was water conservation and flood management project implemented around 256 BC by state of Qin [82]. After the occurrence of flood in 1998, Government of China switched their course to adoption of structural as well as nonstructural approaches for management. Among structural approaches of flood management more or less 86,000 reservoirs have been constructed in past 5 decades (50 years) in China. These reservoirs posed momentous role in managing and mitigating flood risks and associated losses [83]. For the higher achievements in flood management, Chinese governments gathered allied departments under one umbrella to exert holistic action. For effective operation and management of floods, National Flood Control and Drought Defying Chief Headquarters of China has appointed Dalian University of Technology (DUT), Wuhan Hydroelectric University, and Hohai University to develop an integrated management system for flood control of reservoirs (IMSFCR) since 1998, with a duration of 5 years [83]. Moreover, number of initiatives were undertaken to conserve and manage water and water resources. Government implemented Water laws and regulations framed in succession with water sector since 1980s. Some noteworthy laws are Law of Flood Control (1997), Water Law (1998) later on revised in 2001, and Law of Soil and Water Conservation (1991). Some regulations were embarked in water sector of China as well, for instance, regulations of flood proofing, regulations of river course management, and guide to safety building of flood storage and detention basins. China has integrated wide network of technology in assessment and management of flood risk under flood control framework. For instance, Geographic Information System (GIS) is extensively adopted in regional disaster evaluation which exhibited useful outcomes [84–88]. ArcGIS technology also has been used by the Chinese government in mitigation of highway flood risk management [89]. These multiple initiatives are reflection of true zeal of Chinese government toward flood risk management.

3.5. Japan

Japan is another flood prone country of Asia. Extensive rains hit Japan more often in rainy as well as typhoon season which raises the precipitation level across the country. Annual precipitation is around 1500–1600 mm which is very high and fully capable to craft flood threats. Annual precipitation in main city Tokyo is much greater as compared to western countries. On pacific coast of Japan, annual precipitation of 50–60% is recorded from June–October in particular. This situation has caused number of floods in past and casted huge economic damage. In this context, Government of Japan formulated concrete policy to mitigate flood risks.

It is well noteworthy that in Japan about 70% of land is mountainous and covered with forests. Increased area of forests is usually a particular method to combat floods. Due to often occurrence of floods, Japanese utilize their indigenous knowledge to build flood management measures [90]. UNISDR [91] confirmed in connection of Hyogo Framework for Action (2005–2015) that "traditional and indigenous knowledge and cultural heritage" as one of the prime source of innovation, knowledge, and education to bring resilience at all levels.

Flood management history of Japan is extensive most probably came to inception during Yayoi period (300BC-AD 300); hence, Japan build refined and wide ranging disaster management plan. During 1960, efforts were made with focus on hydrologic function of "Green Dams" which acts upon the flow impeding potential of forests to reduce flood vulnerability. Less attention was paid on construction of concrete dams. It may be right to say that 70% mountainous area covered with forest laid foundation of "Green Dams" in country.

Current response system was initiated in short time, but this system enabled government to organize resources and comply in inclusive manner against the prevailed disasters to control damage [92, 93]. Ministry of Land, Infrastructure, and Transport (MLIT) of Japan is mainly responsible for the flood management policy in country at current. Currently, the ministry is managing 109 main river system covering greater than 70% of national land. Rests of the rivers of Japan are under control and management of municipal governments. Japanese Government propelled River Act in 1896. Under the act, each local government is accountable for flood control activities, management of rivers, hydrologic monitoring, and conductance of relevant studies for planning appropriate. Flood Fighting Act was recognized in 1949 with emphasize on systematic forecasting of flood and early warning system with collaboration of meteorological departments, healthy promotion of status and role of flood fighting brigades, and diffusion of flood hazard maps for rivers.

Japanese government has adopted structural as well as nonstructural measures to combat floods. Flood control dams, embankments, flood channels, and regulatory reservoirs were leading structural measures followed by nonstructural measures like forecasting of floods, early warning systems, land use regulations, and disasters and hazards maps adopted at national level. Unfortunately, despite of number of approaches, economic loss associated with floods is increasing due to socioeconomic development and condensed concentration of vastly prized assets. However, the number of deaths due to flood has been reduced significantly since 1960s [94].

3.6. Nepal

Nepal faced devastating floods in Tinao Basin, Koshi River, Tadi River Basin, Sunkoshi Basin, and in Kulekhani area in 1978, 1980, 1985, 1987, and 1993, respectively. The flood prevailed in Kulekhani area during 1993 alone caused 1336 deaths [95]. Floods prevailed in 1985, 1983, and 2004 damaged and destroyed large spheres of land terraces, farm areas, pastures lands, and orchards [96]. Therefore, Nepal is usually anticipated as vulnerable country to natural disasters, and huge proportion of GDP was lost annually in wake of natural disasters. Governmental of Nepal urged need of flood risk management and concentrated on Disaster

Risk Reduction Program with intention to identify, assess, and minimize disaster risk reduction [97–99]. This action facilitated sustainable development process in Nepal with reduction of adverse outcomes of natural disasters.

Disaster risk reduction programs were implemented among communities even in schools [100, 101]. Government of Nepal revised the vision and initiated Disaster Risk Reduction Campaign from 2006–2007. Under the campaign, risk reduction program was incorporated in educational curricula, and disaster management was disseminated on village level and districts level [102]. This integration raised the awareness level of community through involvement of instructors, teachers, and local youth clubs. A number of disaster education-related programs were initiated by government and nongovernment organizations [101, 103] on community level.

Nepali Government also launched National Plan for Disaster Reduction in 1996 in harmony of International Decade of Natural Disaster Reduction [104] to deal disasters on different stages. In 1999, an act named as "Local Self Governance Act" was launched which decentralized the responsibilities to lower level of politicoadministrative.

In 1996, the Government of Nepal (GoN) produced the National Action Plan for Disaster Risk Management [104] in accordance with the International Decade of Natural Disaster Reduction. This plan dealt with different stages of a disaster (pre, during and post) and was, in theory, supported by the 1999 Local Self Governance Act which advocated devolution of responsibilities to lower levels of the politicoadministrative pyramid. However, the impact of act appeared sluggish due to inadequate guidance and insufficient funds [105].

Hyogo Framework for Action (HFA) 2005–2015 documented pivotal significance in fostering agenda of disaster risk reduction [106]. The framework reflected a stimulus of change in field of disaster risk management. Disaster Management Policy and Act and the National Strategy for Disaster Risk Management (hereafter the National Strategy) were initiated in 2006 with aim of disaster risks management with improvement in institutional structures. Oxfam NGO funded both the initiatives. It is noteworthy that major funding of disaster programs was sponsored by NGOs. Similarly, Yoqi [107] presented rapid expansion of NGOs in Nepal. About 40,000 NGOs are registered by Social Welfare Council of Nepal. Heavy reliance on overseas aid and NGOs in absence of fully functioning government could raise issues and hinder disaster risks management policies [108–111]. Governance in Nepal is assumed meager which could be proven by the fact that none of the official mechanism to monitor transparency of NGOs registered under Social Welfare Council is in practice [108]. Jones et al. [112] argued sluggish political situation in Nepal, and progress of Nepal is stalled in framing a legislative context encouraging to effective disaster risk reduction measures. Insufficient public awareness, meager disaster preparedness, feeble governance, political instability, poor coordination among government institutions, agencies, and department followed by scanty financial assistance are notable factors hindering disaster risks reduction policies and measures [113, 114]. It can be said that Nepal is yet to understand the bond between disasters and development practices. Progress can be made only with the enhanced cooperation followed by effective policies to facilitate vulnerable in Nepal where each day about two persons die due to natural disasters [115].

3.7. Vietnam

Vietnam witnesses devastating flood in 1999 evolving economic damage of more than 200 USD, 547 human deaths and damaging 630,000 homes. Coastal flooding is foremost threat to Vietnam due to geographical location and socioeconomic absorption. The coastline is widespread over 3200 kilometers and 70% population is living there [116]. Since then, Vietnam is a flood prone country [117]. Greater observation and management are inevitable to combat vulnerability, and Vietnam government is harnessing different flood management approaches since inception in 1945. First order of Vietnam government (Order no. 70/SL) originated Central Committee for Dyke Maintenance for flood risk reduction, and later on, committee was replaced by Central Committee for Flood and Storm Control (CCFSC). Legislation was ensured, and hundreds of legal documents have been passed comprising laws, decisions, rules and regulations, and legal framework for effective control of floods [118, 119]. Despite of multiple documents, there has been no concrete and comprehensive law and policy on disaster management in Vietnam. For sustainable water resources management, Law in Water Resources was administered in 2012 followed by Law on Dykes (2006) and Ordinance on Flood and Storm Prevention and Control (2000) for flood management in Vietnam. Dykes in Vietnam adhere to dual purpose part, both for irrigation bank and drainage canal (Central Project Office, 2011); hence, law emphasizing on Dykes was of great worth. Ordinance on Flood and Storm Prevention and Control (2000) illustrated the need of further detailed and comprehensive legislation on flood management. This ordinance was replaced by the Law on Natural Disaster Prevention and Mitigation (LNDPM) in 2014. Government of Vietnam also brought a National Strategy for Natural Disaster Prevention, Response and Mitigation to 2020 during 2009. This national strategy was particularly attentive upon integration of disaster management to socioeconomic position of community for effective response and preparedness to combat the disaster. The strategy also tends to raise public awareness.

Flood management strategies revealed inadequate success pertinent to several inefficiencies. Water Resource and Flood Storm Control Division Report (2011) reflected that inadequate funds allocation hindered flood management policies. Improved financial assistance and healthier observation could have enhanced the effectiveness [120], but it does not appear practical. In fact, existing flood management approaches remain failed in combating disasters. Defense system against floods was insufficient in recent incidences [121]. To take floods in account, early warning system could be strengthened [117, 120], which is meager at current. More focus on improving socioeconomic position of community through alleviating inequality and poverty [122, 123] is need of time. This phenomenon may bring a concept of integrated flood management approaches and holistic action [124] for sustainable management of flood risks in Vietnam.

3.8. Sri Lanka

Sri Lanka is a flood prone country of Asian region. The country has faced some noteworthy floods in 1989, 2006, and 2008 damaging huge areas and human killings. Flood is the often hazard faced by Sri Lanka [125]. In last four decades, floods were more often and devastating for highest number of people [126]. However, Sri Lanka has implemented flood mitigation measures from the past. In late nineties, systematic forecasting programs were started in Sri Lanka [127]. Occurrence of Tsunami in 2004 urged government of Sri Lanka to precede tangible mitigation measures against the floods. Soon after Tsunami, Disaster Management Act, No. 13, was established in 2005 with aim of building institutional framework and national policy on disasters management [128]. National Council for Disaster Management (NCDM) is a high-level intern-ministerial body. Disaster Management Act leads to establishment of Disaster Management Center responsible for the implementation and coordinating national as well as subnational programs of disaster management. The country lingers to execute disaster reduction plans with assistance of nongovernmental agencies.

Among various initiatives regarding flood management in Sri Lanka, only two were found with success (Hazard Mapping Project by the National Building Research Organization [1991–1996] and Sri Lanka Urban Multi-Hazard Disaster Mitigation Project [1997–2003]). Incompetence in implementation of structural as well nonstructural measures against floods was significant reason behind poor performance of projects [129]. Local participation and adaptive capacities of local communities were not kept in consideration [130]. The research regarding assessment, planning, preparedness, and evaluation of floods and management techniques is scanty in Sri Lanka [130].

3.9. Indonesia

Indonesia is usually exposed to coastal and riverine flooding [131]. United Nations named Indonesia 38th "at risk" country in 2014. Bengawan Solo and Benanain River are the rivers more often vulnerable to flooding. In past decade, major floods prevailed during 2002, 2007, 2013, and 2014 costing billions of dollars economic damage [132–134]. To combat the vulnerability of floods, government of Indonesia launched different plans and approaches. For instance, in 2008, National Disaster Management Agency was inaugurated in Bahasa with aim to lessen floods. Government of Indonesia has implemented National Disaster Management Coordinating Board in 1979 which provided central coordination with support from 13 ministries of that time and armed forces. This board formulated flood management policy comprising rescue, mitigation, prevention, and reconstruction. Interagency disaster task force was also formulated at provincial, district, and subdistrict level. Flood risk management is also identified in law, 24/2007 and government regulation 21/2008.

Rivers of Indonesia are more often vulnerable to flooding. To combat the vulnerability of floods, government of Indonesia launched different plans. Indonesian Government administered National Disaster Management Plan (2010–2014) in 2012. The plan was more emphasized on flood policies, management strategies, and disaster risk reduction for 5 years span. National Development Planning agency ensured the implementation of plan [135], and technical protection measures were adapted to lower floods risks [136]. Success of flood management plan met with criticism. For instance, local communities were not having adequate environmental awareness to combat risk [137], which reflects that local participation in plan was scanty. Lack of coordination between departments, poor land using pattern and poor awareness, and knowledge followed by forecasting of floods were some other debatable issues not discussed properly. Early warning and forecasting were also criticized by Shrestha et al. [138]. Marfai et al. [139] pointed a dire need of strong institutional linkage between different stakeholders for quick and effective disaster response. Literature on flood risk management in Indonesia is scanty. Flood management aspects are poorly debated and assessed through different research and evaluation techniques.

4. Conclusion

It is myth that Asia is most vulnerable region across the world, and this vulnerability may surpass with the passage of time. Climate of Asian region is already posing abrupt and abnormal variations which are backing the natural disasters. In response, across the region, efforts are in trend with support of concrete policies and huge investment for infrastructure. Each country is documenting viable policy according the topography and geographical location. But, still efforts are inadequate and disasters are occurring frequently. Somehow, killings are downfallen, but economic damage is inverse to killings. Hence, entire Asian region would have to rethink their policies and redefine their actions for sustainability.

Pakistan surely needs to redefine their disaster management policy by replacing "mosque microphone" lead early warning system with integrated modern media. Improving local knowledge on disaster management should be included in policy. Country would have to strengthen their weather forecasting system and water conservation and management. Structural measures (construction of dams) could be fostered. In past, poor storage capacity of water has also been the reason behind occurrence of floods. More research should be ensured on flood risk assessment and management. For the best outcome, country would have to add social research in policy formulation. Social research could embark feasibility, community perceptions, and actions in best entrust.

Disaster management plans of state level in India have been successful but there is need to strengthen inter-departmental coordination and initiation of accountability for those who are expressing poor performers. Floodplain zoning needs more emphasize for accomplishment. For quick and effective risk management, allocation of funds would have been increased. Flood risk management is least discussed in research, hence, more research should be conducted which could help policy makers and practitioners in engineering best models.

Bangladesh being one of the most flood prone regions formulated numerous policies and executed multiple initiatives to mitigate flood risk. Hence, the country would have to shift toward nonstructural measures as well. Bringing climate change into legislation proved remarkable success in flood management. But, current regulatory framework of Bangladesh is mainly operative for response and relief only during times of floods. Government would have to strengthen their framework prior occurrence of flood. For this purpose, weather forecasting and early warning system should be empowered. Local participation should be ensured in policy formulation and planning. More research on different aspects of flood management could harness effective pathways. At present research in Bangladesh on flood management is meager.

Research on flood management is better in China, which is the most flood prone region of Asia. Deforestation has been found one of the major reasons of flooding in China. China would have to put emphasize on increased forestation. This is also nonstructural flood management measure.

Japan has been implementing structural as well as nonstructural measures for flood risk management. Increased forestation has been natural barrier against flood prevalence. However, socioeconomic development is still questionable, and the need is to strengthen socioeconomic condition of communities and enrich them with local indigenous knowledge to tackle floods. Increased focus on research regarding floods management could disclose existing gaps among communities in front.

Nepali Government has been rendering numerous policy efforts in managing floods. The strong aspect of policy was inclusion of disaster management in educational curriculum for the public awareness. Inadequate funds provisions and guidance on flood management were reported as hindering factor. Hence, government would have to sanction more funds and arrange community level trainings for the guidance of community on risk management.

Vietnam is a flood prone country, and several policies and approached were executed. However, still, there is no concrete and comprehensive policy documented on flood management in Vietnam. There is dire need of concrete policy followed by law and legislation on floods management. Inadequate financial assistance has been seen as constraining factor in flood management. Adequate funds provision and boosting early warning system which is already meager, could harness the better returns.

Despite of number of flood management approaches in Sri Lanka, only few (probably two) sowed successful outcomes. Country also remained unable to implement structural as well as nonstructural measures effectively. Moreover, local participation in planning, execution and management of flood management policy was ignored.

Indonesia was declared 38th "at risk" country, and multiple strategies and plans were initiated. Local participation in these strategies was scanty and environmental awareness among community was low. Lack of coordination among departments and poor forecasting of floods were reported major issues yet to be resolved by the government through effective policies.

To render sustainable management of floods across Asia, it is imperative to develop cross-country linkage and coordination in policy formulation and planning. Mutual financial assistance and technology sharing could support flood management initiatives. Asian countries would have to focus on cross country research on policies, assessment and mitigation of flood risks.

Author details

Saleem Ashraf[1]*, Muhammad Luqman[2], Muhammad Iftikhar[3], Ijaz Ashraf[3] and Zakaria Yousaf Hassan[4]

*Address all correspondence to: saleem1828@gmail.com

1 In-Service Agricultural Training Institute, Rahim Yar Khan, Pakistan

2 University College of Agriculture, University of Sargodha, Pakistan

3 Institute of Agricultural Extension and Rural Development, University of Agriculture Faisalabad, Pakistan

4 Directorate of Agricultural Extension and Adaptive Research, Punjab, Pakistan

References

[1] Bates B, Kundzewicz ZW, Wu S, Palutikof JP. Climate Change and Water, Technical Paper of the Intergovernmental Panel on Climate Change. Geneva, Switzerland: IPCC Secretariat; 2008. 210 pp

[2] Brakenridge, GR, Syvitski, JPM, Overeem, I, Higgins, SA, Kettner, AJ, Stewart-Moore, JA and Westerhoff, R. Global mapping of storm surges and the assessment of coastal vulnerability. Natural hazards. 2013; **66**(3):pp. 1295-1312

[3] Kundzewicz, ZW, editors. Changes in Flood Risk in Europe. Wallingford: IAHS Press; 2012

[4] Copsey T, Nguyen Y and Pham PH. How the People of Vietnam Live with Climate Change and What Communication Can Do. 2012. www.bbc.co.uk/climateasia

[5] Mamun AA, Stoll N and Whitehead S. How the People of Bangladesh Live with Climate Change and What Communication Can Do. 2012. www.bbc.co.uk/climateasia

[6] Gambhir V and Kumar P. How the People of India Live with Climate Change and What Communication Can Do. 2012. www.bbc.co.uk/climateasia

[7] Mawilmada N et al. National Climate Change Adaptation Strategy for Sri Lanka, 2011-2016. 2010. Retrieved from: www.climatechange.lk/adaptation/Files/Strategy_Booklet-Final_for_Print_Low_res(1).pd

[8] Zaheer K and Colom A. How the People of Pakistan Live with Climate Change and What Communication Can Do. 2012. www.bbc.co.uk/climateasia

[9] Colom A and Pradhan S. How the People of Nepal Live with Climate Change and What Communication Can Do. 2012. www.bbc.co.uk/climateasia

[10] Copsey T. Dalimunthe S, Hoijtink L and Stoll N.. How the People of Indonesia Live with Climate Change and What Communication Can Do. 2012. www.bbc.co.uk/climateasia

[11] Copsey T, Hoijtink L, Shi X and Whitehead S. How the People of China Live with Climate Change and What Communication Can Do. 2012. www.bbc.co.uk/climateasia

[12] Bouwer, LM, Crompton RP, Faust E, Höppe P, and Pielke RA Jr. Confronting disaster losses. Science. 2007;**318**:753

[13] Dash S, Jenamani R, Kalsi S, and Panda S. Some evidence of climate change in twentieth-century India, Climatic Change. 2007;**85**:299-321

[14] Shen C, Wang WC, Hao Z and Gong W. Characteristics of anomalous precipitation events over eastern China during the past five centuries. Climate Dynamics. 2008;**31**(4):463-476

[15] Douglas I. Climate change, flooding and food security in south Asia. Food Security. 2009;**1**(2):127-136

[16] ADB (Asian Development Bank). Legal Preparedness for Respondingto Disasters and Communicable Disease Emergencies: Study Report (Viet Nam). Kuala Lumpur; 2009. 206 p

[17] Ministry of Environment and Forest. National Adaptation Programme of Action (NAPA). Final Report. Dhaka, Bangladesh: Government of the People's Republic of Bangladesh; 2005

[18] Ranger N, Hallegatte S, Bhattacharya S, Bachu M, Priya S, Dhore K, Rafique F, Mathur P, Naville N, Henriet F, Herweijer C, Pohit S, and Corfee-Morlot J. An assessment of the potential impact of climate change on flood risk in Mumbai. Climatic Change. 2011;**104**:139-167

[19] Kshirsagar N, Shinde R, and Mehta S. Floods in Mumbai: Impact of public health service by hospital staff and medical students. Journal of Postgraduate Medicine. 2006;**52**(4):312

[20] IWR (Institute for Water Resources, US Army Corps of Engineers). Flood Risk Management Approaches As Being Practiced in Japan, Netherlands, United Kingdom, and United States. 2011. IWR Report no. 2011–R–08. 124 pp

[21] Kamiguchi K, et al. Changes in precipitation–based extremes indices due to global warming projected by a global 20-km-mesh atmospheric model. SOLA. 2006;**2**:64-67

[22] Peduzzi P, Dao H, Herold C, and Mouton F. Assessing global exposure and vulnerability towards natural hazards: The Disaster Risk Index. Natural Hazards and Earth System Sciences. 2009;**9**:1149-1159

[23] Atta-Ur-Rehman and Shaw, R. "Flood Risk and Reduction Approaches in Pakistan." Disaster Risk Reduction Approaches in Pakistan. Springer Japan, 2015. 77-100

[24] Zeshan A, and Khan MB. Implementation of Disaster Risk Reduction Policy in Pakistan–An Evidence from Sialkot; ISSRA Papers (The Journal of Government and Public Policy, 2015;**7**(1):85-98

[25] Tariq MAUR, and Giesen NVDFloods and flood management in Pakistan. Physics and Chemistry on the Earth. 2011;**47**:DOI:10.1016/j.pce.2011.08.014

[26] Adnan M. Pakistan's crisis management: Examining proactive and reactive strategies. Journal of Political Studies. 2014;**21**(1):p.161

[27] Khwaja MZ, Jawad S, Khan T. People's Experiences and Perspectives Regarding Disaster Management (Disaster Response, Mitigation, Preparedness and Risk Reduction, Project Report Series # 14 February 2009. Islamabad: Sustainable Development Policy Institute; 2008

[28] ICIMOD. Disaster Preparedness for Natural Hazards – Current Status in Pakistan [Internet]. Kathmandu: International Centre for Integrated Mountain Development (ICIMOD); 2007. Available from: http://lib.icimod.org/record/22456/files/attachment_288.pdf

[29] Bahadur A, Lovell E and Pichon F. Strengthening Disaster Risk Management in India: A Review of Five State Disaster Management Plans. London, UK: Climate & Development Knowledge Network Report. CDKN; 2016

[30] BMTPC. Guidelines: Improving Flood Resistance of Housing [Internet]. New Delhi (India): Building Materials and Technology Promotion Council, Ministry of Housing and Urban Poverty Alleviation, Government of India; 2010. [cited 2015 Jun 25]. Available from: http://www.bmtpc.org/DataFiles/CMS/file/flood-guidelines.pdf

[31] GoI. National Policy on Disaster Management 2009. New Delhi: Ministry of Home Affairs, Government of India; 2009

[32] Prasad E, Mukherjee N. Situation Analysis on Floods and Flood Management. New Delhi: IUCN, International Union for Conservation of Nature; 2014

[33] Tripathi A. Four Decades After Bill, Flood Plain Zoning Still a Far Cry [Internet]. 2014. Deccan Herald. [cited 2015 May 4]. Available from: http://www.deccanherald.com/content/407644/four-decades-bill-flood-plain.html

[34] GoI. Report of Working Group on Flood Management and Region Specific Issues. New Delhi: Planning Commission, Government of India; 2011

[35] Das P. Disaster management in India: policy review and institutional structure. Asiapac Journal of Socical Sciences. 2012;**4**:37-52

[36] Linnerooth-Bayer, Joanne, Reinhard Mechler, and Stefan Hochrainer-Stigler. "Insurance against losses from natural disasters in developing countries: evidence, gaps and the way forward." Journal of Integrated Disaster Risk Management, 2011;**1**(1):59-81

[37] Government of Bangladesh. Country Report 2003 Bangladesh, Online. 2009. Available at http://www.adrc.asia/countryreport/BGD/2003/page2.html

[38] Mozumder P. Exploring flood mitigation strategies in Bangladesh, MS thesis (unpublished). New Mexico: University of New Mexico, Albuquerque, NM; 2005

[39] Paul SK and Routray, JK. Flood proneness and coping strategies: the experiences of two villages in Bangladesh. Disasters. 2010;**34**(2):489-508

[40] WMO. Integrated flood plain management case study: Bangladesh flood management. Geneva: World Meteorological Organization; 2003

[41] Khandakar AT. Participatory water management: Bangladesh model. In: Osti R, Miyake K, editors. Forms of Community Participation in Disaster Risk Management Practices. New York (NY): Nova Science Publishers; 2011. pp. 149-156

[42] Samuels P, Klijn F, Dijkman J. An analysis of the current practice of policies on river flood risk management in different countries. Irrig Drain. 2006;**55**:141-150

[43] Paul BK. Flood research in Bangladesh in retrospect and prospect: A review. Geoforum. 1997;**28**(2):121-131

[44] Paul SK and Hossain MN. People's perception about flood disaster management in Bangladesh: A case study on the Chalan Beel Area. Stamford J Environ Hum Habitat. 2013;**2**:72-86

[45] Custers P. Banking on a flood-free future? Flood management in Bangladesh. Ecologist. 1992;**22**(5):241-247

[46] Wescoat JLJ and Jacobs JW. The Evolution of Flood Hazards Programs in Asia: The Current Situation. Boulder, CO: University of Colorado; 1993

[47] Haque CE and Zaman MQ. Human response to riverine hazards in Bangladesh: A proposal for sustainable floodplain development. World Development. 1993;**21**(1):217-229

[48] Hutton D and Haque CE. Human vulnerability, dislocation and resettlement: Adaptation process of river-bank erosion-induced displaces in Bangladesh. Disasters. 2004;**28**(1): 41-62

[49] Brammer H. Floods in Bangladesh II: Flood mitigation and environmental aspects. The Geographical Journal. 1990;**156**(2):158-165

[50] Townshend T, Fankhauser S, Aybar R, Collins M, Landesman T, Nachmany M, Pavese C. How national legislation can help to solve climate change. Nature Climate Change. 2013;**3**:430-432

[51] IPCC. Managing the Risks of Extreme Events and Disasters to Advance Climate Change Adaptation: A Special Report of Working Groups I and II of the Intergovernmental Panel on Climate Change. Cambridge (UK): Cambridge University Press; 2012

[52] Khan MR, Rahman MA. Partnership approach to disaster management in Bangladesh: A critical policy assessment. Natural Hazards. 2007;**41**:359-378

[53] Sultana P, Thompson P, Green C. Can England learn lessons from Bangladesh in introducing participatory floodplain management?. Natural Hazards. 2008;**22**:357-376

[54] Pal SK, Adeloye AJ, Babel MS, Gupta AD. Evaluation of the effectiveness of water management policies in Bangladesh. International Journal of Water Resources Development. 2011;**27**:401-417

[55] Huq MJ and Bracken L. From risk to opportunity: Climate change and flood policy in Bangladesh. Handbook of Climate Change Adaptation. 2015;pp. 1023-1047

[56] Ali AMMS. Decentralization for development: Experiment in local government administration in Bangladesh. Asian Survey. 1987;**27**(7):787-799

[57] John S. Understanding development: The experience of voluntary agencies in South Asia. Asian Affairs. 1998;**29**(2):152-160

[58] Mirza QMM, Warrick RA, Ericksen NJ and Kenny GJ. Are floods getting worse in the Ganges, Brahmaputra and Meghna Basins? Environmental Hazards. 2001;**3**(2):37-48

[59] Rahman MA, Rahman S. Natural and traditional defense mechanisms to reduce climate risks in coastal zones of Bangladesh. Weather Clim Extremes. 2015;**7**:84-95

[60] Burton I and Kates RW, White GF. The Environment as Hazard. 2nd ed. New York: The Guildford Press; 1993

[61] Cannon T. Vulnerability analysis and the explanation of "natural" disasters. In Varley A, editors. Disasters, Development and Environment. Chapter 2. Wiley; 1994

[62] Cannon T, Twigg J and Rowell J. Social vulnerability, sustainable livelihoods and disasters. Report for the Conflict and Humanitarian Assistance Department, Department for International Development, and the Sustainable Livelihoods Support Office, Natural Resources Institute, University of Greenwich, United Kingdom; 2003

[63] Maskrey, Andrew. "Reducing global disasters." Natural Disaster Management (1999): 84-6

[64] Alcántara-Ayala I. Geomorphology, natural hazards, vulnerability and prevention of natural disasters in developing countries. Geomorphology (Elsevier publications). 2002;**47**(2):107-24

[65] Few R. Flooding, vulnerability and coping strategies: Local responses to a global threat. Progress in Development Studies. 2003;**3**(1):43-58

[66] Farhad AKM. Coping strategies practiced by the farmers in post-flood period in Bangladesh. Bangladesh Res. Pub. J. 2009;**3**(2):922-933

[67] Hoque MM and Siddique MA. Flood Control Projects in Bangladesh: Reasons for Failure and Recommendations for Improvement (unpublished). Dhaka: Institute of Flood Control and Drainage Research, BUET; 1995

[68] Hossain MZ and Sakai T. Severity of flood embankments in Bangladesh and its remedial approach. Agricultural Engineering International: The CIGRE Journal, Manuscript LW 08 004. 2008;Vol. **X**

[69] Abbas A, Amjath-Babu TS, Kächele H, Usman M and Müller K. An overview of flood mitigation strategy and research support in South Asia: Implications for sustainable flood risk management. International Journal of Sustainable Development & World Ecology. 2016;**23**(1):pp. 98-111

[70] Shajahanl A and Reja MY. Towards Sustainable Flood Mitigation Strategies: A Case Study of Bangladesh (unpublished). Dhaka: Department of Architecture, BUET; 2001

[71] Wang JH. The study of flood law in China. Journal of catastrophology. 1999;**14**(3):36-42

[72] Zhang JQ, Wei WQ. The use of microcomputer information systems in assessing flood hazards. International Journal of Hydroelectric Energy. 1995;**13**(3):156-160

[73] Zhang J, Zhou C, Xu K and Watanabe M. Flood disaster monitoring and evaluation in China. Global Environmental Change Part B: Environmental Hazards. 2002;**4**(2):pp. 33-43

[74] Chen SP. Remote Sensing Analysis in Geoscience. Beijing, China: Beijing Mapping Press; 1990

[75] OAS. Disaster, Planning, and Development: Managing Natural Hazards to Reduce Loss. Organization of American States, Washington, DC: Department of Regional Development and Environment; 1990. p. 80

[76] UNDRO. Mitigating Natural Disasters: Phenomena, Effects and Options. United Nations, New York: United Nations Disaster Relief Coordinator; 1991. p. 164

[77] Ding Y and Zhang J, editors. Rainstorm and Flood. Beijing: China meteorological press; 2009. (in Chinese)

[78] Wang B, Yang X, Zhang B. Dynamics of ecosystem services in China during 1973-2003. J Zhejiang Forestry Col. 2009;**26**(5):714-721. (in Chinese)

[79] Deng X-P, Shan L, Zhang H, Turner NC. Improving agricultural water use efficiency in arid and semiarid areas of China. Agricultural Water Management. 2006;**80**(1-3):23-40

[80] Qaiser K, Ahmad S, Johnson W, Batista J. Evaluating the impact of water conservation on fate of outdoor water use: A study in an arid region. Journal of Environment Management. 2011;**92**(8):2061-2068

[81] Gu H. History for Flood Control in China: Irrigation and Water. Power Press; 2006

[82] Cao S, Liu X, Er H. Dujiangyan Irrigation System–a world cultural heritage corresponding to concepts of modern hydraulic science. Journal of Hydro-Environment Research (Elsevier). 2010;**4**(1):3-13

[83] Cheng CT and Chau KW. Flood control management system for reservoirs. Environmental Modelling & Software. 2004;**19**(12):1141-1150

[84] Zhang C, Han J, Sun W. Assessment of geo-hazard danger zoning in Long xian County, Shaanixi, China. Geol. Bull China. 2008;**27**:1795-1801

[85] Xiang X, Huang R. Application of GIS-based artificial Neural Networks on assessment of geo-hazards risk. Chinese Journal of Geological Hazard Control. 2000;**11**:23-27

[86] Chen X, Qi S, Ye H. Fuzzy comprehensive study on seismic landslide hazard based on GIS. Acta Scientiarum Naturalium Universitatis Pekinensis. 2007;**2**:1-6

[87] Zhang L, Li G, Zhou Z. Grey clustering method-based zoning assessment of regional geological disaster. Journal of Natural Disasters. 2009;**18**:164-168

[88] Zhang F, He Z, Yang B. Debris flow hazard assessment system based on GIS and neural network. Computer Engineering. 2009;**35**:205-207

[89] Ou-Yang B, Chu CC, Da YB, Liu XF and Zhang HY. Highway flood disaster risk evaluation and management in China. Natural Hazards. 2015;**75**(2):pp. 381-397

[90] Shaw R, Takeuchi Y, Uy N, Sharma A. Indigenous Knowledge: Disaster Risk Reduction, Policy Note. Bangkok: UNISDR Asia and the Pacific; 2009

[91] UNISDR (United Nations International Strategy for Disaster Reduction). Hyogo Framework for Action 2005-2015: Building the Resilience of Nations and Communities to Disasters. Geneva: UNISDR; 2007

[92] Takara K, Tachikawa Y, Kojima T. Flood Control Function of Mountain Slopes Covered with Forests: Quantitative Assessment of the Effects of So-Called 'green dam' from the Viewpoint of Basin-Scale Hydrology. Disaster Prevention Research Institute Annuals, Kyoto University; 2004. pp. 171-182

[93] Calder IR. Forests and water—ensuring forest benefits outweigh water costs. Forest ecology and management (Elsevier). 2007;**251**(1):110-20

[94] Ikeda T, Yoshitani J, Terakawa A. Flood management under climate variability and its future perspective in Japan. Water Science and Technology. 2005;**51**(5):133-140

[95] Ghatak M, Kamal A, Mishra OP. Background Paper Flood Risk Management in South Asia: SAARC Workshop on Flood Risk Management in South Asia; 2012

[96] Dixit, A. M., Yatabe, R., Dahal, R. K., & Bhandary, N. P. Initiatives for earthquake disaster risk management in the Kathmandu Valley. Natural hazards, Springer Publication. 2013;**69**(1):631-654

[97] Bendimerad F. Disaster risk reduction and sustainable development. World Bank Seminar on The Role of Local Governments in Reducing the Risk of Disasters, Held in Istanbul, Turkey, 28 April–2 May 2003;2003. 57-75

[98] Kameda, H. (2007, March). Networking disaster risk reduction technology and knowledge through Disaster Reduction Hyperbase (DRH). In Proceedings of the disaster reduction hyperbase (DRH) Contents Meeting, Kobe, Japan (pp. 12-13)

[99] Onstada PA, Danesb SM, Hardmanc AM, Olsonc PD, Marczakc MS, Heinsd RK, et al. The road to recovery from a natural disaster: Voices from the community. Community Development. 2012;**3**(5):566-580

[100] Shiwaku K, Shaw R, Kandel RC, Shrestha SN, Dixit AM. Future perspective of school disaster education in Nepal. Disaster Prevention and Management. 2007;**16**(4):576-587

[101] ActionAid. Disaster Risk Reduction through Schools: A Groundbreaking Project. 2011. http://www.actionaid.org/sites/files/actionaid/disaster_risk_reduction_through_schools.pdf. Accessed on June 2013

[102] UNESCO (United Nations Educational, Scientific and Cultural Organization) and UNICEF (United Nations Children's Fund). Disaster Risk Reduction in School Curricula: Case Studies from Thirty Countries. Peris: France United Nations Educational, Scientific and Cultural Organization and United Nations Children's Fund; 2012

[103] MercyCorp. Supporting the Role of Schools in Disaster Risk Reduction (SRSD). 2013. http://nepal.mercycorps.org/projects/disaster-projects/srsd.php

[104] MoHA. National Action Plan on Disaster Management in Nepal. Kathmandu: His Majesty's Government, Ministry of Home Affairs (MoHA); 1996

[105] Pradhan BK. Disaster Preparedness for Natural Hazards: Current Status in Nepal. Kathmandu: International Centre for Integrated Mountain Development; 2007. http://dmc.iris.washington.edu/hq/files/about_iris/governance/ds/docs/PreparednessInNepal.pdf. Accessed 12-07-13

[106] Djalante R, Thomalla F, Sinapoy MS. Building resilience to natural hazards in Indonesia: Progress and challenges in implementing the Hyogo Framework for Action. Natural Hazards. 2012;**62**(2012):pp. 779-803

[107] Yogi B. Role of state and non-state actors in Nepal's Development. Readings on Governance and Development, Vol XIV. Kathmandu: Institute for Governance and Development; 2012

[108] Manandhar N. Nepal: Corruption inside NGOs, 13th March 2013. Telegraph Nepal; 2013. http://www.telegraphnepal.com/views/2013-03-13/nepal:-corruption-inside-ngos. Accessed 28-10-13

[109] Bhatta CD. Nepal: Antinomies of Democracy, Peace and External Influence, 13th June 2013. Telegraph Nepal; 2013. http://telegraphnepal.com/views/2013-06-13/nepal:-antinomies-of-democracy-peace. Accessed 11-10-13

[110] Newar N. Rules of Engagement. Nepal Times; 2013. http://nepaltimes.com/article/nation/Rules-of-engagement,247. Accessed 11-10-13

[111] IRIN. Analysis: Politicians, Donors Question Donor Neutrality in Nepal. IRIN – Humanitarian News and Analysis. 26 February 2013; 2013. http://www.irinnews.org/report/97551/analysis-politicians-donors-question-donor-neutrality-in-nepal. Accessed 27-02-13

[112] Jones S, Oven, KJ, Manyena B and Aryal K. Governance struggles and policy processes in disaster risk reduction: A case study from Nepal. Geoforum. 2014;**57**:pp.78-90

[113] TuladharG, Yatabe R, Dahal RK and Bhandary NP. Disaster risk reduction knowledge of local people in Nepal. Geoenvironmental Disasters. 2015;**2**(1):p. 5

[114] Transparency International. Corruption Threatens Global Economic Recovery. Greatly Challenges Countries in Conflict. Transparency International; 2009. http://transparency.ie/sites/default/files/CPI_2009_Presskit_complete_en.pdf. Accessed 08-10-13

[115] MoHA. National Strategy for Disaster Risk Management. (Unofficial Translation). His Majesty's Government, Ministry of Home Affairs (MoHA); UNDP; and the European Commission for Humanitarian Aid; 2009. http://un.org.np/sites/default/files/report/2010-08-06-nsdrm-in-eng-2009.pdf. Accessed 14-07-13

[116] GFDRR. Country Profile—Vietnam; 2015. https://www.gfdrr.org/sites/gfdrr/files/region/VN.pdf

[117] Valeriano OCS, Koike T, Yang D, Nyunt CT, Van Khanh D and Chau NL. Flood simulation using different sources of rainfall in the Huong River, Vietnam/Simulation d'inondation à l'aide de différentes sources d'information pluviométrique dans le bassin de la Rivière Huong, Vietnam. Hydrological Sciences Journal. 2009;**54**(5):pp. 909-917

[118] CCFSC (Central Committee for Flood and Storm Control). Legal Documents on Flood, Storm Prevention and Natural Disaster Mitigation. Agricultural Publisher; 2011

[119] Goosby S, Mielbrecht S, Sage N, Chiesa C, Askov D. Flood modeling and early warning capacity development: concept of operations (Unpublished). Hanoi, Vietnam. 2010

[120] Hallegatte, Stephane, Bangalore, Mook, Bonzanigo, Laura, Fay, Marianne, Kane, Tamaro, Narloch, Ulf, Rozenberg, Julie, Tréguer, David and Vogt-Schilb, Adrien, Shock Waves, The World Bank, 2016. http://EconPapers.repec.org/RePEc:wbk:wbpubs:22787

[121] Bangalore, Mook; Smith, Andrew; Veldkamp, Ted. Exposure to floods, climate change, and poverty in Vietnam. Policy Research working paper; no. WPS 7765. Washington, D.C.: World Bank Group, 2016. http://documents.worldbank.org/curated/en/928051469466398905/Exposure-to-floods-climate-change-and-poverty-in-Vietnam

[122] Rubin O. Social vulnerability to climate-induced natural disasters: Cross-provincial evidence from Vietnam. Asia Pacific Viewpoint (Wiley Online Library). 2014,1;**55**(1):67-80

[123] Chau VN, Holland J and Cassells S. Institutional structures underpinning flood management in Vietnam. International Journal of Disaster Risk Reduction. 2014;**10**:pp. 341-348

[124] Tran P, Marincioni F, Shaw R and Sarti M. Flood risk management in Central Viet Nam: Challenges and potentials. Natural Hazards. 2008:**46**(1):pp. 119-138

[125] Tissera CH. Natural hazards. In: Somasekeram S, editors. Arjuna's Atlas of Sri Lanka. Colombo: Arjuna Publishers; 1997pp. 76-78

[126] De Silva K and Jayathilaka R. Gender in the context of disaster risk reduction; a case study of a flood risk reduction project in the Gampaha District in Sri Lanka. Procedia Economics and Finance. 2014;**18**:pp. 873-881

[127] Sivakumar SS. Flood mitigation strategies adopted in Sri Lanka a review. International Journal of Scientific and Engineering Research. 2015;**3**:p. 6

[128] Fernando RLS and Muthulingam A. Effectiveness of Administrative preparedness: A Case Study on Flooding Conditions in Ambagamuwa Korale Division of Nuwara Eliya District in Sri Lanka. In Strategic Disaster Risk Management in Asia. India: Springer; 2015. pp. 87-98

[129] Idris S and Dharmasiri LM. Flood risk inevitability and flood risk management in urban areas: A review. Journal of Geography and Regional Planning. 2015;**8**(8):pp. 205-209

[130] Dilhani KAC and Jayaweera N. A study of flood risk mitigation strategies in vernacular dwellings of Rathnapura, Sri Lanka. Built-Environment Sri Lanka. 2016;**12**(1)

[131] Caljouw M, Nas PJM, and Pratiwo L. Flooding in Jakarta. Towards a blue city with improved water management. Bijdragen tot de Taal-, Land- en Volkenkunde (BKI). 2005;**161**:454-484

[132] Bappenas: Laporan Perkiraan Kerusakan dan Kerugian Pasca Bencana Banjir Awal Februari. di Wilayah JABODETABEK, (Jakarta, Bogor, Depok, Tangerang, dan Bekasi), Kementrian Negara Perencanaan Pembangunan Nasional (Bappenas). Jakarta; 2007

[133] Ward PJ, Budiyono Y and Marfai MA. Flood risk in Jakarta. In: Munich R, Munich R, Munich. Severe Weather in Eastern Asia, Perils, Risks, Insurance (Munich Re Knowledge Series Natural Hazards). Germany; 2013. 106-107

[134] Sagala S, Lassa J, Yasaditama H, Hudalah D. The evolution of risk and vulnerability in Greater Jakarta: contesting government policy. Working Paper, Institute for Resource Governance and Social Change; 2013. Irgsc.org

[135] Bappenas: National Action Plan for Disaster Risk Reduction. 2010-2012. State Ministry for National Development Planning (Bappenas), National Board for Disaster Management (BNPB): Jakarta; 2010

[136] Texier P. Floods in Jakarta: When the extreme reveals daily structural constraints and mismanagement. Disaster Prevention and Management. 2008;**17**:358-372. DOI: 10.1108/09653560810887284

[137] Neolaka A. Flood disaster risk in Jakarta, Indonesia. WIT Transactions on Ecology and the Environment. 2012;**159**:pp. 107-118

[138] Shrestha M, Goodrich C, Udas P, Rai D, Gurung M and Khadgi V. Flood Early Warning Systems in Bhutan: A Gendered Perspective. International Centre for Integrated Mountain Development (ICIMOD); 2016

[139] Marfai MA, Sekaranom AB, Ward P. Community responses and adaptation strategies toward flood hazard in Jakarta, Indonesia. Natural hazards (Springer Publications). 2015,**75**(2):1127-44

Chapter 10

Analysis of the Influence of the 2007–2008 La Niña Events, Land Use, and Dam Management Modes on the 2008 Spring Freshet Characteristics in Quebec, Canada

Ali Arkamose Assani

Additional information is available at the end of the chapter

http://dx.doi.org/10.5772/intechopen.68646

Abstract

The most intense spring freshet observed since 1950 in many regions of southern Quebec took place in 2008. The goal of the chapter was to examine the influence of natural (La Niña) and human (land use and dam management) factors on the characteristics (magnitude, duration, timing, and flow variability) of this freshet. As far as natural factors are concerned, a positive correlation was found between La Niña events (both moderate and strong) and flood peaks in natural rivers. Despite its high intensity, however, the 2008 freshet was produced by a relatively moderate 2007–2008 La Niña event. The influence of land use, for its part, resulted in a higher flood peak but of relatively shorter duration in the agricultural watershed (L'Assomption River) than in the forested watershed (Matawin River) due to greater runoff in the former watershed. Finally, dam management mode affected the timing, duration, and flow variability of the freshet, as well as the number of days with zero flow. The greatest changes were observed downstream from the Matawin dam, which causes an inversion of the natural annual cycle of flow.

Keywords: spring freshet, La Niña, land use, dam management modes, Quebec

1. Introduction

Flood characteristics (magnitude, duration, timing, frequency, and flow variability) are affected by complex interactions between natural (climate and physiography) and human (land use and dams) factors. The most intense spring freshet observed in many regions of southern Quebec since 1950 occurred in 2008. This freshet led to numerous floods and substantial material damage (e.g. [1]). However, no study has looked at the climatic causes of this freshet or the influence of human activity on its characteristics. The main goal of this chapter was to fill this gap by constraining the respective influence of climatic and human factors.

As far as natural climatic factors are concerned, Assani et al. [2] showed that annual maximum daily flows were significantly correlated with the Atlantic Multi-decadal Oscillations (AMO), Artic Oscillation (AO), and Southern Oscillation Index (SOI) climate indices in Quebec. However, this study did not look at the specific impacts of El Niño and La Niña events on these flows. From a climate standpoint, the 2007–2008 season was characterized by a moderate La Niña event [3]. One of the goals of the present chapter is to test for a link between La Niña events and the magnitude of spring maximum daily flows in southern Quebec, which, to date, has not been analyzed. Because the 2008 spring freshet was associated with the occurrence of a moderate La Niña event, the underlying hypothesis is that the characteristics of spring freshets are independent of the intensity of La Niña events in southern Quebec.

As far as human factors are concerned, many dams have been built in Quebec since the nineteenth century to develop natural resources and fulfill the demand for both domestic and industrial hydroelectric power [4]. Many studies have looked at the impacts of these works on streamflow [5–20] and have shown that the extent of changes in streamflow downstream from these dams depends, among other things, on the dam management mode. Four management modes have been identified and described, each corresponding with a specific regulated hydrological regime downstream from dams. The first regulated hydrological regime is characterized by an increase in winter flows and a decrease in spring flows compared to flows in natural rivers. However, the annual cycle of flows downstream from dams is preserved: minimum flows occur in winter, despite their increase, and maximum flows occur in the spring, during snowmelt. This regime is called the "natural-type" regulated hydrological regime because of its similarity with hydrological regimes observed in pristine rivers. The second regulated hydrological regime is characterized by nearly constant flows throughout the year, with much lower month-to-month flow variability than that observed in natural rivers. This is called the "homogenization-type" regulated hydrological regime. The third regulated hydrological regime is characterized by a significant increase in winter flows and a commensurate decrease in spring flows, resulting in a complete inversion of the annual cycle of flow downstream from the dams: maximum flows occur in winter in the absence of runoff (with precipitation falling mainly as snow), and minimum flows occur in springtime during snowmelt. This regime is called the "inversion-type" regulated hydrological regime due to the inversion of the annual cycle of flows. The fourth and final regulated hydrological regime is observed downstream from dams diverting water from one watershed to another. This regime, called the "diversion-type" regulated hydrological regime, is characterized by a significant decrease in minimum flows, with maximum flows being conserved.

Several studies looking at the impacts of land use (deforestation and agriculture) on flows in rivers in Quebec have shown that land use exerts a greater influence on minimum flows (decrease) than on maximum flows (e.g., [21–23]) with little effect on flood flows. Recently, Assani et al. [13] showed that in the agricultural L'Assomption River watershed, flood peaks are higher than in the forested Matawin River watershed. These authors did not look at the impacts of land use on other flood characteristics, however. The second specific goal of the present chapter is to analyze the impacts of dams and land use on the magnitude, duration, timing, and flow variability of the 2008 spring freshet by comparing these flows in the two watersheds. The underlying hypothesis is that the characteristics of the 2008 spring freshet were strongly influenced by land use and dam management modes.

2. Methods

2.1. Choice of watersheds

The analysis of the impacts of La Niña was carried out in two steps. The first step involved comparing the 2008 spring freshet characteristics with those of the 2007 and 2009 spring freshets under natural conditions at the Joliette station (1340 km^2) in the L'Assomption River watershed and the Saint-Michel-des-Saints station (1390 km^2) in the Matawin River watershed, on the one hand, and downstream from the Rawdon dam (1240 km^2), on the Ouareau River in the L'Assomption River watershed, the Matawin dam (4070 km^2) on the Matawin River, and the Manouane dam (3686 km^2) on the Manouane River, on the other hand. The second step involved correlating the magnitude of spring maximum daily flows at stations located in natural rivers (Joliette and Saint-Michel-des-Saints stations) and downstream from the Ouareau and Matawin dams with quarterly seasonal mean values for summer (from July to September), fall (from October to December), and winter (from January to March) of SOI indices corresponding with strong and moderate La Niña events since 1950 [3]. The location of the five flow-gauging stations analyzed is shown in **Figure 1**.

To analyze the impacts of land use, the characteristics of the 2008 freshet in the agricultural L'Assomption River watershed (Joliette station) and forested Matawin River watershed (Saint-Michel-des-Saints station) were compared. These two rivers originate in the Canadian Shield, and the whole of the Matawin River watershed falls within this geological unit, whereas only two-thirds of the L'Assomption River watershed is within this unit, the last third being in the St. Lawrence River Lowlands, where all agricultural activity is concentrated. There is no agriculture in the Matawin River watershed [7]. Because both watersheds have similar physiographic and climatic characteristics (e.g. [14, 20]), a comparison of spring freshet characteristics as a function of land use between the two watersheds is straightforward. Moreover, the watershed surface areas at their respective flow-gauging stations (Joliette station on the L'Assomption River and Saint-Michel-des-Saints station on the Matawin River) are also similar.

To analyze the impacts of dam management modes, a comparison of the characteristics of the 2007, 2008, and 2009 spring freshets downstream from four dams was carried out. These four dams are the Rawdon (natural-type regulated hydrological regime characterized by maximum flows in springtime and minimum flows in winter), Matawin (inversion-type regulated hydrological regime characterized by maximum flows in winter and minimum flows in springtime during snowmelt), and Manouane (diversion-type regulated hydrological regime characterized by a significant decrease in spring flows and minimum flows in all seasons) dams. The Manouane

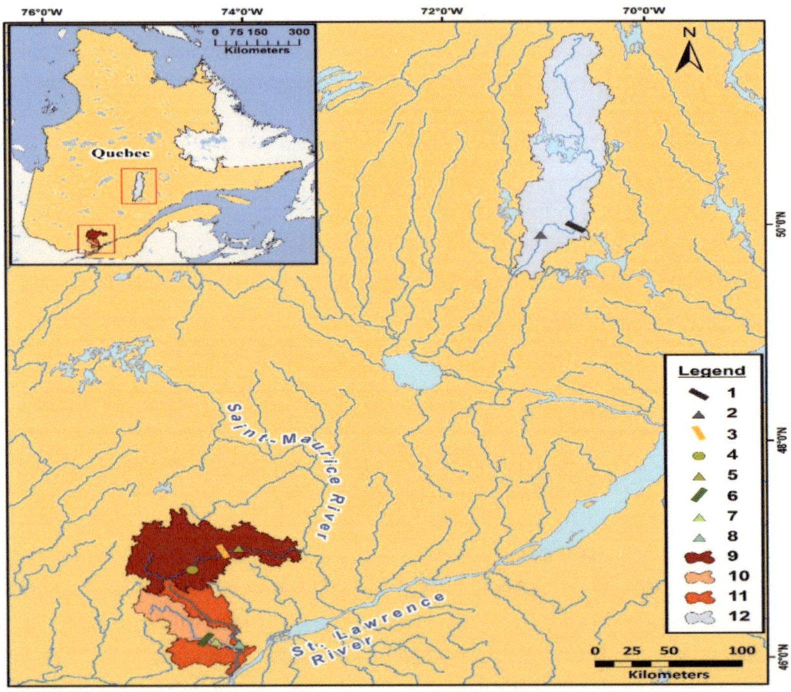

Figure 1. Location of dams and flow-gauging stations. 1 = Manouane dam, 2 = Manouane station, 3 = Matawin reservoir, 4 = Saint-Michel-Des-Saints station (upstream from Matawin reservoir), 5 = Matawin station (downstream from reservoir), 6 = Rawdon dam, 7 = Rawdon station (downstream from dam), 8 = Joliette station (L'Assomption river), 9 = Matawin watershed, 10 = Ouareau watershed, 11 = L'Assomption watershed, and 12 = Manouane watershed.

River flows entirely within the Canadian Shield and is the main tributary of the Péribonka River, which flows out into Lac Saint-Jean. This is no agricultural activity in the Manouane River watershed. The characteristics of these three dams are shown in **Table 1**, and their monthly flow coefficients are presented in **Figure 2**.

Dams	Type	ID	Height (m)	Watershed surface area (km²)	Year of construction	Total storage capacity (m³)	Reservoir surface area (ha)
Taureau (Matawin River)	Reservoir	X0004459	25	4070	1930	946,000,000	9505
Rawdon (Ouareau River)	Hydroelectric power plant	X0004205	14.6	1259	1913	5,976,417	194
Manouane (Manouane River)	Diversion dam	X2015243	9.5	3600	2003	70,000,000	–

ID, identification number.

Table 1. Dam characteristics.

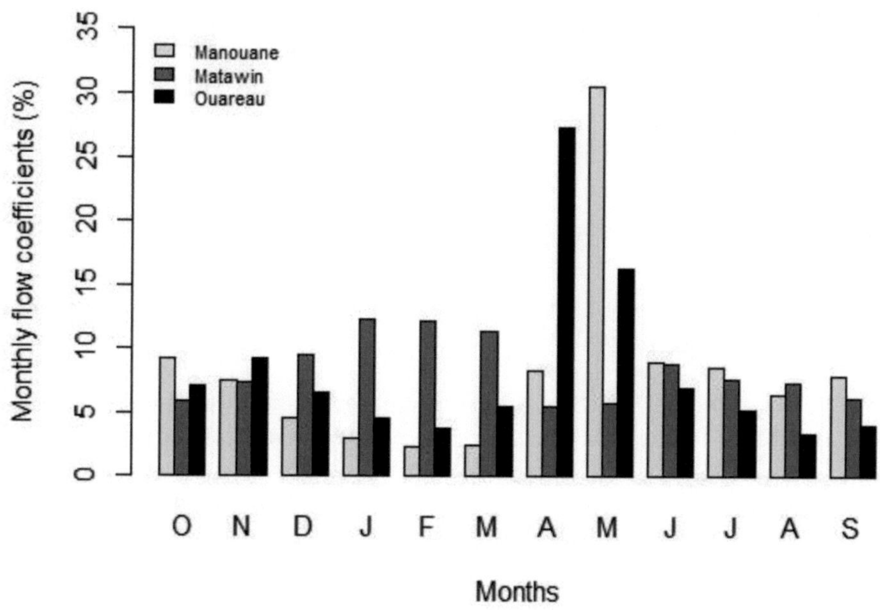

Figure 2. Monthly flow coefficients downstream from the three dams.

2.2. Sources of data and definition of spring freshet characteristics

Daily flow data downstream from the Rawdon and Manouane dams were taken from the website of the CEHQ [24]. As for flow data measured downstream from the Matawin dam, they were kindly provided by Hydro-Québec, the dam manager since 1963. SOI index data were taken from the National Oceanic and Atmospheric Administration (NOAA) website [25]. The characteristics of the 2007, 2008, and 2009 spring freshets used in the comparisons are as follows:

- Maximum flow magnitude (Q_{max}, spring maximum daily flow for a given year) during the spring freshet, expressed in $l/s/km^2$ in order to compare magnitude values between watersheds of different sizes.
- Timing (TQ_{max}) of these maximum flows, expressed in Julian days.
- Total duration in days (DQ_{max}) of the spring freshet, which is the total number of days that the freshet lasted.
- Spring freshet flow variability, measured using two indices [12]: the coefficient of variation (CV), expressed as a percentage, and the coefficient of immoderation (CI). The former index is a measure of the inter-day variability of flows during the freshet and the latter, which is the ratio of maximum (Qmax) and minimum (Qmin) flows during the freshet, is a measure of the maximum amplitude of the variability of extreme flows during the freshet.
- Total number of days with zero flow (NDZF) during the spring season (from April to June). These are the days during which flows become nil (total cessation of flow in channels).

For precipitation, data were taken from a Quebec scale map of the amount of snow (snowfall) measured during the winter of 2007–2008 (from November 2007 to March 2008) by the Ministère du Développement durable, de l'Environnement et de la Lutte contre les changements climatiques (Mddelc) [26, 27].

3. Results

3.1. Analysis of the influence of La Niña events and land use on the characteristics of spring freshets in natural rivers

The characteristics of the 2007, 2008, and 2009 spring freshets are presented in **Table 2**, and the inter-day variability of freshet daily flows is shown in **Figures 3** (Joliette station) and **4** (Saint-Michel-des-Saints station). In both watersheds, the magnitude of the freshet (Q_{max}) was higher in 2008 (La Niña year) than in 2007 and 2009. However, during the 3 years, this magnitude is higher in the agricultural watershed (L'Assomption River) than in the forested watershed (Matawin River). As for freshet duration, the 2008 freshet lasted longer than the 2009 freshet but not as long as the 2007 freshet. The 2008 freshet occurred nearly synchronously in both watersheds. In the agricultural watershed, the 2008 spring freshet occurred almost on the same date as the 2007 freshet but later than in 2009 (**Figure 3**). In contrast, in the forested Matawin River watershed, the 2008 freshet occurred slightly earlier than in 2007 and 2009 (**Figure 4**). The two flow variability indices (CV and CI) reveal that the variability of flows during the 2008 freshet was stronger than in 2009 but not as high as in 2007. In both watersheds, values of both flow variability indices are similar. Finally, no days with zero flow were observed in either watersheds during the 3 years.

Table 3 compares values of the magnitude of spring maximum daily flows as a function of moderate and strong La Niña events since 1950. The highest magnitude value since 1950 was

	L'Assomption River (agricultural watershed)			Matawin River upstream from dam (forested watershed)		
	2007	2008	2009	2007	2008	2009
Q_{max} (l/s/km^2)	187.3	206.7	143.3	123.7	157.1	99.7
Duration (days)	31	24	15	43	31	24
Timing (JD)	115	114	96	117	115	119
CV (%)	65.1	48.4	46	69.4	51.4	32.3
CI	12.1	5.7	3.6	8.3	5.4	2.8
NZFD	0	0	0	0	0	0

JD, Julian day; timing of Q_{max} 2008, La Niña year; NZFD, number of zero flows days.

Table 2. Comparison of the characteristics of spring freshets under natural conditions.

http://dx.doi.org/10.5772/intechopen.68646

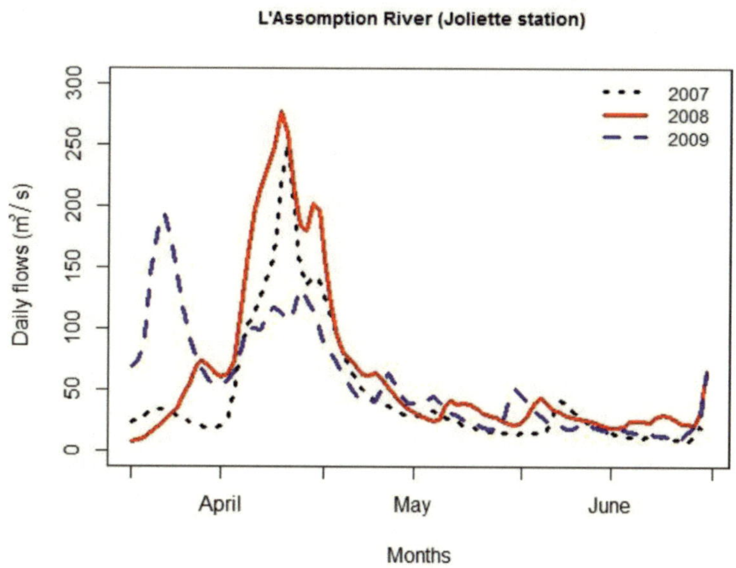

Figure 3. Comparison of the inter-day variability of spring daily flows in 2007 (blue or fair grey dotted curve), 2008 (red or grey full curve, La Niña year), and 2009 (black or dark grey broken or discontinuous curve) at the Joliette station on the L'Assomption River.

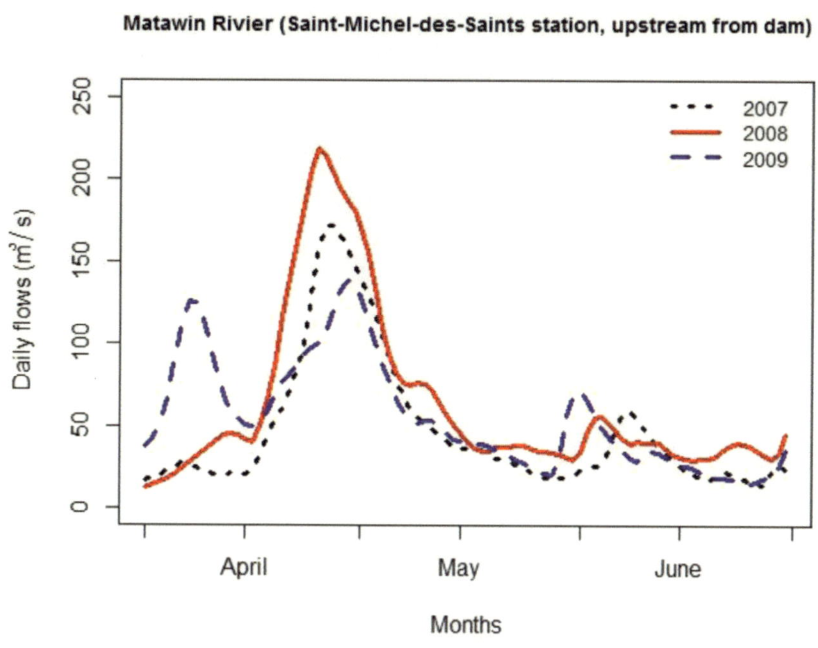

Figure 4. Comparison of the inter-day variability of spring daily flows in 2007 (blue or fair grey dotted curve), 2008 (red or grey full curve, La Niña year), and 2009 (black or dark grey broken or discontinuous curve) at the Saint-Michel-Des-Saints station upstream from Matawin dam (Matawin River).

Year of occurrence of La Niña events	Intensity of events*	Maximum flows	
		Joliette station (L'Assomption river)	Saint-Michel-Des-Saints station (Matawin River)
1955–1956	M	87	113
1970–1971	M	140	149
1973–1974	S	214	212
1975–1976	S	262	173
1988–1989	S	121	122
1998–1999	M	115	118
1999–2000	M	160	135
2007–2008	**M**	**277**	**218**
2010–2011	M	229	175

*Null [3].

M, moderate event; S, strong event.

The informations about 2007-2008 La Niña event are showed in bold.

Table 3. Comparison of maximum daily flows (m^3/s) for spring flood peaks in pristine rivers as a function of La Niña event intensity since 1950.

measured during the 2007–2008 event in both watersheds, despite the moderate intensity of this La Niña event. There is, however, a significant positive correlation between La Niña event intensities, expressed by values of the SOI indices, and the magnitude of spring freshets in both watersheds (**Table 4**). This correlation is only significant for winter SOI indices.

River stations	JAS	OND	JFM
Joliette station (L'Assomption River)	0.0014	0.3411	**0.6472**
Saint-Michel-Des-Saints (Matawin River)	−0.1885	0.3317	**0.7728**

JAS, from July to September; OND, from October to December; JFM, from January to March. Statistically significant coefficients of correlation at the 5% level are shown in bold.

Table 4. Coefficients of the correlation between spring flood peaks and mean quarterly seasonal indices for the nine moderate and strong La Niña events (SOI indices values) since 1950 in natural rivers.

3.2. Analysis of the influence of dam management modes on the characteristics of spring freshets

As in natural rivers, the magnitude of freshet flows is higher in 2008 than in 2007 and 2009 downstream from the three dams (**Table 5** and **Figures 5–7**). A comparison of the dams shows that magnitude is lower downstream from the Matawin dam, characterized by an inversion-type flow regime, than being downstream from the other two dams. The freshet lasted longer in 2008 than in 2007 and 2009 downstream from the Ouareau and Manouane dams, but the opposite is true downstream from the Matawin dam, where the duration of the 2008 spring

	Ouareau (natural-type regime)			Matawin (inversion-type regime)			Manouane (diversion-type regime)		
	2007	2008	2009	2007	2008	2009	2007	2008	2009
Q_{max} (l/s/km^2)	235.4	259.4	183.1	29.2	60.4	53.2	50.6	108.3	81
Duration (days)	30	41	15	23	14	28	39	58	41
Timing (JD)	114	114	95	152	179	156	129	121	136
CV (%)	68.4	70	49.3	37.7	36.7	59.4	61.2	71.6	59.3
CI	12.2	10.7	4.4	119	246	216.6	12.2	19.4	14
NZFD	0	0	0	48	13	29	0	0	0

JD, Julian Days; Timing of Q_{max} 2008, La Niña year; NZFD, number of zero flows days.

Table 5. Comparison of the characteristics of spring freshets downstream from three dams.

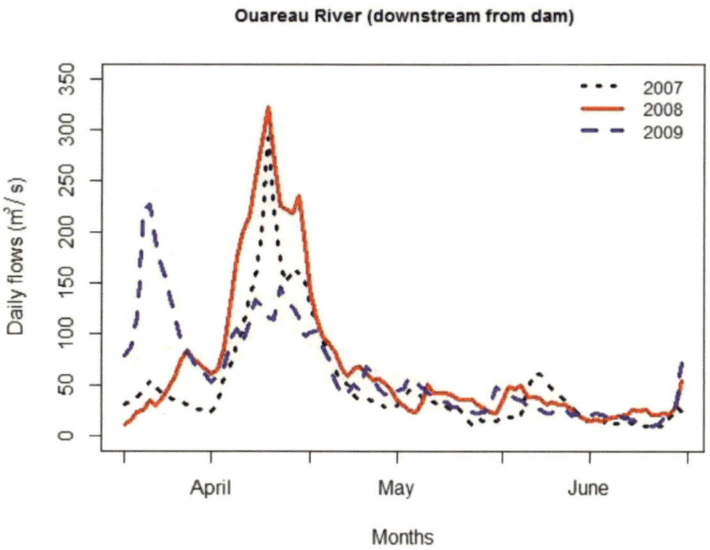

Figure 5. Comparison of the inter-day variability of spring daily flows in 2007 (blue or fair grey dotted curve), 2008 (red or dark grey full curve, La Niña year), and 2009 (black broken or discontinuous curve) downstream from Ouareau dam.

freshet was roughly half as long as those of the 2007 and 2009 freshets. Downstream from the Matawin dam, the freshet occurred later in the season in 2008 than in 2007 and 2009, whereas from the Manouane dam, it occurred early in the season. The inter-day variability of flows (CV) for the 2008 freshet was lower downstream from the Matawin dam than downstream from the other two dams. However, the maximum amplitude of extreme daily flow variations during the freshet, expressed as the coefficient of immoderation (CI), was stronger downstream from the Matawin dam than from the other two dams. Finally, despite the occurrence

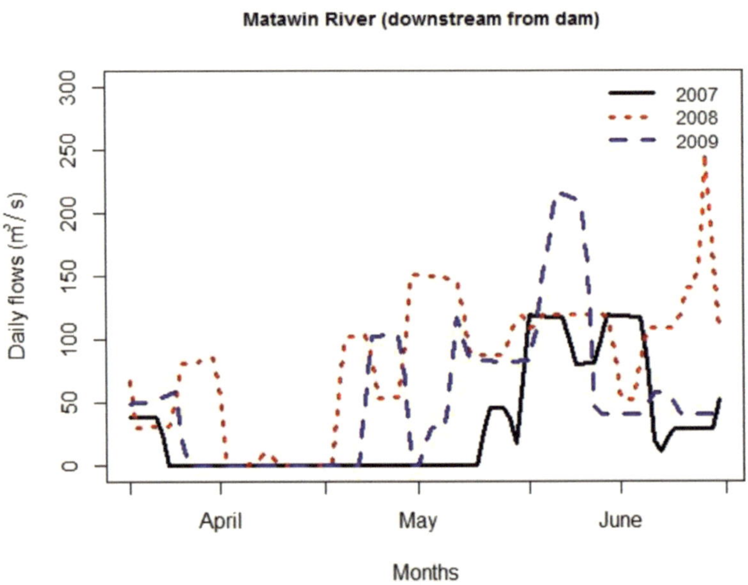

Figure 6. Comparison of the inter-day variability of spring daily flows in 2007 (blue or fair grey dotted curve), 2008 (red or dark grey full curve, La Niña year), and 2009 (black broken or discontinuous curve) downstream from Matawin dam.

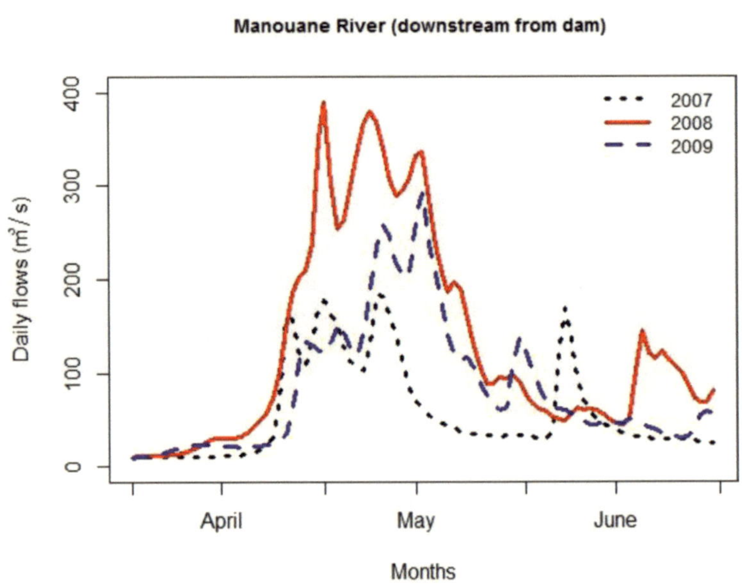

Figure 7. Comparison of the inter-day variability of spring daily flows in 2007 (blue or fair grey dotted curve), 2008 (red or dark grey full curve, La Niña year), and 2009 (black broken or discontinuous curve) downstream from Manoaune dam.

of the freshet, 13 days with zero flow were observed downstream from the Matawin dam, unlike the other two dams. However, this frequency of days with zero flow downstream from the Matawin dam was lower in 2008 than in 2007 and 2009. Finally, **Table 6** reveals the lack of any statistically significant correlation between SOI indices, which are associated with La

http://dx.doi.org/10.5772/intechopen.68646

Stations	JAS	OND	JFM
Downstream from Ouareau dam station	−0.1774	0.1466	0.5557
Downstream from Matawin dam station	−0.0614	0.3559	0.3762

JAS, from July to September; OND, from October to December; JFM, from January to March. No coefficient of correlation is statistically significant at the 10% level.

Table 6. Coefficients of correlations between spring flood peaks and mean quarterly seasonal indices for the nine moderate and strong La Niña events (SOI indices values) since 1950 downstream from the dams.

Niña events, and spring maximum daily flows downstream from the dams, contrary to what is observed for natural rivers, indicating that dams alter the link between climate factors and freshet flows.

4. Discussion

Very few studies in the literature have looked at the relationship between freshet characteristics and La Niña events. As far as the magnitude of freshets at the global scale is concerned, Ward et al. [27] observed that there are more areas in which annual floods intensify with La Niña and decline with El Niño than vice versa. This observation is consistent with results of the present study. Thus, a positive correlation was observed between the magnitude of spring freshets and values of winter quarterly (JFM) indices associated with moderate and strong La Niña events. This correlation, however, is only observed for natural rivers and is lacking downstream from dams. Despite this positive correlation in natural rivers, the 2008 spring freshet, which was the most intense since 1950 in most regions of southern Quebec, was caused by a moderate La Niña event. Two factors may account for the exceptionally high intensity of the 2008 spring freshet [26–28]:

- Abnormally high snowfalls during the winter of 2007–2008 (from November to March). In most regions of Quebec, the highest snowfalls since 1950 were recorded. In all regions of Quebec, the amount of snow that fell during that season was at least twice as large as the normal seasonal amount.
- Abundant rainfall from April 28 to May 2, which accelerated melting of this abundant snow, causing flooding and landslides in several regions of southern Quebec.

As far as land use is concerned, a comparison of the characteristics of the 2008 spring freshet between the agricultural L'Assomption River watershed and the wholly forested Matawin River watershed showed that spring freshets are of higher magnitude and longer duration in the former watershed than in the latter. These differences in hydrological behavior of the freshets are accounted for higher runoff in the agricultural watershed due to reduced plant cover and enhanced soil compaction (decreased porosity) caused by farm machinery [13]. However, very little difference was observed in the timing and amplitude of freshet flow variability between the two watersheds.

Downstream from the dams, the magnitude of the 2008 freshet was also higher than in 2007 and 2009, regardless of management mode. Therefore, this factor did not influence the magnitude of flows downstream from the dams compared to natural rivers during extreme hydrological events such as the 2008 spring freshet. Flows during this freshet were higher both in natural rivers and downstream from dams. In contrast, the other characteristics of the 2008 spring freshet were strongly affected by differences in dam management modes. Thus, the other characteristics of the 2008 freshet downstream from the Matawin dam, characterized by an inversion-type regulated regime, are significantly different from those observed downstream from the other two dams. The inversion-type management mode is characterized by storage of water in the reservoir during springtime (snowmelt and rainfall) and summer (rainfall) and its evacuation in winter. There are two reasons for this practice:

- To supply water to hydroelectric power plants built downstream from the reservoirs to produce electricity in winter. Thus, water stored in spring and summer is released in winter to supply power plants built downstream because precipitation falls mainly as snow (there is no water input from runoff). In the case of the Matawin dam, its reservoir is used to supply water in winter to hydroelectric power plants located downstream on the Saint-Maurice River. This substantial water storage accounts for the occurrence of days with zero flow downstream from this dam despite the fact that the freshet is actually taking place in natural rivers (see **Figure 6**).
- To control flood downstream from the reservoirs. To limit flooding caused by natural tributaries downstream from reservoirs, large amounts of water are stored in these reservoirs during strong spring freshets, such as the 2008 freshet. This flood control accounts for the relatively short and late nature of the 2008 freshet downstream from the Matawin dam. Releasing water at the start of the freshet in May would have enhanced the effects of flooding caused by natural tributaries downstream from the reservoir and on the Saint-Maurice River. To limit these effects, all the water from the freshet was first stored in the reservoir (occurrence of days with zero flow downstream from the reservoir), to be released downstream from the reservoir later in June (late occurrence of the freshet downstream from the reservoir) over a relatively short period (short duration of the freshet downstream from the reservoir).

This practice also resulted in a significant decrease in the magnitude of the freshet downstream from the Matawin dam, its late timing (in June rather than in May), its relatively short duration, and the strong amplitude of its flow variability due to the occurrence of days with zero flow resulting from the storage of all water derived from snowmelt and rainfall at the end of April and beginning of May 2008.

5. Conclusion

This chapter highlighted a significant positive correlation between the magnitude of spring flood peaks and the intensity (moderate and strong) of La Niña events in Quebec since 1950 in natural rivers. Thus, in Quebec, high La Niña intensities are associated with higher magnitudes of flood peaks. Despite this relationship, however, the magnitude of the 2008 spring freshet, the strongest to have occurred in Quebec since 1950, is associated with a moderate 2007–2008 La Niña event. Abnormally high snowfall in winter (from November 2007 to March 2008) and very high rainfall in the spring (from April 28 to May 2) account for the exceptionally high intensity of this freshet. Thus, both in natural rivers and downstream from dams, the magnitude of the 2008 spring freshet was the highest on record since 1950. This magnitude was much higher in the agricultural L'Assomption River watershed than in the forested Matawin River watershed. In addition, the freshet was of shorter duration in the former watershed than in the latter. This difference in land use, however, did not affect the other characteristics of the 2008 freshet (timing of flood peak, variability of flows). Downstream from the dams, differences in management modes had a significant influence on the characteristics of this freshet. Thus, downstream from the Matawin dam, characterized by the storage of snowmelt- and rainfall-derived water in spring and summer and the release of this water the following winter for hydroelectric generation (inversion-type management mode), the 2008 spring freshet was characterized by a relatively short duration, a late occurrence of flood peaks, a high amplitude (variability) of extreme flows, and, above all, the occurrence of days with zero flow, contrary to what was observed downstream from the other two dams. The chapter shows that the same extreme hydroclimate event, namely the spring freshet, does not induce the same hydrological impacts downstream from different dams due to their different flow management modes. This must be considered when assessing reserved flows required for the restoration and conservation of the ecological integrity of fluvial ecosystems in Quebec.

Author details

Ali Arkamose Assani

Address all correspondence to: Ali.Assani@uqtr.ca

Department of Environmental Sciences, University of Quebec at Trois-Rivières, Trois-Rivières, Québec, Canada

References

[1] Environment Canada [Internet]. 2008. Available from: https://http://www.ec.gc.ca/meteo-weather/default.asp? [Accessed: January 1, 2010]

[2] Assani AA, Charron S, Matteau M, Mesfioui M, Quéssy JF. Temporal variability modes of floods for catchments in the St. Lawrence watershed (Quebec, Canada). Journal of Hydrology. 2010;**385**:292-299

[3] Null J. El Niño and La Niña years and intensities based on oceanic Niño index (ONI) [Internet]. 2016. Available from: http://ggweather.com/enso/oni.html [Accessed: December 14, 2016]

[4] Astrade L. La gestion des barrages-réservoirs au Québec: exemples d'enjeux environnementaux. Annales de Géographie. 1998;**604**:590-609

[5] Assani AA. Comparaison de la variabilité interannuelle des débits en fonction de régimes hydrologiques artificialisés dans un complexe de barrages sur la rivière Saint-Maurice (Québec, Canada). Bulletin de la Société Géographique de Liège. 2004;**44**:87-96

[6] Assani AA. The usefulness of the Lombard method for analyzing the hydrological impacts of dams: The case of the Manouane River diversions dam, Quebec, Canada. Water. 2016;**8**:410. DOI: 10.3390/w8090410

[7] Assani AA, Buffin-Bélanger T, Roy AG. Analyse d'impacts d'un barrage sur le régime hydrologique de la rivière Matawin (Québec, Canada). Revue des Sciences de l'Eau. 2002;**15**:557-574

[8] Assani AA, Gravel É, Buffin-Bélanger T, Roy AG. Impacts des barrages sur les caractéristiques des débits annuels minimums en fonction des régimes hydrologiques artificialisés au Québec (Canada). Revue des Sciences de l'Eau. 2005;**18**:103-127

[9] Assani AA, Stichelbout É, Roy AG, Petit F. Comparison of impacts of dams on the annual maximum flow characteristics in the three regulated regimes in Québec. Hydrological Processes. 2006;**20**:3485-3501

[10] Assani AA, Lajoie F, Laliberté C. Impacts des barrages sur les caractéristiques des débits moyens annuels en fonction du mode de gestion et de la taille des bassins versants au Québec. Revue des Sciences de l' Eau. 2007;**20**:127-146

[11] Assani AA, Landry R, Daigle J, Chalifour A. Reservoirs effects on the interannual variability of winter and spring streamflow in the St-Maurice watershed (Québec, Canada). Water Resources Management. 2011;**25**:3661-3675

[12] Assani AA, Delisle F, Landry R, Muma M. Effects of land use on flow rate change indices. Forests. 2015;**6**:4349-4359

[13] Assani AA, Landry R, Kinnard C, Azouaoui O, Demers C, Lacasse K. Comparison of the spatiotemporal variability of temperature, precipitation, and maximum daily spring flows in two watershed in Quebec characterized by different land use. Advances in Meterology. 2016;**2016**:1-8. DOI: 10.1155/2016/3746460

[14] Aubry L, Assani AA, Biron S, Gratton D. Comparison of the hydromorphological evolution of the L'Assomption and Ouareau river channels (Quebec, Canada). River Research and Applications. 2013;**29**:979-990

[15] Fortier C, Assani AA, Mesfioui M, Roy AG. Comparison of interannual and interdecadal variability of heavy floods characteristics between upstream and downstream from Matawin dam. River Research and Applications. 2011;**27**:1277-1289

[16] Lajoie F, Assani AA, Matteau M, Mesfioui M, Roy AG. Comparaison entre les débits réservés écologiques et les débits lâchés en aval des barrages au Québec. Influence du mode de gestion des barrages, de la taille des bassins versants et de la saison. Water Quality Research Journal of Canada. 2006;**47**:263-274

[17] Lajoie F, Assani AA, Roy AG, Mesfioui M. Impacts of dams on monthly flow characteristics. The influence of watershed size and seasons. Journal of Hydrology. 2007;**334**:423-439

[18] Landry R, Assani AA, Biron S, Quéssy J-F. The management modes of seasonal floods and their impacts on the relationship between climate and streamflow downstream from dams in Quebec (Canada). River Research and Applications. 2014;**30**:287-298

[19] Matteau M, Assani AA, Mesfioui M. Application of multivariate statistical in analysis of the hydrologic impact of dams. Journal of Hydrology. 2009;**371**:120-128

[20] Sylvain J-M, Assani AA, Landry R, Quessy J-F, Kinnard C. Comparison of the spatio-temporal variability of annual minimum daily extreme flow characteristics as a function of land use and dam management mode in Quebec, Canada. Water. 2015;**7**:1232-1245

[21] Lavigne M-P, Rousseau AN, Turcotte R, Laroche A-M, Fortin J-P, Villeneuve J-P. Validation and use of a semidistributed hydrological modelling system to predict short-term effects of clear-cutting on a watershed hydrological regime. Earth Interactions. 2004;**8**:1-19

[22] Quilbé R, Rousseau AN, Moquet J-S, Savary S, Ricard S, Garbouj MS. Hydrological responses of a watershed to historical land use evolution and future land use scenarios under climate change conditions. Hydrology and Earth Systems Sciences.2008;**12**:101-110

[23] Muma M, Assani AA, Landry R, Quessy JF, Mesfioui M. Effects of the change from forest to agricultural land use on the spatial variability of summer extreme daily characteristics in southern Quebec (Canada). Journal of Hydrology.2011;**407**:153-163

[24] Centre d'Expertise Hydrique (Cehq). [Internet]. 2014. Available from: http://www.cehq.gouv.qc.ca/. [Accessed: May 20, 2014]

[25] National Oceanic and Atmospheric Administration (NOAA). [Internet]. 2017. Available from: http://www.cpc.ncep.noaa.gov/products/precip/CWlink/MJO/enso.shtml [Accessed: January 18, 2017]

[26] Ministère du Développement durable, de l'Environnement et de la Lutte contre les changements climatiques du Québec (Mddelcc) [Internet]. 2008. Available from: http://www.mddelcc.gouv.qc.ca/climat/surveillance/precipitation-neige2007-2008.pdf [Accessed: January 14, 2017]

[27] Ward PJ, Eisner S, Flörker M, Dettinger MD, Kummu M. Annual flood sensitivities to El Niño-southern oscillation at the global scale. Hydrology and Earth Systems Sciences. 2014;**18**:47-66

[28] Ici-Radio-Canada. [Internet]. 2008. Available from: http://ici.radio-canada.ca/regions/est-quebec/2008/05/04/003-transport-ferroviaire_n.asp. [Accessed: January 20, 2017]

Chapter 11

One- and Two-Dimensional Hydrological Modelling and Their Uncertainties

Mohd Talha Anees, Khiruddin Abdullah,
Mohd Nawawi Mohd Nordin,
Nik Norulaini Nik Ab Rahman,
Muhammad Izzuddin Syakir and
Mohd. Omar Abdul Kadir

Additional information is available at the end of the chapter

http://dx.doi.org/10.5772/intechopen.68924

Abstract

Earth processes, which occur in land, air and ocean in different environment and at different scales, are very complex. Flooding is also a part of the complex processes, which need to be assessed accurately to know the accurate spatial and temporal changes of flooding and their causes. Hydrological modelling has been used by several researchers in river and floodplain modelling for flood analysis. In this chapter, factors affecting flash flood, possible options of basic input parameters in one- and two-dimensional hydrological models in data sparse environment, some case studies and uncertainty in hydrological modelling were discussed. This discussion will help the readers to understand the flooding factors, selection of input parameters in data sparse environment, a brief insight of one- and two-dimensional hydrological models and uncertainties in their input and model parameters and model structures.

Keywords: flooding, floodplain, modelling, river, uncertainties

1. Introduction

Natural hazards are inevitable, unfortunate events resulting from combination of natural, geological and anthropogenic disturbances. Flooding is one of the main natural hazards and occurs frequently all over the world [1], especially in Asia and Africa than other countries. Every year, flooding incurs loss of life, economy, environment and agriculture. Economic loss includes damage to businesses, residential properties, roads, bridges, buildings and automobiles. As an example, environmental damage includes contamination of water supplies and destruction to natural habitats by chemical waste and oil spills from vehicles and industrial facilities. According to the United Nation's report [2], floods accounted for 47% of all weather-related disasters since 1995, affecting 2.3 billion people, killing 1.57 lakhs and incurring damages of about US$19.3 billion and US$0.83 billion in Asia and Africa, respectively [3].

Flooding is mainly caused as a result of increased settlement along levees, unexpected high rainfall, deforestation, sediment deposition and river channel changes [4–7]. Due to heavy rainfall in a short period of time, rivers are unable to transport the increased volume of water and other materials along its course that results in over-bank flow causing inundation of neighbouring lands. Once the river water has overflowed or breached the levees, water races through the almost level flood plain submerging the cultivated fields and villages along its way causing enormous damage to lives and properties. In-depth exploration into the causes and effects of flooding would lead to improved flood risk monitoring, prediction, mitigation and relief operations. However, poor disaster management practices, limited financial resources and high population pressure are some common characteristics of less developed countries such as Ethiopia, which affects the monitoring, mitigation and relief operations [8].

Hydrological model, a simplified and conceptual representation of complex hydrological processes, provides spatial and temporal changes over large areas and simplifies the complex reality. It is very important to find accurate causes and effects and to understand the behaviour of flooding according to their specific locations. The complex processes of river and floodplain due to combination of natural and anthropogenic activities have been assessed by several researchers [9, 10]. However, hydrological modelling has some uncertainties either in input or model parameter, which affects accuracy and efficiency of a model [11, 12]. The poor spatial distribution of basic input and model parameters data in hydrological modelling such as precipitation, evapotranspiration, infiltration and runoff can affect the model accuracy. For example, in complex topography, precipitation has uncertainty in its spatial distribution due to uplifting air masses by the wind. Some studies [13, 14] considered precipitation's spatial discontinuity and used different occurrence/non-occurrence estimation approach to improve the spatial distribution of precipitation. The spatial distribution of input and model parameters also affects the accuracy of river's geometry such as selection of their spacing [15] and channel shape [16].

With the advancement of computational technology, many one-dimensional (1D) and two-dimensional (2D) hydrological models have been developed for both river and floodplain modelling to analyse the behaviour of flooding and to identify the causes and effects of flooding. Several researchers used 1D and 2D models for floodplain modelling [17] and for river modelling [18]. The use of mixed approach of 1D and 2D numerical models increases the quality of results [17] and also saves time and computer memory, which can be limiting factors for the application of 2D models [19]. Bladé et al. [19] studied the conservation of mass and momentum by coupling of 1D and 2D models for river channels and floodplain, respectively. However, results of these models also affected by the complexity and quality of topographic and input data [20, 21].

In this chapter, first, the role of factors affecting flash flood and their use in hydrological modelling will be discussed. Secondly, in data sparse environment, the possible options for basic

http://dx.doi.org/10.5772/intechopen.68924

input variables for 1D and 2D modelling techniques would be considered. Finally, some case studies based on different 1D and 2D modelling techniques and uncertainties of hydrological modelling will be discussed in brief.

2. Factors affecting flash flood

Precipitation is the most important factor in the occurrence of flash flood, but many other contributing factors such as natural factors such as catchment characteristics, soil type and land use cover or anthropogenic activities at the river and floodplain also increase the flood frequency. Therefore, in this section, the role of these factors and their application in hydrological modelling will be discussed.

2.1. Precipitation

The dynamic behaviour and spatial distribution of precipitation due to climate changes is a major factor of concern. In general, higher precipitation can result in more runoff but the surface condition sometimes will be of greater importance. In hydrological modelling, spatial distribution of precipitation without any discontinuity should be considered in complex topography. Castro et al. [14] studied the effect of precipitation discontinuity in complex topography where orographic and meteorological parameters are effecting the spatial distribution of precipitation. Therefore, seasonal shifting of precipitation, extreme precipitation, orographic effect and behaviour of meteorological parameters such as wind characteristics, temperature and relative humidity are very important to understand the behaviour of precipitation in any region to get accurate results in hydrological modelling.

2.2. Catchment characteristics

A drainage basin is an area having a common outlet for its surface runoff. When rain falls on a drainage basin, the movement of water towards common outlet depends on the size and shape of the area. The size of the contributing area of precipitation in a basin has a direct influence on the total volume of runoff that drains from that basin. Runoff starting from most upstream point of the larger basin will take longer to reach the basin outlet than runoff travelling from the farthest point in the smaller basin, since it needs to travel a longer distance. If the basin is circular in shape, the precipitation will enter the river at roughly the same time because all points in the basin are equidistant from one another and vice versa (**Figure 1**). This will produce a high peak discharge and can lead to flash floods. Basin shape also has an influence on magnitude and timing of the peak flow at the basin outlet. **Figure 2** shows two basins of equal area with different size in which runoff is more likely to arrive at same time to the outlet in small size.

Stream density is another important characteristic of a drainage basin which is the length of all channels within the basin divided by the area of the basin. A drainage basin with a large number of tributaries has a higher stream density than a basin with very few tributary streams. Higher stream density allows the terrain to drain more runoff and vice versa. Moreover, slope

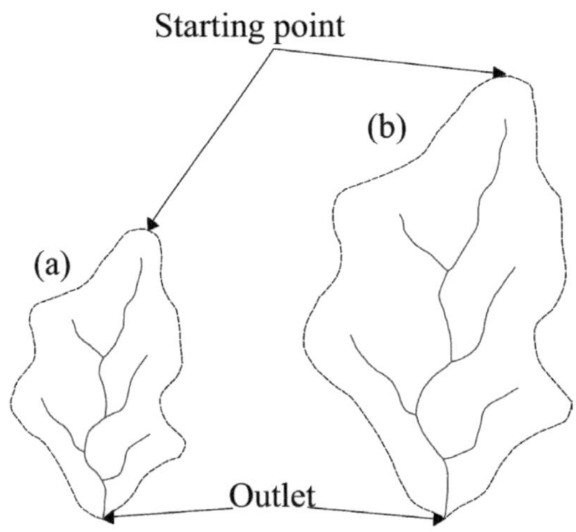

Figure 1. In basin (a), time taken from starting point to the outlet is less as compared to basin (b).

maintains moisture, which increases river's discharge and triggers flash flood. The above discussions indicates that the basins of the study area should be prioritized on the basis of flood risk and then hydrological modelling has to be carried out in prioritized basins, which will save simulation time and give better results.

2.3. Soil characteristics

Precipitated water first encounters intercepting surfaces such as foliage and man-made structures, then infiltration starts when surface water interacts with soil or bed rock. It first restores the soil moisture deficiency and then percolates downward by the force of gravity and reaches the water

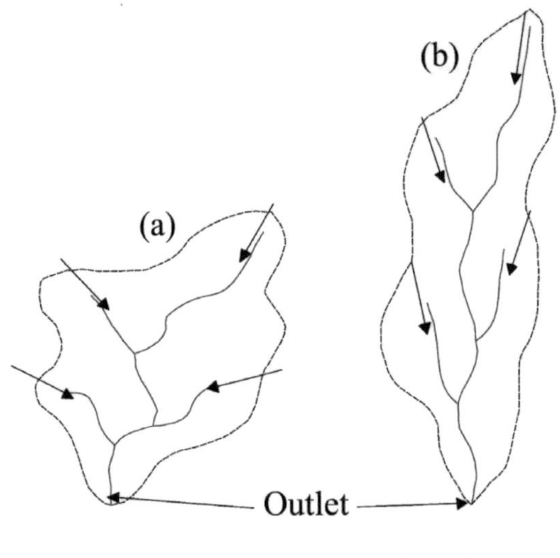

Figure 2. In (a) basin, water is more likely to come at same time than the basin (b).

http://dx.doi.org/10.5772/intechopen.68924

table [22]. During this process, soil properties and bed rock properties play an important role in the movement of water. Soil texture refers to the relative proportion of sand, silt and clay-sized particles, which controls the infiltration capacity of the basin. Clay and silt soils have the ability to retain high soil moisture while sandy soil ability to retain soil moisture is poor. De Lannoy et al. [23] explained in detail the control of soil properties on the spatial-temporal variability of infiltration and soil moisture processes, while Ref. [24] explained the influence of land use on soil moisture. Apart from soil properties, decomposing plant material on the surface also affects infiltration and runoff considerably. Litter (fresh leaves) and duff (fermented humus) are the two layers on the forest floor [25]. These two layers, especially the duff layer, aids in water repellence and affects the rate of infiltration [26]. **Table 1** shows different hydrological properties for different soil material. These properties should be included in hydrological modelling for accurate results.

2.4. Land use and land cover

Land cover and land use are essential influences on runoff. If a basin has very dense vegetation cover, the vegetation will intercept precipitation and store it, reducing the volume of

Classification	Hydraulic conductivity (in/h)	Porosity	Soil suction	Volumetric moisture deficiency	
				Dry (% diff)	Normal (% diff)
Sand and loamy sand	2.41–8.27	0.437	1.9–2.4	35	30
Sandy loam	1.02	0.437	4.3	35	25
Loam	0.52	0.463	3.5	35	25
Silty loam	0.27	0.501	6.6	40	25
Silt	–	–	7.5	35	15
Sandy clay loam	0.17	0.398	8.6	25	15
Clay loam	0.09	0.464	8.2	25	15
Silty clay loam	0.06	0.471	10.8	30	15
Sandy clay	0.05	0.430	9.4	20	10
Silty clay	0.04	0.479	11.5	20	10
Clay	0.02	0.475	12.5	15	5
Very slow	<0.06[3]				
Slow	0.06–0.20[3]				
Moderately slow	0.20–0.63[3]				
Moderate	0.63–2.0[3]				
Rapid	2.0–6.3[3]				
Very rapid	>6.3[3]				

Table 1. Soil properties that affect infiltration [27].

water entering a river. It also helps in binding the soil. In vegetation-sparse basin, the soil is highly susceptible to mass wasting which can cause huge volumes of sediment deposition on the riverbed which affect hydraulic geometry of a river. The hydraulic geometry refers to the interrelationship between water and sediment discharge, stream width, depth, velocity and river platform [28]. In channel, vegetation slows the velocity of the water flowing in it. The slower the water moves, the higher the water level, and the greater extent to which the floodplain surrounding the river will be inundated. This, however, can reduce downstream flood levels and flows. Vegetation also supports river banks in decreasing erosion and increasing the deposition of sediment. Urbanisation near the river or in floodplain also enhances the flooding frequency by reducing infiltration and increasing surface runoff into a river. Urbanisation generally involves the laying down of tarmac and concrete, impermeable substances that will increase surface runoff into the river and therefore increase the river's discharge.

In hydrological modelling, man-made structure along the river such as bridges, culverts, embankments, spillways and weirs should be included which gives accurate results for runoff. The runoff coefficient (C) is a dimensionless coefficient in hydrological modelling which is related to the amount of runoff to the amount of precipitation received. It has a larger value for areas with low infiltration and high runoff and lower for permeable, well-vegetated areas. A high C value may indicate flash flooding areas during storms as water moves fast overland on its way to a river channel or a valley floor (**Table 2**).

Area description	*C*	**Area description**	*C*
Business:		*Agricultural land:*	
Downtown areas	0.70–0.95	*Bare packed soil*	
Neighbourhood areas	0.50–0.70	Smooth	0.30–0.60
Residential:		Rough	0.20–0.50
Single-family areas	0.30–0.50	*Cultivated rows*	
Multi-units, detached	0.40–0.60	Heavy soil, no crop	0.30–0.60
Multi-units, attached	0.60–0.75	Heavy soil, with crop	0.20–0.50
Suburban	0.25–0.40	Sandy soil, no crop	0.20–0.40
Apartment	0.50–0.70	Sandy soil, with crop	0.10–0.25
Industrial:		*Pasture*	
Light areas	0.50–0.80	Heavy soil	0.15–0.45
Heavy areas	0.60–0.90	Sandy soil	0.05–0.25
Parks, cemeteries	0.10–0.25		
Playgrounds	0.20–0.35	Woodlands	0.05–0.25
Railroad yard areas	0.20–0.40	***Streets:***	
Lawns:		Asphaltic	0.70–0.95
Sandy soil, flat, 2%	0.05–0.10	Concrete	0.80–0.95
Sandy soil, avg., 2-7%	0.10–0.15	Brick	0.70–0.85

http://dx.doi.org/10.5772/intechopen.68924

Area description	C	Area description	C
Sandy soil, steep, 7%	0.15–0.20	Unimproved areas	0.10–0.30
Heavy soil, flat, 2%	0.13–0.17	Drives and walks	0.75–0.85
Heavy soil, avg., 2–7%	0.18–0.22	Roofs	0.75–0.95
Heavy soil, steep, 7%	0.25–0.35		

Table 2. Runoff coefficient (*C*) values [29, 30].

The roughness coefficient is also an important factor in terms of energy loss in both river and floodplain flowing water. The Manning's equation is a well-known equation used by almost all hydrological models to calculate channel roughness, which is given by:

$$Q = \frac{c}{n} AR^{\frac{2}{3}} S^{\frac{1}{2}} \quad (1)$$

where c is the constant, Q is the flow (length^3/time), A is the cross-sectional area (length^2), R is the hydraulic radius (length) = A/P, P is the wetted perimeter (length), S is the slope (length/length) and n is Manning's roughness coefficient. The selection of Manning's n values is important for accurate results because of its high variability and dependence on the number of factors such as surface roughness, vegetation, channel irregularities, channel alignment, scour and discharge, seasonal changes, temperature and suspended material and bedload [31]. **Table 3** shows some basic natural and man-made channel and floodplain conditions with their n values which can be used in hydrological modelling.

Type of channel and description	Range
A. Natural streams	
1. Main channels	
a. Clean, straight, full, no rifts or deep pools	0.025–0.033
b. Clean, winding, some pools and shoals	0.033–0.060
c. Sluggish reaches, weedy, deep pools	0.050–0.080
d. Very weedy reaches, deep pools, or floodways with heavy stands of timber and brush	0.070–0.150
2. Floodplain	
a. Pasture no brush	0.025–0.035
b. Cultivated areas	0.020–0.050
c. Brush	0.035–0.160
d. Trees	0.030–0.160
3. Mountain streams, no vegetation in channel, banks usually steep, with trees and brush on banks submerged	
a. Bottom: gravel, cobbles and few boulders	0.030–0.050
b. Bottom: cobbles with large boulders	0.040–0.070

Type of channel and description	Range
B. Lined or built-up channels	
1. Concrete	0.011–0.025
2. Concrete bottom float finished with different types of sides	0.015–0.035
3. Gravel bottom with different types of sides	0.017–0.036
4. Brick	0.011–0.018
5. Metal	0.011–0.030
6. Asphalt	0.013–0.016
7. Vegetal lining	0.030–0.500
C. Excavated or dredged channels	
1. Earth, straight and uniform	0.016–0.033
2. Earth, winding and sluggish	0.023–0.050
3. Dragline-excavated or dredged	0.025–0.060
4. Rock cuts	0.025–0.050
5. Channels not maintained, weeds and brush	0.040–0.140

Table 3. Manning's values for different types of channels and floodplains [31].

3. Basic input variables in data sparse environment

In order to achieve accurate flow magnitudes and water levels in 1D and 2D hydrological models, it is needed to use parameters that accurately describe the channel and flood plain geometries [31]. Digital elevation model (DEM) is an input in any hydrological modelling and its resolution significantly affects the accuracy of results. Although high resolution DEM such as airborne LIDAR (light imaging, detection, and ranging) bathymetry (ALB) technology (up to 2.5 m resolution topographical data) is available, the freely available space-borne Thermal Emission and Reflection Radiometer Global Digital Elevation Model (ASTER GDEM) of 30 m resolution and Shuttle Radar Topographic Mission Digital Elevation Model (SRTM DEM) of 90 m resolution can also be used for channel and floodplain geometries. However, ASTER GDEM and SRTM DEM has some limitations such as low penetrating power into a water column in dark or dirty water (**Figure 3**) but it can be enhanced by geographic information system (GIS) by selecting different river cross-section shape.

Several studies have been done for the selection of cross-section shape such as the trapezoid and rectangular (horizontal bottom), semicircle, parabolic, catenary, semi-cubic parabolic, egg and circular sections (curved bottom). These are generally used in different situations such as the hydraulic, economical, hydrogeological and seepage situations [16]. Curved channel, especially the parabolic shape, has the advantages such as slope stability and lacking of sharp edges [32]. Vatankhah [33] mentioned that the best hydraulic sections are those which have the least wetted perimeter for a given cross-sectional area and have the maximum hydraulic

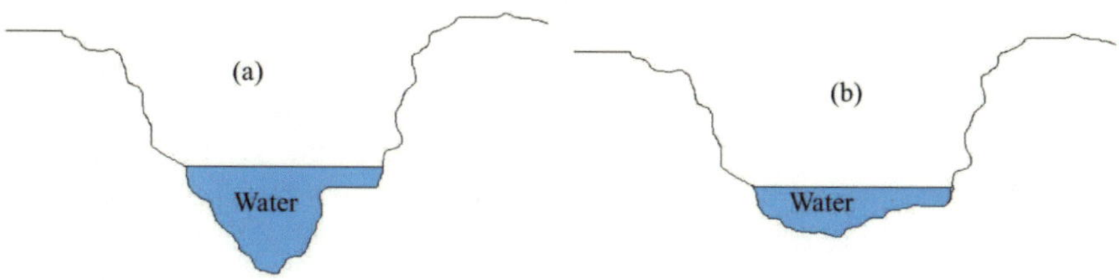

Figure 3. Two river cross sections. Cross section (a) is derived from LIDAR-based DEM while (b) is derived from ASTER GDEM.

radius. As far as channel cross-section spacing is concerned, Ref. [34] presented several guidelines for the choice of cross section and distance between them on the basis of river geometry and flows. Ref. [15] investigated the guidelines and their results confirmed the validity of these rules on the optimal spacing between cross sections. They also mentioned different equations for cross-section spacing based on different situations.

Land use and land cover are also an important input parameter in hydrological modelling because they affect the runoff frequency. The user has to assign values for different land cover classes as given in **Tables 1** and **2**. In high-resolution data sparse environment such as SPOT 5 (2.5 m resolution), IKONOS (4–1 m), QuickBird (2.4–0.61 m) and WorldView-2 (2.4–0.46 m), freely available Landsat data of 30 m resolution play a crucial role in the classification of land use and land cover. This data is useful because Landsat sensors record blue, green and red light along with the near-infrared, mid-infrared and thermal-infrared light. Landsat data has been used to monitor water quality, glacier recession, sea ice movement, invasive species encroachment, coral reef health, land use change, deforestation rates and population growth. Landsat is also helpful in assessing the damage from natural disasters such as fires, floods and tsunamis, and subsequently help in planning disaster relief and flood control programs [35]. Many studies have investigated the spatial-temporal changes of the earth surface, which could be used in place of fieldwork data [36–39]. In 2D hydrological modelling, user has to assign several floodplain and river man-made structures. So these structures can be assigned from Landsat data with the help of Google Earth information because of the difficulty in identifying objects in 30 m resolution.

4. One-dimensional model

A flood occurs due to bursting of river's banks and the water spills onto the floodplain. The nature of the terrain around a river will tell how quickly rainwater reaches the channel. Upstream anthropogenic activities and topographic changes result in increased rates of sedimentation, which modifies the river geometries and results in downstream flooding. A river flood simulation helps us to understand the spatial-temporal changes in the nature of the river and its effects on the surrounding environment [28]. Hence, for river modelling, 1D modelling is essential to know the behaviour of river flooding. In 1D modelling, it is assumed that all water flows in the longitudinal direction. One-dimensional model represents the landscape as a sequence of cross sections and simulate flow to estimate the average velocity and water depth at each cross section [20].

4.1. Overview of HEC RAS

Hydrologic Engineering Centre's River Analysis System (HEC-RAS), well-known and widely used freely available software for 1D simulation with GIS compatibility, was developed by Ref. [40]. HEC-RAS can perform steady and unsteady flow hydraulics, sediment transport/mobile bed computations and water temperature modelling. The implementation of HEC-RAS requires huge amount of data. Several studies have been done using HEC-RAS, such as unsteady flow and sediment modelling [41], steady and unsteady flow simulations [42], discharge estimation [43], determination of freedom space for rivers [44], bedload transport computations [45], backwater height calculation [46], flood analysis [18, 44, 47], bedrock channel morphology variations [48], sediment transport simulation [49, 50], cross-section spacing testing [51], hydrologic simulation [52] and river water surface profile simulation [53].

4.2. Overview of MIKE 11

MIKE 11 is another widely used, but not free, 1D hydrological model. It can perform flood analysis, dam break analysis, water quality analysis, sediment transport analysis, optimization of reservoir and river structure operations, river salinity intrusion analysis, integrated flood and catchment modelling and wetland restoration studies [54]. Some of the studies found in the literature are river stage simulation [55], groundwater response investigation to overland flow and topography [56], lowland wet grassland modelling [57], water resource management [58], organic carbon loading simulation [59], dam impact on water and sediment [60] and runoff estimation [61].

4.3. Case studies of 1D model

Several case studies have been conducted using 1D models to show their capabilities in flood analysis. The estimation and selection of river cross section have been discussed in the above section. Apart from river cross-section selection, spacing between cross sections and in-channel vegetation effects on river geometry are also assessed by some researchers. Ali et al. [51] tested different cross section using LIDAR-based DEM in HEC-RAS without including any hydraulic structural across the river and found negligible difference between different cross-section spacing. Castellarin et al. [15] retrieved different number of cross section from high-resolution Digital Terrain Model (DTM) for their study. They found that once the satisfactory spacing is achieved depending on riverbed geometry, physical wave flood and tide shape, the inclusion of additional cross sections into the 1D hydraulic model does not necessarily improves the model accuracy. Ali et al. [51] studied the influence of downstream in-channel vegetation on cross-sectional geometry and found that with decreasing slope at the downstream, the velocity also decreases due to in channel vegetation and hence widening of river takes place. Mohammadi et al. [47] conducted a simulation of flood by suing HEC-RAS to estimate flood damages. Pietsch and Nanson [62] compared the performance of the artificial neural network (ANN) technique with MIKE 11 for hydrologic simulation and found better results by ANN in terms of goodness-of-fitness and Nash–Sutcliffe index while MIKE 11 showed good results in terms of RMSE. One-dimensional flow is the limitation of 1D model and hence it cannot give causes of flooding from floodplain.

http://dx.doi.org/10.5772/intechopen.68924

5. Two-dimensional models

Two-dimensional hydraulic models are either based on finite element or finite volume to solve steady or unsteady flow equations in which water flows both longitudinally and laterally. The finite element method is a numerical procedure for solving differential equations in which continuous quantities are approximated by sets of variables at discrete locations forming a network [63]. Some of the widely used 2D models are HEC-RAS 2D, LISFLOOD-FP and FLO-2D.

5.1. Overview of HEC-RAS 2D

HEC-RAS 2D is a newly developed model and it has the ability to perform both in 1D and 2D hydrodynamic routing with unsteady flow separately as well as the combination of 1D and 2D unsteady flow by solving Saint-Venant or diffusion wave equation which allows user to work on larger river systems. For more details of this model, readers can pursue to model's reference manual [64].

5.2. Overview of LISFLOOD-FP

LISFLOOD-FP is a two-dimensional hydrodynamic model specifically designed to simulate floodplain inundation in a computationally efficient manner over complex topography. It is capable of simulating grids up to 10^6 cells for dynamic flood events and can take advantage of new sources of terrain information from remote sensing techniques such as airborne laser altimetry and satellite interferometric radar. The model predicts water depths in each grid cell at each time step, and hence can simulate the dynamic propagation of flood waves over fluvial, coastal and estuarine floodplains [65].

5.3. Overview of FLO-2D

FLO 2D is a volume conservation, flood routing and grid-based 2D model (economical as compared to MIKE 11) which uses a dynamic-wave momentum equation and a finite-difference routing scheme [66] and routes precipitation-runoff and flood hydrographs over unconfined surfaces and channels using either kinematic, diffusive or dynamic wave approximation to the momentum equation [67]. It can be used to delineate flood hazards and regulate floodplain zoning. While FLO 2D can simulate over-bank flows, it can be used to solve other flood-related problems, such as unconfined flows over complex plains and split channels, mud or debris flows and urban flooding [27].

5.4. Overview of MIKE 21

MIKE 21 is a 2D simple and faster Cartesian grid-based hydrological model for free surface flow, waves, sediment transport and environmental processes. The flood screening tool (FST) option introduces an additional numerical solver for it. A diffusive wave approach brings a simpler and thus faster numerical solution that enables even more rapid flood screening outputs [54].

5.5. Overview of TUFLOW

TUFLOW is also a grid-based 2D hydrodynamic model for free-surface flow. It has two products, viz., TUFLOW and TUFLOW FV in which TUFLOW FV has both 2D and three-dimensional (3D) flexible mesh solver. TUFLOW solves the full two-dimensional, depth averaged, momentum and continuity equations for free-surface flow using a second-order semi-implicit matrix solver. According to Ref. [68], the difference between TUFLOW and other 2D model is the inclusion of the viscosity or sub-grid-scale turbulence term that other mainstream software omits. A powerful feature of TUFLOW is its ability to dynamically link between their 1D and 2D networks.

5.6. Overview of XPSWMM

XPSWMM is a fully dynamic hydraulic and hydrologic modelling software that combines 1D calculations for upstream to downstream flow with 2D overland flow calculations which will help in understanding what truly happens to storm water system, foul water system or floodplain when waters flow, populations increase or catastrophic events hit. It allows integrated analysis of flow, pollutant transport and sustainable design measures in engineered and natural systems including ponds, rivers, lakes, overland floodplains and the interaction with groundwater [69].

5.7. Case studies based on 1D and 2D models

In 2D modelling, topographic data and its resolution are utilized in accurate modelling. However, the computational cost rises exponentially as the resolution goes finer [11]. Horritt [70] conducted 2D finite volume model with mesh resolution ranges from 2.5 to 50 m and found better results in high-resolution mesh. Chen et al. [11] proposed an approach by using multiple layers on coarse grid cell and found more accuracy with fine grid cell in same processing time which will reduce the cost of modelling. Chen et al. [71] built features in coarse grids by using the building coverage ratio (BCR) and the conveyance reduction factor (CRF) parameters in a 2D model to simulate flooding in urban areas. They found that the proposed model can minimise the errors due to terrain averaging and provide a much better accuracy of modelling results at a marginally increased computing cost. Liu et al. [56] analysed the influence of groundwater and topography on the response characteristics of overland flow by using combined 1D and 2D model in arid region. They successfully used fractal-wavelet approach to analyse the temporal characteristics of groundwater. Neal et al. [21] used three 2D explicit hydraulic models (defined as simulating diffusive, inertial or shallow water waves) to know the physical complexity needed in flood inundation simulation. They found that the diffusive-type model required much longer simulation times while inertia model was the quickest. Differences in simulated velocities and depths due to physical complexity were within 10% and simpler models were unable to simulate supercritical flows accurately. Quiroga et al. [72] used the application of 2D model in flood analysis and found good performance when compared with satellite image of the flood event. They found that the simulation provide information such as water depth, flow velocity, temporal variation of the flood and specific locations where water begins to overflow. Vozinaki et al. [73] compared the combined

1D and 2D models by using high-resolution DEM to check the results accuracy and found success in estimation of accurate flood hazard area. Some other recent studies of 1D modelling are debris flow simulation [74–76], dam break simulation [77], hydrodynamic modelling [78] and mudflow simulation [79].

6. Three-dimensional model

One- and two-dimensional hydrological models have some limitations such as the assumption of hydrostatic pressure, more horizontal length than vertical length, viscous shear stresses and bed friction on fluid components, roughness implementation on grid plane and macro-effects of changes of channel shape and direction but these limitations can be improved by 3D models [80]. Some of 3D hydrological models are as follows.

6.1. Overview of FLOW 3D

FLOW-3D is a powerful and highly accurate computational fluid dynamics (CFD) software that gives engineers valuable insight into many physical flow processes. It allows flexible gridding system, which is referred to as free-gridding because grids or geometry can be freely changed each independent of the other. FLOW-3D accurately predicts the detail of severe storm and tsunami wave run-up on coastal structures and is used for flash flood and critical structures flood and damage analysis. It is also used in large hydroelectric power projects and small municipal wastewater treatment systems, which is helpful in finding accurate results for testing design options, reduction in complexity and focussing efforts on optimized solutions [81].

6.2. Overview of MIKE 3

MIKE 3 is developed on the same module of MIKE 21, which provides the simulation tools to model 3D free surface flows and associated sediment or water quality processes. It also has a flexible gridding system. The ideal applications of MIKE 3 are mostly associated with coastal and marine hydrological modelling, lake hydrodynamic, ecology and environmental impact assessment of marine infrastructures.

Some studies used 3D models in the modelling of river width vegetated floodplain [82], simulation of curved open channel flows [83] and modelling for small debris flows [84]. Various 1D, 2D and 3D models are listed in **Table 4** with their case studies references.

S. No.	Model	References of related studies
1	HEC RAS	[18, 41–53]
2.	MIKE 11	[55–61]
3.	HEC RAS 2D	[72, 73, 85]
4.	LISFLOOD-FP	[17, 86–89]
5.	FLO 2D	[66, 67, 90, 91]

S. No.	Model	References of related studies
6.	TUFLOW	[92–95]
7.	MIKE 21	[96, 97]
8.	XPSWMM	[98–102]
9.	FLOW 3D	[103, 104]
10.	MIKE 3	[105–107]

Table 4. Hydrological models with their related study references.

7. Uncertainty in hydrological modelling

Earth processes in which changes occur in land, air and ocean in different environment and at different scales are very complex. River and floodplain processes have been studied through hydrological modelling which simplify the complex reality of Earth surface. However, hydrological model have some uncertainties, which should be included to understand the accurate causes and effects of flooding in both river and floodplain. River and floodplain modelling are characterized by uncertainties in input and model parameters, model structures and model calibration. Precipitation, one of the most basic input parameters in rainfall-runoff modelling, is taken as a parameter with uncertainty by some recent studies. The discontinuity in precipitation has already been discussed in the previous sections. McMillan et al. [108] highlights the dependency of rainfall error on the data time step in hydrological modelling. Benke et al. [109] analysed the impact of parameter uncertainty on predictions of streamflow for a water-balance hydrological model and found that the shape (skewness) of the distributed parameter's uncertainty had a significant effect on model output uncertainty. Poulin et al. [110] investigated the effects of model structure and parameter results from different events on the uncertainty related to hydrological modelling in climate change impact studies and concluded that the impact of hydrological model structure uncertainty is more significant than the effect of parameter uncertainty. Dobler et al. [111] studied the uncertainty of model parameter and hydrological projection in different models and found that the hydrological projection uncertainty varied with the choice of models affected by choice of model parameters. Engeland et al. [12] explore the effect of input uncertainty and poor observation quality on hydrological model calibration and predictions. They found that insufficient information of input parameters affects model parameters and hence results in poorer calibration and prediction. Therefore, from the above discussion it can be concluded that model uncertainty should be given utmost importance for prediction of accurate results.

8. Conclusions

Flood is the most devastating natural hazard in the world. It could be a cause of complex topographic and climatic changes on the Earth. Hydrological modelling is a useful tool to

http://dx.doi.org/10.5772/intechopen.68924

determine dynamic behaviour of flooding, its causes and effects. Based on the above discussion, this chapter concludes with the following key points:

- Precipitation is most important factor in flash flood modelling which should be used without any discontinuity. Basin shape, size, slope and stream density, spatio-temporal land use and land cover changes are important factor in controlling runoff frequency.
- Soil characteristics and the presence of decomposing plant material on the surface affect infiltration rate, which should be considered in hydrological modelling for accurate results.
- In-channel vegetation should also be considered in hydrological modelling which effects in changing downstream channel geometry.
- Basins of any study area should be prioritized on the basis of flood risk prior to hydrological modelling and then it has to be carried out in prioritized basins, which will save simulation time and will give better results.
- In spite of the use of high-resolution topographic data, which increases processing time and computational cost, freely available ASTER GDEM can be used by including multiple layers of parameters, which will give accurate results, less processing time and low computational cost.
- In data sparse environment, cross section can be selected as per the above-mentioned methods, and land use and land cover classification can be carried out from freely available Landsat data with the help of Google Earth information.
- Overview of widely used 1D, 2D and 3D hydrological models with their case studies are also helpful in knowing the usage of these models for different flood-related studies.
- At last, uncertainties of hydrological modelling were should be taken into account while modelling to enhance the results.

Acknowledgements

We would also like to acknowledge Universiti Sains Malaysia for providing the financial support through grants 203/PJJAUH/6764002 and 1001/PTEKIND/8011021 to carry out this work.

Author details

Mohd Talha Anees[1]*, Khiruddin Abdullah[1], Mohd Nawawi Mohd Nordin[1], Nik Norulaini Nik Ab Rahman[2], Muhammad Izzuddin Syakir[3] and Mohd. Omar Abdul Kadir[3]

*Address all correspondence to: talhaanees_alg@yahoo.in

1 School of Physics, Universiti Sains Malaysia, Minden, Penang, Malaysia

2 School of Distance Education, Universiti Sains Malaysia, Minden, Penang, Malaysia

3 School of Industrial Technology, Universiti Sains Malaysia, Minden, Penang, Malaysia

References

[1] International Strategy for Disaster Reduction. Global Assessment Report on Disaster Risk Reduction: Risk and poverty in a changing climate Invest today for a safer tomorrow, United Nations Geneva, 2009.

[2] UNISDR. The Human Cost of Weather Related Disasters 1995-2015. Geneva: UNISDR; 2015. Available: http://www.unisdr.org/2015/docs/climatechange/COP21_WeatherDisastersReport_2015_FINAL.pdf

[3] Nkwunonwo UC, Malcolm W, Brian B. Flooding and flood risk reduction in Nigeria: Cardinal gaps. Journal of Geography & Natural Disasters, 2015;**5**:136. DOI: 10.4172/2167-0587. 1000136

[4] Ghani AHA, Lihan T, Rahim SA, Musthapha MA, Idris WMR, Rahman ZA. Prediction of sedimentation using integration of RS, RUSLE model and GIS in Cameron Highlands, Pahang, Malaysia. In: The 2013 UKM FST Postgraduate Colloquium: Proceedings of the Universiti Kebangsaan Malaysia, Faculty of Science and Technology 2013 Postgraduate Colloquium, AIP Publishing. 2013;**1571**(1):543-548. DOI: 10.1063/1.4858711

[5] Sidek LM, Nor MM, Rakhecha PR, Basri H, Jayothisa W, Muda RS, Ahmad MN, Razad AA. Probable Maximum Precipitation (PMP) over mountainous region of Cameron Highlands-Batang Padang Catchment of Malaysia, IOP Conference Series: Earth and Environmental Science. IOP Publishing. 2013;**16**(1):012049. DOI: 10.1088/1755-1315/16/1/012049

[6] Ismail MH, Othman CKACK, Malek IAA, Abdullah SA. Land Use Trends Analysis Using SPOT 5 Images and Its Effect on the Landscape of Cameron Highlands, Malaysia, Designing Low Carbon Societies in Landscapes. Springer; 2014. pp. 223-238. DOI: 10.1007/978-4-431-54819-5_13

[7] Kamarudin MKA, Toriman ME, Rosli MH, Juahir H, Aziz NAA, Azid A, Zainuddin SFM, Sulaiman WNA. Analysis of meander evolution studies on effect from land use and climate change at the upstream reach of the Pahang River, Malaysia. Mitigation and Adaptation Strategies for Global Change. 2015;**20**(8):1319-1334. DOI: 10.1007/s11027-014-9547-6

[8] Koriche SA, Rientjes TH. Application of satellite products and hydrological modelling for flood early warning. Physics and Chemistry of the Earth, Parts A/B/C. 2016;**93**:12-23. DOI: 10.1016/j.pce.2016.03.007

[9] Chang TJ, Wang CH, Chen AS. A novel approach to model dynamic flow interactions between storm sewer system and overland surface for different land covers in urban areas. Journal of Hydrology. 2015;**524**:662-679. DOI: 10.1016/j.jhydrol.2015.03.014

[10] Massari C, Brocca L, Moramarco T, Tramblay Y, Lescot JFD. Potential of soil moisture observations in flood modelling: Estimating initial conditions and correcting rainfall. Advances in Water Resources. 2014;**74**:44-53. DOI: 10.1016/j.advwatres.2014.08.004

[11] Chen AS, Evans B, Djordjević S, Savić DA. A coarse-grid approach to representing building blockage effects in 2D urban flood modelling. Journal of Hydrology. 2012;**426**:1-16. DOI: 10.1016/j.jhydrol.2012.01.007

[12] Engeland K, Steinsland I, Johansen SS, Petersen-Øverleir A, Kolberg S. Effects of uncertainties in hydrological modelling. A case study of a mountainous catchment in Southern Norway. Journal of Hydrology. 2016;**536**:147-160. DOI: 10.1016/j.jhydrol.2016.02.036

[13] Hwang Y, Clark M, Rajagopalan B, Leavesley G. Spatial interpolation schemes of daily precipitation for hydrologic modeling. Stochastic Environmental Research and Risk Assessment. 2012;**26**(2):295-320. DOI: 10.1007/s00477-011-0509-1

[14] Castro LM, Gironás J, Fernández B. Spatial estimation of daily precipitation in regions with complex relief and scarce data using terrain orientation. Journal of Hydrology. 2014;**517**:481-492. DOI: 10.1016/j.jhydrol.2014.05.064

[15] Castellarin A, Di Baldassarre G, Bates PD, Brath A. Optimal cross-sectional spacing in Preissmann scheme 1D hydrodynamic models. Journal of Hydraulic Engineering, 2009;**135**(2):96-105. DOI: 10.1061/(ASCE)0733-9429(2009)135:2(96)

[16] Han YC. Horizontal bottomed semi-cubic parabolic channel and best hydraulic section. Flow Measurement and Instrumentation. 2015;**45**:56-61. DOI: 10.1016/j.flowmeasinst.2015.04.001

[17] Dimitriadis P, Tegos A, Oikonomou A, Pagana V, Koukouvinos A, Mamassis N, Koutsoyiannis D, Efstratiadis A. Comparative evaluation of 1D and quasi-2D hydraulic models based on benchmark and real-world applications for uncertainty assessment in flood mapping. Journal of Hydrology. 2016;**534**:478-492. DOI: 10.1016/j.jhydrol.2016.01.020

[18] Merkuryeva G, Merkuryev Y, Sokolov BV, Potryasaev S, Zelentsov VA, Lektauers A. Advanced river flood monitoring, modelling and forecasting. Journal of Computational Science. 2014;**10**:77-85. DOI: 10.1016/j.jocs.2014.10.004

[19] Bladé E, Gómez-Valentín M, Dolz J, Aragón-Hernández JL, Corestein G, Sánchez-Juny M. Integration of 1D and 2D finite volume schemes for computations of water flow in natural channels. Advances in Water Resources. 2012;**42**:17-29. DOI: 10.1016/j.advwatres.2012.03.021

[20] Cook A, Merwade V. Effect of topographic data, geometric configuration and modeling approach on flood inundation mapping. Journal of Hydrology. 2009;**377**(1):131-142. DOI: 10.1016/j.jhydrol.2009.08.015

[21] Neal J, Schumann G, Bates P. A subgrid channel model for simulating river hydraulics and floodplain inundation over large and data sparse areas. Water Resources Research. 2012;**48**(11): 1-16. DOI: 10.1029/2012WR012514

[22] Raghunath HM. Hydrology: Principles, Analysis and Design. New Age International, New Delhi, 2006.

[23] De Lannoy GJ, Verhoest NE, Houser PR, Gish TJ, Van Meirvenne M. Spatial and temporal characteristics of soil moisture in an intensively monitored agricultural field (OPE 3). Journal of Hydrology. 2006;**331**(3):719-730. DOI: 10.1016/j.jhydrol.2006.06.016

[24] Zucco G, Brocca L, Moramarco T, Morbidelli R. Influence of land use on soil moisture spatial–temporal variability and monitoring. Journal of Hydrology. 2014;**516**:193-199. DOI: 10.1016/j.jhydrol.2014.01.043

[25] Keith D, Johnson E, Valeo C. Moisture cycles of the forest floor organic layer (F and H layers) during drying. Water Resources Research. 2010b;**46**(7):W07529. DOI: 10.1029/2009WR007984

[26] Neris J, Tejedor M, Rodríguez M, Fuentes J, Jiménez C. Effect of forest floor characteristics on water repellency, infiltration, runoff and soil loss in Andisols of Tenerife (Canary Islands, Spain). Catena. 2013;**108**:50-57. DOI: 10.1016/j.catena.2012.04.011

[27] O'Brien J. FLO-2D Reference manual, Nutrioso, Arizona. version 2009; 2011.

[28] Anees MT, Abdullah K, Nawawi MNM, Rahman NNNA, Piah ARM, Zakaria NA, Syakir MI, Omar AM. Numerical modeling techniques for flood analysis. Journal of African Earth Sciences. 2016;**124**:478-486. DOI: 10.1016/j.jafrearsci.2016.10.001

[29] Fetter CW, Fetter C. Applied Hydrogeology. Upper Saddle River: Prentice Hall, New Jersey, 2001.

[30] http://water.me.vccs.edu/courses/CIV246/table2b.htm

[31] Gichamo T, Popescu I, Jonoski A, Solomatine D. River cross-section extraction from the ASTER global DEM for flood modeling. Environmental Modelling & Software. 2012;**31**:37-46. DOI: 10.1016/j.envsoft.2011.12.003

[32] Easa SM. Improved channel cross section with two-segment parabolic sides and horizontal bottom. Journal of Irrigation and Drainage Engineering. 2009;**135**(3):357-365. DOI: 10.1061/(ASCE)IR.1943-4774.0000002

[33] Vatankhah AR. Semi-regular polygon as the best hydraulic section in practice (generalized solutions). Flow Measurement and Instrumentation. 2014 **38**:67-71. DOI: 10.1016/j.flowmeasinst.2014.05.016

[34] Samuels PG. Cross-section location in one-dimensional models. In 2nd International Conference on River Flood Hydraulics. Chichester: Wiley, at Wallingford, UK. 17 -20 September, 1990. pp. 339-350

[35] http://landsat.gsfc.nasa.gov/?page_id=6724

[36] Hutchinson J, Jacquin A, Hutchinson S, Verbesselt J. Monitoring vegetation change and dynamics on US Army training lands using satellite image time series analysis. Journal of Environmental Management. 2015;**150**:355-366. DOI:10.1016/j.jenvman.2014.08.002

[37] Scullion JJ, Vogt KA, Sienkiewicz A, Gmur SJ, Trujillo C. Assessing the influence of land-cover change and conflicting land-use authorizations on ecosystem conversion on the forest frontier of Madre de Dios, Peru. Biological Conservation. 2014;**171**:247-258. DOI: 10.1016/j.biocon.2014.01.036

[38] Al-Bakri JT, Duqqah M, Brewer T. Application of remote sensing and GIS for modeling and assessment of land use/cover change in Amman/Jordan. Journal of Geographic Information System. 2013;**5**(5):509. DOI: 10.4236/jgis.2013.55048

[39] Eisakhani M, Malakahmad A. Water quality assessment of Bertam River and its tributaries in Cameron Highlands, Malaysia. World Applied Sciences Journal. 2009;**7**(6): 769-776. DOI: 10.1.1.388.3202

[40] Brunner GW. HEC-RAS River Analysis System: User's Manual. US Army Corps of Engineers, Institute for Water Resources, Hydrologic Engineering Center, Davis, California, 2001.

[41] Shelley J, Gibson S, Williams A. Unsteady Flow and Sediment Modeling in a Large Reservoir Using HEC-RAS 5.0, Federal Interagency Sediment Conference, 2015

[42] D'Oria M, Mignosa P, Tanda MG. An inverse method to estimate the flow through a levee breach. Advances in Water Resources. 2015;**82**:166-175. DOI: 10.1016/j.advwatres.2015.05.002

[43] Yang TH, Ho JY, Hwang GD, Lin GF. An indirect approach for discharge estimation: A combination among micro-genetic algorithm, hydraulic model, and in situ measurement. Flow Measurement and Instrumentation. 2014;**39**:46-53. DOI: 10.1016/j.flowmeasinst.2014.07.003

[44] Buffin-Bélanger T, Biron PM, Larocque M, Demers S, Olsen T, Choné G, Ouellet MA, Cloutier CA, Desjarlais C, Eyquem J. Freedom space for rivers: An economically viable river management concept in a changing climate. Geomorphology. 2015;**251**:137-148. DOI: 10.1016/j.geomorph.2015.05.013

[45] Pearson AJ, Pizzuto J. Bedload transport over run-of-river dams, Delaware, U.S.A. Geomorphology. 2015;**248**:382-395. DOI: 10.1016/j.geomorph.2015.07.025

[46] Wu T, Zheng Z, Ma W. Application of HEC-RAS in floating bridge Calculation of the backwater height. 3rd International Conference on Mechatronics, Robotics and Automation (ICMRA 2015), San Francisco, USA, 7-8 June 2015, Published by Atlantis Press. pp. 205-208

[47] Mohammadi S, Nazariha M, Mehrdadi N. Flood damage estimate (quantity), using HEC-FDA model. Case study: The Neka River. Procedia Engineering. 2014;**70**:1173-1182. DOI: 10.1016/j.proeng.2014.02.130.

[48] Hopkins MC, Dawers NH. Changes in bedrock channel morphology driven by displacement rate increase during normal fault interaction and linkage. Basin Research. 2015;**27**(1):43-59. DOI: 10.1111/bre.12072

[49] Pietroń J, Jarsjö J, Romanchenko AO, Chalov SR. Model analyses of the contribution of in-channel processes to sediment concentration hysteresis loops. Journal of Hydrology. 2015;**527**:576-589. DOI: 10.1016/j.jhydrol.2015.05.009

[50] Ali H, Boualem R, Mohamed R. Numerical modelling of solid transport caused by an extreme flood: Case of the Hamiz dam failure (Algeria). Journal of King Saud University-Engineering Sciences. 2014. in press. DOI: 10.1016/j.jksues.2014.08.003

[51] Md Ali A, Di Baldassarre G, Solomatine DP. Testing different cross-section spacing in 1D hydraulic modelling: A case study on Johor River, Malaysia. Hydrological Sciences Journal. 2015;**60**(2):351-360. DOI: 10.1080/02626667.2014.889297

[52] Eirini Vozinaki A, Alexakis D, Pappa P, Tsanis I. Hydrologic simulation in Mediterranean flood prone watersheds using high-resolution quality data. EGU General Assembly Conference Abstracts. 2015;**17**:10258

[53] Wang CH. Application of HEC-RAS Model in Simulation of Water Surface Profile of River, Applied Mechanics and Materials. Vol. 641. Trans Tech Publications; 2014. pp. 232-235. DOI: 10.4028/www.scientific.net/AMM.641-642.232

[54] DHI, MIKE. 11. A Modelling System for Rivers and Channels. Reference Manual. Denmark: DHI Water & Environment; 2011

[55] Panda RK, Pramanik N, Bala B. Simulation of river stage using artificial neural network and MIKE 11 hydrodynamic model. Computers & Geosciences. 2010;**36**(6):735-745. DOI: 10.1016/j.cageo.2009.07.012

[56] Liu HL, Chen X, Bao AM, Wang L. Investigation of groundwater response to overland flow and topography using a coupled MIKE SHE/MIKE 11 modeling system for an arid watershed. Journal of Hydrology. 2007;**347**(3):448-459. DOI: 10.1016/j.jhydrol.2007.09.053

[57] Thompson J, Sørenson HR, Gavin H, Refsgaard A. Application of the coupled MIKE SHE/ MIKE 11 modelling system to a lowland wet grassland in southeast England. Journal of Hydrology. 2004;**293**(1):151-179. DOI: 10.1016/j.jhydrol.2004.01.017

[58] Doulgeris C, Georgiou P, Papadimos D, Papamichail D. Ecosystem approach to water resources management using the MIKE 11 modeling system in the Strymonas River and Lake Kerkini. Journal of Environmental Management. 2012;**94**(1):132-143. DOI: 10.1016/j.jenvman.2011.06.023

[59] Kanda EK, Kosgei JR, Kipkorir EC. Simulation of organic carbon loading using MIKE 11 model: A case of River Nzoia, Kenya. Water Practice & Technology. 2015;**10**(2):298-304. DOI: 10.2166/wpt.2015.035

[60] Vinh V, Ouillon S, Thanh T, Chu L. Impact of the Hoa Binh dam (Vietnam) on water and sediment budgets in the Red River basin and delta. Hydrology and Earth System Sciences. 2014;**18**(10):3987-4005. DOI: 10.5194/hess-18-3987-2014

[61] Wang L, Liu HL, Bao AM, Pan XL, Chen X. Estimating the sensitivity of runoff to climate change in an alpine-valley watershed of Xinjiang, China. Hydrological Sciences Journal. 2014;**61**(6):1069-1079. DOI: 10.1080/02626667.2014.964718

[62] Pietsch TJ, Nanson GC. Bankfull hydraulic geometry; the role of in-channel vegetation and downstream declining discharges in the anabranching and distributary channels of the Gwydir distributive fluvial system, southeastern Australia. Geomorphology. 2011;**129**(1):152-165. DOI: 10.1016/j.geomorph.2011.01.021

[63] Cook AC. Comparison of one-dimensional HEC-RAS with two-dimensional FESWMS model in flood inundation mapping [PhD diss.]. Purdue University West Lafayette, Indiana, 2008.

[64] http://www.hec.usace.army.mil/software/hec-ras/documentation/HEC RAS%205.0%20 Users%20Manual.pdf

[65] Bates P, Trigg M, Neal J, Dabrowa A. LISFLOOD-FP User's Manual. School of Geographical Sciences, University of Bristol, UK, 2013.

[66] Cesca M, D'Agostino V. Comparison between FLO-2D and RAMMS in debris-flow modelling: a case study in the Dolomites. WIT Transactions on Engineering Sciences. 2008;**60**: 197-206. DOI: 10.2495/DEB080201

[67] Hübl J, Steinwendtner H. Two-dimensional simulation of two viscous debris flows in Austria. Physics and Chemistry of the Earth, Part C: Solar, Terrestrial & Planetary Science. 2001;**26**(9):639-644. DOI: 10.1016/S1464-1917(01)00061-7

[68] WBM B. TUFLOW User Manual. GIS Based, BMT WBM, Australia. 2010

[69] Solutions XP. XP-SWMM Stormwater and Eastewater Management Model: Getting Started Manual. Newbury, UK; 2013

[70] Horritt MS. A methodology for the validation of uncertain flood inundation models. Journal of Hydrology. 2006;**326**(1):153-165. DOI: 10.1016/j.jhydrol.2005.10.027

[71] Chen AS, Evans B, Djordjević S, Savić D. A. Multi-layered coarse grid modelling in 2D urban flood simulations. Journal of Hydrology. 2012;**470**:1-11. DOI: 10.1016/j.jhydrol.2012.06.022

[72] Quiroga VM, Kure S, Udo K, Mano A. Application of 2D numerical simulation for the analysis of the February 2014 Bolivian Amazonia flood: Application of the new HEC-RAS version 5. RIBAGUA-Revista Iberoamericana del Agua. 2016;**3**(1):25-33. DOI: 10.1016/j.riba.2015.12.001

[73] Vozinaki AEK, Morianou GG, Alexakis DD, Tsanis IK. Comparing 1D and combined 1D/2D hydraulic simulations using high-resolution topographic data: A case study of the Koiliaris basin, Greece. Hydrological Sciences Journal. 2016;1-15. DOI: 10.1080/02626667.2016.1255746

[74] Lanni C, Mazzorana B, Macconi P, Bertagnolli R. Suitability of Mono-and Two-Phase Modeling of Debris Flows for the Assessment of Granular Debris Flow Hazards: Insights from a Case Study. Engineering Geology for Society and Territory. Springer International Publishing, 2015;**2**:537-543. DOI: 10.1007/978-3-319-09057-3_89

[75] Adegbe M, Alkema D, Jetten VG, Agbor AT, Abdullahi IN, Shehu OU, Unubi AS. Post seismic debris flow modelling using Flo-2D: Case study of Yingxiu, Sichuan Pronvince, China. Journal of Geography and Geology. 2013;**5**(3):101-115

[76] Llanes F, Ferrer P, Realino V, Aquino DT, Eco RC, Lagmay A. Debris flow simulation using FLO-2D on the 2004 landslide area of real, General Nakar, and Infanta, Philippines. AGU Fall Meeting Abstracts. 2013;**1**:1620

[77] Lee K, Kim S, Kim J. Simulation of inundation zone triggered by dam failure using FLO-2D. AGU Fall Meeting Abstracts. 2014;**1**:3859

[78] Thomas D. Mussetter RA, Boberg SA. FLO-2D Modeling to Support the Rio Grande Bosque Restoration Feasibility Study, Albuquerque Reach of the Middle Rio Grande, New Mexico, World Environmental and Water Resources Congress 2012@ sCrossing Boundaries. ASCE, 2012;2593-2602. DOI: 10.1061/9780784412312.259

[79] Peng SH, Lu SC. FLO-2D simulation of mudflow caused by large landslide due to extremely heavy rainfall in southeastern Taiwan during Typhoon Morakot. Journal of Mountain Science, 2013;**10**(2):207-218. DOI: 10.1007/s11629-013-2510-2

[80] Toombes L, Chanson H. Numerical limitations of hydraulic models. In Proceedings of the 34th World Congress of the International Association for Hydro-Environment Research and Engineering: 33rd Hydrology and Water Resources Symposium and 10th Conference on Hydraulics in Water Engineering, Engineers Australia, 2011. pp. 2322-2329

[81] Hirt CW, Nichols B. Flow-3D User's Manual. Flow Science Inc; 1988

[82] Stoesser T, Wilson CAME, Bates PD, Dittrich A. Application of a 3D numerical model to a river with vegetated floodplains. Journal of Hydroinformatics. 2003;**5**(2):99-112

[83] Ye J, McCorquodale JA. Simulation of curved open channel flows by 3D hydrodynamic model. Journal of Hydraulic Engineering. 1998;**124**(7):687-698. DOI: 10.1061/(ASCE)0733-9429(1998)124:7(687)

[84] Bertolo P, Wieczorek GF. Calibration of numerical models for small debris flows in Yosemite Valley, California, USA. Natural Hazards and Earth System Science. 2005;**5**(6):993-1001. DOI: 10.5194/nhess-5-993-2005

[85] Wang W, Gao Y, Anacona PI, Lei Y, Xiang Y, Zhang G, Li S, Lu A. Integrated hazard assessment of Cirenmaco glacial lake in Zhangzangbo valley, Central Himalayas. Geomorphology. 2015. In press. DOI: 10.1016/j.geomorph.2015.08.013

[86] Horritt MS, Bates PD. Evaluation of 1D and 2D numerical models for predicting river flood inundation. Journal of Hydrology. 2002;**268**(1):87-99. DOI: 10.1016/S0022-1694(02)00121-X

[87] Fernández A, Najafi MR, Durand M, Mark BG, Moritz M, Jung HC, Neal J, Shastry A, Laborde S, Phang SC, Hamilton IM. Testing the skill of numerical hydraulic modeling to simulate spatiotemporal flooding patterns in the Logone floodplain. Cameroon. Journal of Hydrology. 2016;**539**:265-280. DOI: 10.1016/j.jhydrol.2016.05.026

[88] Wood M, Hostache R, Neal J, Wagener T, Giustarini L, Chini M, Corato G, Matgen P, Bates P. Calibration of channel depth and friction parameters in the LISFLOOD-FP hydraulic model using medium-resolution SAR data and identifiability techniques. Hydrology and Earth System Sciences. 2016;**20**(12):4983. DOI: 10.5194/hess-20-4983-2016

[89] Seenath A, Wilson M, Miller K. Hydrodynamic versus GIS modelling for coastal flood vulnerability assessment: Which is better for guiding coastal management?. Ocean & Coastal Management. 2016;**120**:99-109. DOI: 10.1016/j.ocecoaman.2015.11.019

[90] Haltas I, Tayfur G, Elci S. Two-dimensional numerical modeling of flood wave propagation in an urban area due to Ürkmez dam-break, İzmir, Turkey. Natural Hazards. 2016;**81**(3):2103-2119, DOI: 10.1007/s11069-016-2175-6

[91] Samela C, Manfreda S, Paola FD, Giugni M, Sole A, Fiorentino M. DEM-based approaches for the delineation of flood-prone areas in an ungauged basin in Africa. Journal of Hydrologic Engineering. 2015;**21**(2):06015010. DOI: 10.1061/(ASCE)HE.1943-5584.0001272

[92] Kaase CT, Kupfer JA. Sedimentation patterns across a Coastal Plain floodplain: The importance of hydrogeomorphic influences and cross-floodplain connectivity. Geomorphology. 2016;**269**:43-55. DOI: 10.1016/j.geomorph.2016.06.020

[93] McCoy A, Schubert M, Stafne B. Using 2D hydraulic models for planning and designing more resilient infrastructure. In: River Flow. CRC Press; Taylor & Francis Group, London. 2016. pp. 1896-1903. DOI:10.1201/9781315644479-297

[94] Nelson EJ, Jones R. Two-dimensional hydrodynamic dam break simulation on the Camú river, Dominican Republic. Aquananotechnology: Global Prospects. 2014;319

[95] Bertram DG. Developing an integrated urban drainage model and proxy-model methodologies [doctoral dissertation]. University of Glasgow, Scotland, UK. 2015.

[96] Parvathy KG, Gopinath DI, Noujas V, Thomas KV. Wave Transformation along Southwest coast of India using MIKE 21. The International Journal of Ocean and Climate Systems. 2014;**5**(1):23-34. DOI: 10.1260/1759-3131.5.1.23

[97] VishnuRadhan R, Vethamony P, Zainudin Z, Kumar KV. Waste Assimilative capacity of coastal waters along Mumbai Mega City, West Coast of India using MIKE-21 and WASP simulation models. CLEAN–Soil, Air, Water. 2014;**42**(3):295-305. DOI: 10.1002/clen.201200549

[98] Toriman ME, Hassan AJ, Gazim MB, Mokhtar M, Mastura SS, Jaafar O, Karim O, Aziz NAA. Integration of 1-d hydrodynamic model and GIS approach in flood management study in Malaysia. Research Journal of Earth Sciences. 2009;**1**(1):22-27

[99] van der Sterren M, Rahman A, Ryan G. Modeling of a lot scale rainwater tank system in XP-SWMM: A case study in Western Sydney, Australia. Journal of Environmental Management. 2014;**141**:177-189. DOI: 10.1016/j.jenvman.2014.02.013

[100] Kwak D, Kim H, Han M. Runoff control potential for design types of low impact development in small developing area using XPSWMM. Procedia Engineering. 2016;**154**: 1324-1332. DOI: 10.1016/j.proeng.2016.07.483

[101] Akram F, Rasul MG, Khan MMK, Amir MSII. Comparison of different hydrograph routing techniques in XPSTORM modelling software: A case study. World Academy of Science, Engineering and Technology International Journal of Environmental, Ecological, Geological and Mining Engineering. 2014;**8**(3):208-218

[102] Coremans D, Basnayaka AP. 1D-2D integrated flood modelling of Strelley brook catchment, Western Australia. In Hydrology and Water Resources Symposium, 2014; 905. Engineers Australia.

[103] https://www.flow3d.com/spillway-hydraulics-assessments/

[104] https://www.flow3d.com/indianapolis-storm-water-system/

[105] Bolaños R, Sørensen JVT, Benetazzo A, Carniel S, Sclavo M. Modelling ocean currents in the northern Adriatic Sea. Continental Shelf Research. 2014;**87**:54-72. DOI: 10.1016/j.csr.2014.03.009.

[106] https://www.dhigroup.com/upload/publications/mike21/Karlsson_2010.pdf

[107] https://www.dhigroup.com/upload/publications/mike21/Liungman1_2010.pdf

[108] McMillan H, Jackson B, Clark M, Kavetski D, Woods R. Rainfall uncertainty in hydrological modelling: An evaluation of multiplicative error models. Journal of Hydrology. 2011;**400**(1):83-94. DOI: 10.1016/j.jhydrol.2011.01.026

[109] Benke KK, Lowell KE, Hamilton AJ. Parameter uncertainty, sensitivity analysis and prediction error in a water-balance hydrological model. Mathematical and Computer Modelling. 2008;**47**(11):1134-1149. DOI: 10.1016/j.mcm.2007.05.017

[110] Poulin A, Brissette F, Leconte R, Arsenault R, Malo JS. Uncertainty of hydrological modelling in climate change impact studies in a Canadian, snow-dominated river basin. Journal of Hydrology. 2011;**409**(3):626-636. DOI: 10.1016/j.jhydrol.2011.08.057

[111] Dobler C, Hagemann S, Wilby RL, Stötter J. Quantifying different sources of uncertainty in hydrological projections in an Alpine watershed. Hydrology and Earth System Sciences. 2012;**16**(11):4343-4360. DOI: 10.5194/hess-16-4343-2012

Multicriteria Decision Analysis for Flood Risk Management: The Case of the Mapai Dam at the Limpopo River Basin, Mozambique

Avelino I. Mondlane

Additional information is available at the end of the chapter

http://dx.doi.org/10.5772/intechopen.68978

Abstract

In this article, we present evaluations of the different scenarios for building the Mapai Dam in the Limpopo River Basin of Mozambique as an integral part of the flood risk management in the basin. The study used various decision analysis software packages such as V.I.S.A and Decide IT, which are basically instances of decision support systems. In order to perform the evaluation, we used sources available to assess the need and feasibility to build the Dam, which provides the basis for our analysis. We first structured the problem in order to make it more understandable through a value measurement tree, then built a decision making tree to develop different scenarios, whereby we compared their outputs based on a probabilistic analysis, according to their storage capacities. The results prove that the system is helpful in decision-making process, particularly when we face multiple choices.

Keywords: decision support system, flood risk management, Mapai Dam, multicriteria decision making

1. Introduction

Mozambique is one of the least developed countries in the Southern African development community (SADC) region with 2700 km of coastline [1]. The country is already experiencing the devastating effects of droughts, floods, and cyclones on agricultural production such as crops, livestock, forest, and fisheries both in rural and in coastal areas from climate change [2]. This is true in particular with respect to some shared international basins with hinterland countries [3], and moreover within the Limpopo River Basin as a result of the extreme weather and the lack of flood control mechanisms. With an area of about 400,000 km^2, the Limpopo Basin occupies between 11 and 16% of the area of the four riparian countries, namely Botswana, Mozambique, South Africa, and Zimbabwe, and the river flows into the Indian Ocean at Mozambique Channel [4] (**Figure 1**). The population of the basin is approximately 14 million, of which half is in Botswana and less than 10% is in Mozambique and Zimbabwe [5].

Together with its tributaries, the Limpopo River is expected to be significantly affected by the climate change [6] and as a consequence there is an urgent need to address the high vulnerability of the population within the basin. The government of Mozambique identified the District of Chicualacuala to start the United Nations joint programme (UNJP) on Environmental Mainstreaming and Adaptation to Climate Change for the period between 2008 and 2011 with a total of USD 7 million funding [7]. According to Ref. [8], climate change is expected to affect the precipitation, potential evaporation, and runoff in Limpopo Basin, which may bring changes in water supply, leading to power shortages due to the reduced rainfall.

In 2000, floods devastated Mozambique with the cost estimated at 20% of the gross domestic product (GDP), slowed down the yearly economic growth to 2% [9]. The development of Mapai Dam is one of the strategies identified by the government of Mozambique to minimize the impact of floods in the Limpopo River Basin, since Mozambique is the only country in the basin with limited capacity of water storage among the riparian countries as indicated in **Figure 1**. The statistical data from [4] show that Mozambique has only one dam with a large storage capacity, which is the Massingir Dam in the Elephants River with around 2800 Mm^3

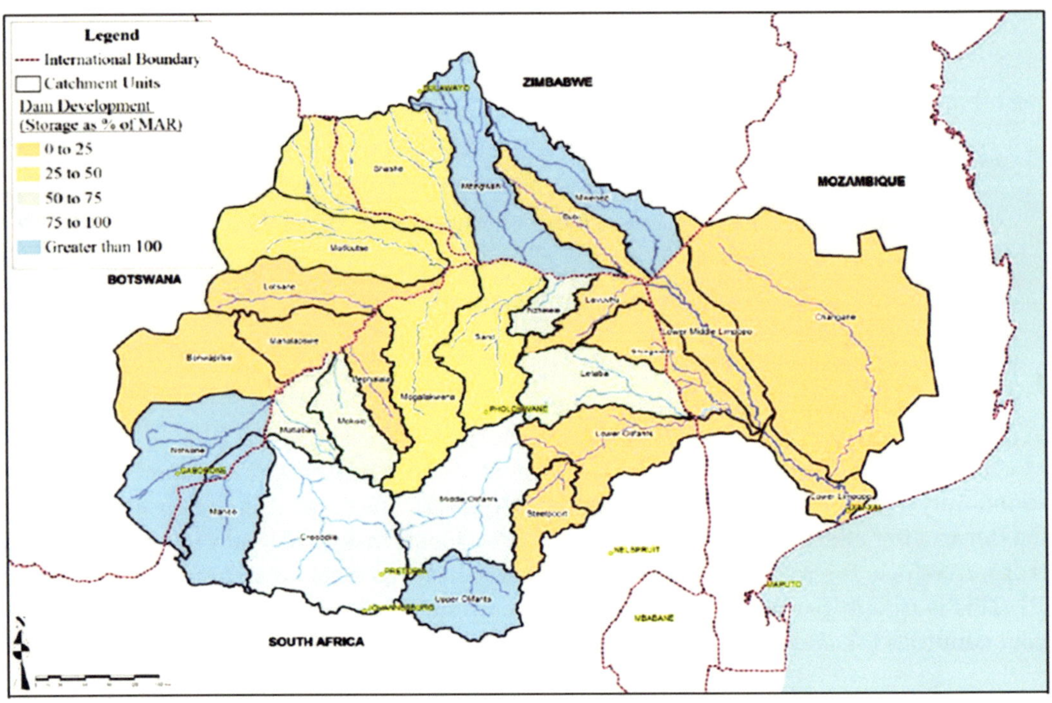

Figure 1. Distribution of water demand (as % of mean annual runoff (MAR)) for sub catchments of the Limpopo River Basin. *Source:* Ref. [11].

of capacity. However, this dam operates below its capacity after going through a technical accident in the 1980s. An additional infrastructure is the Macarretane Dam on the Limpopo River in Chókwe with approximately 4 Mm^3 storage capacity. Botswana has four large dams and plans to build additional three in the near future, while South Africa has a quite large number of dams, with 160 classified as large storage. In Zimbabwe, there are 12 dams ranging from medium to large capacity [10]. The limitation of water storage in the Mozambican section worsens the impact of floods, which has negative economic, social, and environmental implications. Herein, we analyze the different scenarios and impact of this structural coping mechanism.

Limpopo River is of vital importance for the country not only for economic and social systems but also for the functioning of natural ecosystems. An illustrative example is Chókwe District in the downstream area with strong potential for agriculture based on the irrigation system, which is called the granary of the nation, due to the irrigated agricultural practice.

According to Ref. [12], the Master Plan for Prevention and Mitigation of Natural Disasters established in March 2006 by the Government of Mozambique (GoM) outlined the emergency procedures for government agencies, the structure and organization of the national emergency operations center, and its regional branches. The Master Plan also highlights disaster preparedness as an essential part of government concerns to be included in the poverty reduction strategy (PARPA) as a crosscutting agenda. The set of actions includes the flood early warning system, introduction of drought-resistant crops, and the information and communication management systems for national emergencies. PARPA constitutes one of the guidelines for the water resources assistance strategies in the country [2].

One of the objectives of this article is to develop simultaneously an analytical procedure using multicriteria decision analysis (MCDA) in order to accommodate different and conflicting objectives by the involved stakeholders based on a decision support system (DSS) as a part of an MCDA process. Initially, we structure the problem, using value three measurements, using V.I.S.A,[1] aiming to make the problems more understandable. Lastly, we use Decide IT,[2] a decision-making support system to develop a decision tree to structure the problem and develop our analysis.

The outline of this article is as follows: the introduction (present section) gives the background information about Mozambique and the formulation of the problem. The second section summarizes the applied methodology, which is followed by the main water resources strategies for the country. The main steps and the key concepts related to MCDA and multicriteria decision making (MCDM) are presented in Section 3, while in Section 4, we structure the problem and present the basic assumptions toward scenario building in decision making. Section 5 brings the solution to the study problem, by applying a DSS based

[1]V·I·S·A stands for Visual Interactive Sensitivity Analysis. It is a software for decisions with multiple, tough to balance, factors; for decisions where no option matches all of the criteria perfectly; or for decisions where more than one person has a say in how the decision is made. It does not tell you the "right answer," it lets everyone involved see for themselves what the best overall decision is, weighing up all the factors using a considered and sound process. (http://www.visadecisions.com/visa_more.php.)

[2]Decide IT is among the software used at the Department of Computer and Systems Science, developed by Preference-Consulting Developer of Software for Risk and Decision Analysis (http://www.preference.bz).

on Decide IT software, where we developed all possible outputs, provided the decision and willingness of the stakeholders to abide by the decision. We provided the analysis in Section 6 along with the discussion of the results, and finally in Section 7, we present the conclusions and future work.

2. Methodology and materials

Flood risk management as a multidisciplinary activity is subjected to different and multiple choices in a conflicting situation for the involved stakeholders. In Mozambique, flood risk management is under the responsibilities of National Institute for disaster management (INGC), but different stakeholders are actively involved in the process. Among these are the Mozambique Red Cross, the National Institute of Meteorology, United Nations agencies, and the National Defense Forces.

The authors [13] address sensitive analysis as one of the decisive steps in building models, and by adopting such concept, we evaluate different options for the development of Mapai Dam, taking into consideration different scenarios based on storage capacities, bearing in mind two different perspectives: the *de facto* situation and alternative(s) for building the Dam with a specific and optional storage capacity.

Our study was based on a field survey and analysis of documents in order to get insight on whether to build or not the Mapai Dam. Moreover, a fieldwork was performed in Chókwe District between 8 and 22 April 2013 aiming to assess the impact of floods in the basin and evaluate to what extent the Mapai Dam could influence the management of the basin. About 90% of the interviewees indicated that there is a strong need to develop the Mapai Dam in order to minimize the impact of floods and also to bring better control on water management for agriculture.

The data collected during the fieldwork were aggregated for different probabilities according to responses of the interviewees. Furthermore, the probabilities are incorporated into a chosen decision support system, the Decide IT, which is part of the tools used for our research.

We used the DSS module of Decide IT to build a decision tree for scenario analysis. This can be linked to multiple purpose analysis, given the different options that are generated, hence is linked to MCDA. The literature survey of different sources of information among governmental documents in regard to strategies for flood risk management over Limpopo River Basin also shows to what extent there is need of an effective infrastructure mechanism, see **Table 1**. The Mapai Dam is also listed in the AQUASTAT database for irrigation with a capacity of 11.2 Mm^3 [14], as part of major African basin infrastructures.

Table 1 shows the selection and sources of the material used, which stress the need for the development of the Mapai Dam in the Limpopo River Basin in order to minimize the impact of floods downstream. The World Bank has shown some commitment to support the water resources strategies in Mozambique. In 2007, the organization developed the country's water resources assistance strategy, which congregates different approaches and strategies already in use in Mozambique and includes the inputs from consultations with the government, the

Title	Year	Source	Type of document
Strategic plan for the development of Gaza Province	2007	[15]	Strategic plan
Mozambique Country water resources assistance strategy: Making water work for sustainable growth and poverty reduction	2007	[16]	Technical consulting document
Limpopo Basin permanent technical committee – LBPTC Joint Limpopo River Basin Study Scoping Phase	2010	[4]	Technical report
Centro de Promoção De Investimentos CPI	2012	[17]	Advocacy for investment
Profile of the Limpopo Basin in Mozambique, a contribution to the challenge program on water and food project 17 livelihoods: Managing risk, mitigating drought and improving water productivity in the water scarce Limpopo Basin"	2009	[18]	Research project
FCEC-CPLP. Fórum Para a Cooperação Económica E Comercial Entre a China E Os Países de Língua Portuguesa	2012	[19]	International forum advocating investment
Water-use accounts in CPWF basins: Simple water-use accounting of Limpopo Basin	2010	[14]	Working paper, basin focal project series
Sustainable land use planning for integrated land and water management for disaster preparedness and vulnerability reduction in the Limpopo Basin	2007	[20]	Limpopo Basin strategic plan
Mozambique: Government looking for funding for Mapai Dam	2013	[21]	State House news
Republic of Mozambique: Poverty reduction strategy paper—economic and social plan for 2005	2005	[22]	IMF country report (No. 05/312)
Official development assistance to Mozambique Database: Project fact sheet for 8 ACP MOZ 024-19	2005	[23]	Project fact sheet for 8 ACP MOZ 024-19
Limpopo River awareness kit	2014	[24]	Official home page

Table 1. List of main references of Mapai Dam through various studies.

stakeholders, and the donors as shown in **Figure 2**. Here the Mapai Dam project development is addressed as a part of the strategies for the main water sectors.

In Mozambique and the Limpopo River Basin, there is a need to establish agreements to share water resources with the neighbors and build the Mapai Dam on the Limpopo River for irrigation purposes, flood control, control of saltwater intrusion, and ensure minimum flows for ecological reasons. Based on Ref. [18], there is also a need to promote the development of small dams for water supply for livestock and the rural population. According to the National Directorate of Water in Mozambique [18], the main source of water supply in urban centers in the Mozambique Limpopo Basin (Chókwe and Xai-Xai) is ground water, whose characteristics are covered by **Figure 2**. It seems that the agriculture sector has increased the use of surface water in the basin. The potential irrigable area of the Limpopo River basin in the Mozambican territory is estimated around 148,000 ha, of which about 50% is equipped with irrigation infrastructure and 27% is operational [18].

The development of Mapai Dam fits within the country's water resources assistance strategy, and it will bring changes in different sectors such as irrigation, urban and rural water supply,

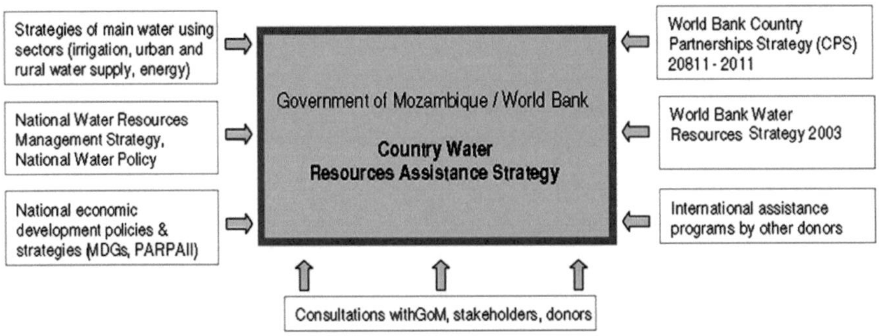

Figure 2. Development of the country water resources assistance strategy.

and energy, as well as to reinforce the national water resources and national economic developments. To meet these challenges Mozambique counts on cooperation of international partnership such as with the World Bank and other international assistance as shown in **Figure 2.**

The development of the Mapai Dam in the Limpopo River Basin is of interest to many of the interviewed parties such as the water management authority of the local government, farmers, other water management institutions such as administração regional de aguas do sul (ARA Sul) and national water administration (ANE), the Government of Gaza, the Center for Promotion of Investment, the INGC, the National Institute of Meteorology, the UN agencies, and the Mozambique Red Cross. It is clear that in many ways the stakeholders play a decisive role for the development of the Mapai Dam.

3. MCDA/(M) methods

According to Ref. [5] by the 1970s, the research on MCDM focused on the theoretical foundations of multiple mathematical programming procedures and algorithms for solving multiple objective mathematical programming problems. The author [25] argues that in multicriteria analysis problems several criteria are simultaneously optimized in a feasible set of finite number of given choices. The author [26] defines multicriteria decision making (MCDM) as a collection of methods to compare, select, or rank multiple alternatives that typically involve incommensurate attributes. It is well suited for eliciting and modeling the flood preferences of stakeholders and for improving the coordination among the flood agencies, various organizations, and the affected citizens.

Multi-attribute methods follow two schools of thought: American/Anglo-Saxon and European/French. Multi-attribute utility or value theory (MAUT or MAVT) and the analytical hierarchy process (AHP) are the main methods used in the United States, while the French school developed the outranking methods of ELECTRE (*elimination et choix traduisant la realité*) and preference ranking organization method for enrichment of evaluations (PROMETHEE) that are dominant in Europe. In Ref. [27], one can find more details about the advantages and disadvantages of both approaches.

Decision support system architecture for flood modeling under consideration integrates the latest advances in MCDM, remote, GIS, hydrologic models, and real-time flood information systems. The author [28] defines a decision support system (DSS) as "a computer-based information system used to support decision-making activities in situations where it is not possible or desirable to have an automated system performing the entire decision process." Based on Ref. [5], by the 1980s, the emphasis within the MCDM, through the increased use of computers, shifted to DSS. Furthermore, they define the multicriteria decision support system (MCDSS) as simply a DSS that implements MCDM and/or MAUT/MAVT models.

The specific characteristics of MCDSS that, in a way, differentiate them from the ordinary DSS is that they include analysis of multiple criteria, involvement of MCDM methods, and the integration of the user in the input during the modeling process [5]. A DSS is customized, interactive computing environment that integrates model/analytical tools, databases, graphical user interfaces, and other systems. In addition, Levy [26] notes that DSSs are designed to help decision makers use data and models to evaluate unstructured problems that require the management judgment. A significant number of authors and their work with respect to DSS and MCDSS are extensively discussed in Ref. [26].

As in many hydrological hazards within flood risk management, the transformation of qualitative factors such as environmental quality, social impact, ecological concerns, and political issues into quantitative, financial, values constitute a significant drawback to cost-benefit analysis approach. Both qualitative and quantitative issues can be handled within the process of decision making through the MCDM because it provides a systematic procedure to the decision makers to identify desirable alternatives under uncertainty. According to Ref. [29], the process of decision making where multiple conflicting criteria are involved is classified into two main stages: (1) multiple objective problems, which have an infinite number of feasible alternatives and (2) multiple attribute problems, which have a finite set of feasible alternatives. Given the importance of this analysis for our research we performed an MCDA analytical process for the dam installation in the Mapai village.

4. Assumptions and problem structuring

Within the sector of water and sanitation, the Strategic Plan for the Development of Gaza Province [15] defines the construction of Mapai Dam as one of the top priorities. The main objective is to improve the territorial waters and to guarantee the flows of international rivers of the Limpopo River Basin. The Mapai Dam site is foreseen between Phafuri and Combomune [15] along the Limpopo River, and it has been identified as one of the most strategic location for the estimated USD 450 million cost dam [19].

The government of Mozambique (GoM) has been advocating among international partners and stakeholders for a fundraising process for the Mapai Dam. The Mozambique Investment Promotion Centre [17] is one of the national entities that have led the process. Recently, at the Forum for Economic and Trade Cooperation between China and Portuguese-Speaking Countries [19], the Mapai Dam was one of the top issues that put forward by the GoM.

According to Ref. [10], the earlier research done by the end of the 1960s showed that the main benefits derived from the development of Mapai Dam are:

(a) Availability of water for irrigation of about 40,000 ha at the low and medium part of Limpopo Valley;

(b) Power Center with a capacity to produce 40 MW electricity;

(c) Reduction of migration by providing job opportunities;

(d) Supply of drinking water to the population (presently, the inhabitants are drinking water from boreholes); and

(e) Flood protection for Xai-Xai and Chókwe at the Low Limpopo Valley in coordination with the Massingir Dam.

In the late 1970s, an additional hydrological study performed in regard to building the Mapai Dam showed a reservoir with an estimated 11,200 Mm^3 of water at retention level of 180 m above Mean Sea Level (MSL), or 6600 Mm^3 at 170 m above Mean Sea Level (MSL). In order to estimate the impact of the usage of these reservoirs, in the scope of flood control, a brief analysis was conducted by ANE in 2006. This analysis, as show in **Figure 3**, focused on the potential changes within Combomune hydrographic designs downstream.

The advantage of value tree measurement illustration is that it can reflect the outcome of the brainstorming sessions by the stakeholders with respect to the problem and concerning a specific order or a course of action. A study for ANE by Ref. [10] shows that for each size of the reservoir and a series of hydrographical analysis, the maximum optimum flow from the reservoir was calculated as the maximum flow that allows the reservoir to support its full

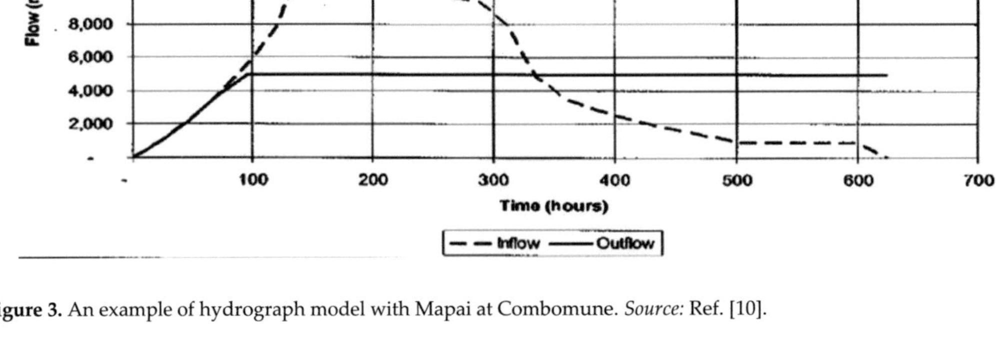

Figure 3. An example of hydrograph model with Mapai at Combomune. *Source:* Ref. [10].

http://dx.doi.org/10.5772/intechopen.68978

capacity, while avoiding an over flow. The calculations, rather simple to perform, require just a perfect knowledge of hydrographic effluent. Being an ideal case, it is rather not applicable in reality. Nevertheless, it is a good guide for an achievable flood mitigation process when a reservoir is used.

To understand the complexity of this project we present in **Figure 4** an overview of the value tree measure, which provides better understanding of the project.

To demonstrate how different scenarios can be illustrated we used a value measurement tree, **Figure 4**, to structure the different possible outcomes. Based on their profiles, we can evaluate the behavior of different outputs in the score profile across the tree. Three potential "active-volumes" reservoirs used to control the floods were assumed as follows:

(a) 2000 Mm^3, close to the capacity of a reservoir initially used for other purpose (for example supply of water for irrigation and hydropower) but with separate capacity for flood control;

(b) 5000 Mm^3, close to the capacity of a reservoir initially used for flood control;

(c) 10,000 Mm^3, close to a reservoir of larger capacity with a sophisticated system of forecasting and administration, in order to maximize the primary flood control function.

The value tree in **Figure 4** shows how different scenarios can be interpreted and helps to make a decision. Here, we can see how the optimal changes in the node of a dam with 10,000 m^3 may bring positive impacts on health facilities, water supply, agriculture, and power supply.

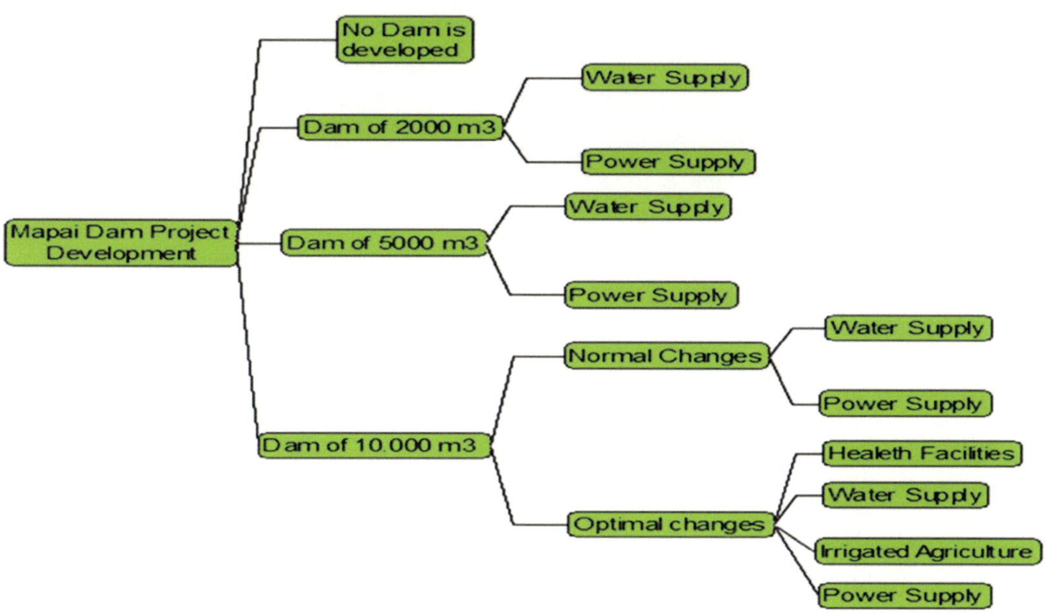

Figure 4. The value tree structure for the Mapai Dam project.

4.1. Decision making and scenario planning

As discussed before, the issue of the Mapai Dam is a strategic decision for the development of the water and sanitation system in Gaza Province [15] with a particular impact along the Limpopo River Basin. Prevailing the uncertainty on the topic, due to the absence of data that could be used for this particular research, going through academic exercise and in order to validate the application used, from the three scenarios defined in this research, we may generate the following assumptions. Initially, we assume the two alternatives of constructing or not constructing the dam, given possible different and conflicting stakeholder opinions, having equal probability of 50% each. Alternative one will be to construct the dam at the cost of USD 450 million according to Ref. [19], and alternative two will be not to construct the dam, meaning to continue with the current situation.

As a result of the fieldwork performed in Chókwe District between 8 and 22 April 2013, we selected several probabilistic indicators, which were used to simulate and analyze the data in a decision support system, or more specifically, to build an analytical decision tree in Decide IT (**Figure 5**).

This decision three is built based on the three main scenarios (**Figure 5**), where alternative one is branched at Nodes E1, E3, and E5, with probabilities $P = 0.4$, $P = 0.35$, and $P = 0.25$, respectively, for building dams with 2000, 5000, and 10,000 Mm^3 storage capacity [10]. Based on the literature (**Table 1**) and interviews with the stakeholders in Chókwe District, we summarize that:

(1) With alternative one, the following three outcomes are expected, based on the information in [31]:

  (a) The 40% probability of installation of a dam with 2000 Mm^3 storage capacity will bring basic changes to the living standard, with probabilities between 10 and 24% for water supply and between 16 and 30% for power supply (node E1);

  (a) The 35% probability of the installation of a dam with 5000 Mm^3 storage capacity will create medium changes to the living standard, with probabilities between 0 and 35% for both the water and the power supplies (node E3);

  (a) Lastly, the 25% probability for installation of a dam with 10,000 Mm^3 storage capacity should bring significant changes to the living standard, with probabilities between 10 and 30% for providing drinking water and irrigation facilities, power supply, and job creation (node D1).

(2) Node D1 in **Figure 5** illustrates the critical scenario for our research. Regrettably, this is not what happens with the node E4, which although provides all qualities given in D1, has limitations.

(3) The availability of water for irrigation of about 40,000 ha farmland at the low and middle part of the Limpopo Valley is to some extent expected to boost both the agriculture and the livestock production capabilities of the area.

(4) Power station with a capacity of 40 MW will increase the capacity of industrial production and the access to clean energy.

http://dx.doi.org/10.5772/intechopen.68978

(5) Reduction of migration by creating job opportunities that will ensure the decline of unemployment rate and rise of social and economic development within the region with some positive impact on gender equality. Guarantee to supply potable water to the population, which will enable access to clean water and guarantee that school aged girls focus on their study requirements and daily responsibilities rather than spending time to fetch water for the families.

(6) Flood protection for Xai-Xai and Chókwe at the Low Limpopo Valley in coordination with the Massingir Dam. This will provide security to thousands of people and the economic assets that are highly exposed to floods risk along the basin, especially to the cities that are frequently flooded.

In the next section, we use two decision support software—Decide IT and V.I.S.A—to provide more analytical options based on the assumed probabilities for each scenario. The main motivation on choosing the Decide IT software is that flexibility that it offers on the analysis of different options based on a decision tree.

5. The application of DSS

The main purpose of using Decide IT is to accommodate the probabilities discussed in Section 4.1 in order to build the decision tree model for the Mapai Dam in Gaza Province, taking into account different proposals and priorities from various stakeholders.

The output of this process is shown in **Figure 5**, which illustrates the outcomes derived from discussions with the stakeholders interviewed during the fieldwork. This decision tree integrates

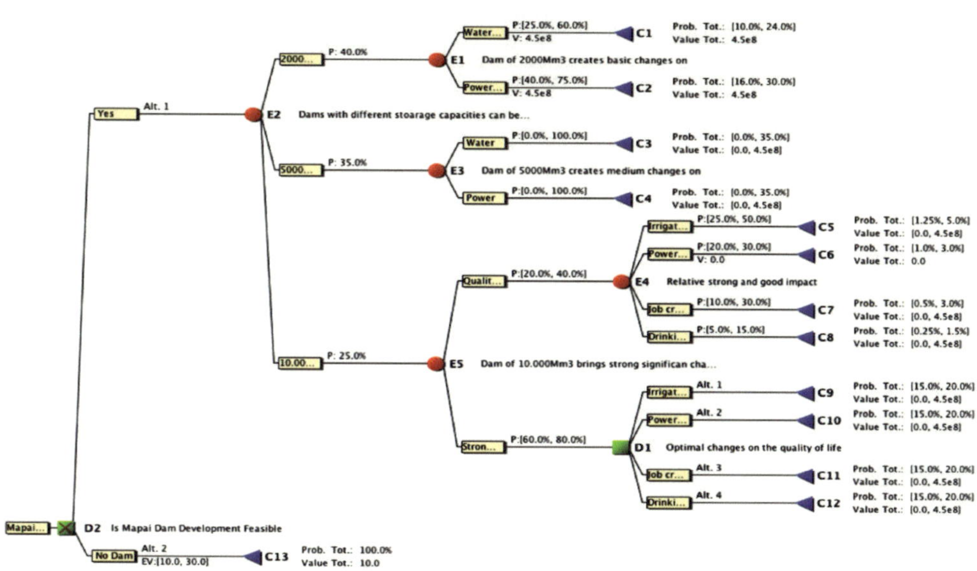

Figure 5. Decision making process for the Mapai Dam project.

multiple objective managements reflected in different scenarios. Hierarchical and network models are used to understand the relationships among multiple and competing objectives. Nodes E1, E2, and E5 represent different stakeholders' options and opinions.

Figure 5 depicts different outputs of stakeholders' options on possible conflicting objectives, which represent the scenarios derived from trade-offs and specific benefits that each variant of the dam may deliver. Modeling processes can also be built on V.I.S.A, another DSS extensively used to build MCDA scenarios, through comparative analysis of the behavior of score profile across tree from the present situation that ranges from not building a dam to the development of the optimal alternative of a dam with the maximum storage capacity, as shown in **Figure 6**.

Figure 6 illustrates the profile of different scenarios that stem from the installation of the dam with different storage capacities. The scale varies from worst, which is equivalent to zero improvement, to best that corresponds to 100 units. The actual or the current situation, where no dam is being built, shows zero units, while a dam of 2000 Mm^3 will provide power and water up to 50 units of the scale.

A dam of 5000 Mm^3 shows some improved results, while the last one even brings changes on agriculture and health, which are reflected in a well positioning of all dimensions, namely, economic, social, environmental, and coping mechanisms. Usually decision making is based on a cost-benefit analysis to determine the optimal alternative. We develop this analysis based on the impact of different dam alternatives or options (types and storage capacities), as shown in **Figure 7**, where we can see the effects that different storage capacities have on the economic, social, environmental, and particularly, health systems.

As shown in **Figures 6** and **7**, apart from the financial constraints, the construction of the Mapai Dam covers different areas of interest including, but not limited to, demography, water and sanitation, flood management, environmental issues, the risk of floods damage, and geology. Different types of real life problems in management practice can be formulated as a multicriteria

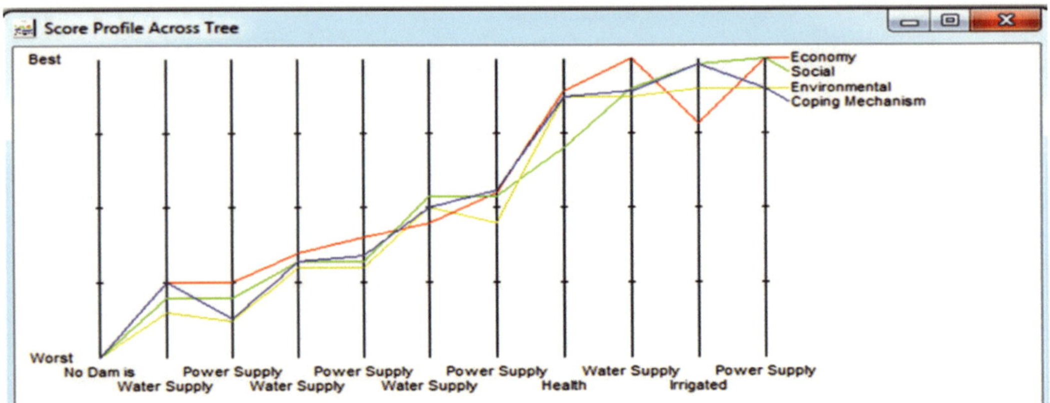

Figure 6. Score profile behavior across Tree.

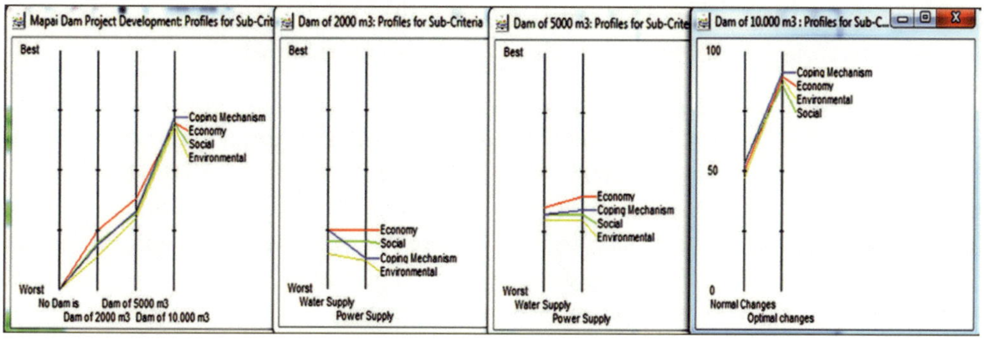

Figure 7. Profile of each sub-criterion according to the storage capacity.

analysis problem, when one has to involve the evaluation and choice of resources, strategies, offers, policies, products, innovations, designs, costs, profits, portfolios, and tools [25].

6. Findings and analysis

The development of the Mapai Dam is considered of vital importance for the basin management, particularly in the mitigation of floods and irrigation in Mozambique. On one hand, we applied V.I.S.A software to build the value measurement tree in order to understand the complexity of the problem whose resolution may lead to improvements in the economy, social, environmental, and coping mechanisms as shown in **Figure 5**. In **Figure 5**, we can see the progress achievable with respect to water supply, power supply, health issues, and the irrigation facilities for agriculture, which are in direct correlation to the different alternatives for the dam storage capacity, and which range from 2000 to 10,000 Mm^3.

The analysis of profiles for different sub-criteria by storage capacity also can be addressed in another chart, where we can aggregate them all together in the first section of the chart at most left of **Figure 6**. This is followed by a progressive analysis of scenarios with a dam of 2000 Mm^3 storage capacity where the environmental issues ranks last from worst to best, while the economy leads the process, this might result from new capacity of job creation and other social facilities. On the second scenario with a dam of 5000 Mm^3 storage capacity the economy leads followed by the coping mechanisms and social issues, while the environmental aspects ranks last in the process. The scenario of 10,000 Mm^3 storage capacity dam brings more reliable and consistent changes with coping mechanisms on the top, followed by economic and environmental issues while the social area ranks last. Notice that in this scenario, all four dimensions are above 75 units in the scale of the worst being 0 to the best being 100 units, respectively. The involved stakeholders should agree on the scale and the units to be used in advance.

In addition, some research done was based on the Decide IT software, with the main result represented in the decision tree in **Figure 5**. The tree shows two main branches for the feasibility of Mapai Dam project: two main outcomes are highlighted, alternative one with "Yes" for the development of the dam and alternative two with "No," which means no action or not developing the project, hence no further developments in this branch.

Alternative one with the main sub-decision node E2 "dams of different storage capacities," which shows three main branches, namely for dams with storage capacities of 2000 Mm^3 with 40% probability to be developed, 5000 Mm^3 with 35% probability, and 10,000 Mm^3 with 25% probability, respectively. In the first two scenarios, E1 and E2, we have possibilities to perform basic and moderate improvements of water and power supply facilities, while the third scenario with 25% of chances might bring profound and significant changes in the water supply, power supply, irrigation for agriculture, and the job creation.

In short, **Figure 5** shows two main decisive moments: D1 and D2, which are represented by two green squares. D2 represents the preliminary step where the stakeholders choose whether to build the dam (alt one) or keep the actual situation (alt two). D1 represents a more developed stage, where different options have been developed and evaluated, see nodes E1, E2, E3, E4, and E5, which are part of key milestones to the process, nonetheless their results were not much feasible to meet the overall objectives, such as shown in the decisive node D1.

7. Conclusion and future work

This research shows how complex problems can be solved through the application of sophisticated tools and techniques, particularly a combination of them as we developed in this article. For a real case, using the support of both V.I.S.A and Decide IT we built different scenarios and simulated different outcomes; here the value measurement tree helped to understand and structure all features and details of the problem, while the decision tree enumerated all the possible outputs according to parameters such as the dam storage capacity. For this purpose we used probabilities of the stakeholders, even when those have different preferences. Here, we combined the inputs from the reviewed documents with those from the interviewed stakeholders in Chókwe. Additional analysis based on monetary values and probabilistic simulation could be developed as a part of sensitivity analysis, but this is beyond the scope of the present research, and, provided that our main aim was to evaluate the possible outputs and implications of building the dam, we limited our analysis at this stage. Given the limitations of data and time to perform more detailed research, for the present study, we will validate it in further research as part of the future work.

Author details

Avelino I. Mondlane

Address all correspondence to: si-aim@dsv.su.se

Department of Computer and Systems Science (DSV), Stockholm University Kista, Sweden and Center for Informatics, Eduardo Mondlane University, Maputo, Mozambique

References

[1] Pereira A. The incomati and limpopo international river basins : A view from downstream. Water Policy. 2000;**2**:99-112

[2] World Bank. Municipal ICT Capacity and Its Impact on the Municipal ICT Capacity and Its Impact on the Climate-Change Affected Urban Poor The Case of Mozambique; 2012

[3] Vaz A. Coping with floods – the experience of Mozambique 1 (November): 27-28; 2000

[4] Bigcon C. Joint Limpopo River Basin Study Scoping Phase; Final Report. On Behalf of the Limpopo Basin Permanent Technical Committee–LBPTC; 2010

[5] Dyer JS, Peter C, Ralph E, Jyrki W, Stanley Z. Multiple Criteria Decision Making, Multiattribute Utility Theory: The Next Ten Years . Management Science -INFORMS, Vol. 38, No. 5. (May, 1992), pp. 645-654

[6] Chilundo M, Kelderman P. Design of a water quality monitoring network for the Limpopo River basin in Mozambique. Physics and Chemistry of the Earth. 2008;**33**:655-665. doi:10.1016/j.pce.2008.06.055

[7] Midgley S, Dejene A, Mattick A. Adaptation to Climate Change in Semi-Arid Environments Experience and Lessons from Mozambique. Edited by FAO - Food and Agriculture Organization of the United Nations Rome; 2012

[8] Twena M. Climate Change and Poverty in Mozambique Realities and Response Options for CARE November 2006: 0-42

[9] Foley C. Mozambique : A Case Study in the Role of the Affected State in Humanitarian Action – September; 2007

[10] Parsons S. Estudo de Propagacao de Caudais de Cheia no Rio Limpopo: Relatorio Final; 2006

[11] USAID, Fact Sheet 1- Regional Centre for Southern Africa (RCSA) 1;2002a. www.sardc.net/imercsa/li,popo

[12] Amin S, Markus G. Data Against Natural Disasters: Establishing Effective Systems for Relief. Recovery and Reconstruction. Washington, DC, USA: The World Bank; 2008. ISBN 978-0-8213-7452-8 — eISBN 978-0-8213-7453-5. DOI: 10.1596/978-0-8213-7452-8

[13] Clemen R, Reilly T. Making Hard Decisions, with "Decision Tools." Edited by Duxbury Thomson Learning. 2nd ed. Duxbury Thomson Learning; 2001

[14] Mohammed M, Mac K, Judy E, and Mark T. Water-use Accounts in CPWF Basins: Simple Water-use Accounting of Limpopo Basin; Working Paper, Basin Focal Project Series. CSIRO Land and Water, Canberra, Australia; 2010

[15] PEPG. Plano Estratégico Da Província de Gaza - Government working document; 2007

[16] World Bank; Mozambique Country Water Resources Assistance Strategy : Currency Unit - Metical International Federation of Accountants Country Director; 2007

[17] GdM, The Government of Mozambique - Centro de Promocao Den Investimentos CPI. CPI_Mozambique, C.P. 4635; 2012

[18] Brito R, Famba P, Munguambe N, Ibraimo C. Profile of the Limpopo Basin in Mozambique a Contribution to the Challenge Program on Water and Food Project 17 Livelihoods : Managing Risk, Mitigating Drought and Improving water productivity in the water scarce Limpopo Basin. Faculdade De Agronomia, and Universidade Eduardo Mondlane; 2009

[19] FCEC-CPLP, Fórum Para a Cooperação Económica E Comercial Entre a China E Os Países de Língua Portuguesa; 2012. FCEC. http://www.forumchinaplp.org.mo/pt/announce.php?id=976.

[20] Limppopo Strategic Plan 2015/16 - 2019/20. LIMPOPO PROVINCE PR338/2014 ISBN: 978-0-621-43204-6

[21] http://africanenvironmentalnews.com/news/mozambique-government-looking-for-funding-for-mapai-dam

[22] Republic of Mozambique: Poverty Reduction Strategy Paper—Economic and Social Plan for 2005 - IMF country report (No. 05/312); August 2005

[23] Official Development Assistance to Mozambique Database: Project Fact Sheet for 8 ACP MOZ. 024-19 - Project fact sheet for 8 ACP MOZ 024-19. http://www.odamoz.org.mz/projects/726

[24] http://www.limpoporak.com/en/management/water+demand/power+generation.aspx

[25] Vassil V, Krasimira G, Mariyana V. A multicriteria analysis support system. Journal, International Theories, Information. 2004;**13**:103-111

[26] Levy J. Multicriteria decision making and decision support systems for flood risk management. Stochastic Environmental Research & Risk Assessment. 2005;**19**:438-447. DOI: Jo10.1007 s00477-005-0009-2. Published online

[27] Hermans C, Jon D. Advances in the Economics of Environmental Resources Emerald Book Chapter : Multicriteria Decision Analysis : Overview and Implications for Environmental Decision Making Multicriteria Decision Analysis : Overview And Implications For Environmental Decision; 2011. doi:10.1016/S1569-3740(07)07010-1. http://www.preference.bz

[28] Ginzberg, M.J. and Stohr, E.A. (1982). "Decision Support Systems: Issues and Perspectives," in Decision Support Systems, M.J. Ginzberg, W. Reitman and E.A. Stohr (eds.), North-Holland, Amsterdam – Center for Digital Economy Research Stem School of Business. Working paper IS-82-12

[29] Cheng S, Christine W and Guo H. Part A : Toxic/hazardous substances and environmental engineering using multiple criteria decision analysis for supporting decisions of solid waste management. Journal of Environmental Science and Health. 2012;37-41

Chapter 13

Flood Risk Management in Mexico

Felipe I. Arreguín-Cortés and
Claudia Elizabeth Cervantes-Jaimes

Additional information is available at the end of the chapter

http://dx.doi.org/10.5772/intechopen.69834

Abstract

Mexico receives an average annual rainfall of 740 mm, which are distributed in the hydrological cycle as follows: 72% evapotranspiration, 21% becomes runoff and 6% as aquifer recharge. Within the Mexican territory, exist a great diversity of climates and high spatial and temporal variability in water resources availability. In the period 2000–2015, damages from hydrometeorological phenomena in Mexico represented between 60 and 99% of total damages and losses at national level due to natural and socioorganizational events. Considering global climate change impact on the selection, design and implementation of flood control measures, represents a major challenge, since the level of certainty regarding its influence on the variables involved, remains insufficient. This chapter provides a description of the main elements directly linked to flooding in México, such as a high spatial and temporal variability in water resources availability and presence of tropical cyclones in both coasts and climate change. A brief summary of the main disasters caused by hydrometeorological phenomena, the annual cost of the damages, the main non-structural measures for flood control and the intervention from the Mexican Institute of Water Technology in the use, development and spread of technology focused on flood risk management are also included.

Keywords: flood risk, flood risk management, water vulnerability, resilience

1. Introduction

Mexico covers an area of 1.964 million km^2 [14], and its annual average rainfall (1981–2010) is 740 mm, equivalent to 1449 km^3, which are distributed in the hydrological cycle as follows: 72% evapotranspiration, 21% becomes runoff and 6% as aquifer recharge. In addition, every year Mexico receives from the United States and Guatemala 48 km^3 through transboundary watersheds and exports 0.43 km^3 to the United States, based on the 1944 U.S.-Mexico Water Treaty [3].

The population, estimated to be around 119.5 million in 2015, ranks 11th in the world, with a growth rate of 1.4%, and 77% of the Mexican habitants live in urban areas [15]. Population forecast for 2030 shows this trend will be accentuated in the next years. Therefore, it will be an increasing and of water concentrated in metropolitan regions.

Water withdrawals from surface and ground water sources represent 52 and 32.9 km^3/year, respectively. In total, 77% of the water from rivers and groundwater goes into irrigation, whereas 14% is used in domestic applications and 9% for industry [3].

It should be note that 77% of the population that needs to be supplied is located in the north and central region of the country, where only 33% of the water resources are found, which leads to overexploitation of basins and aquifers [3].

Because of its geographic location, varied orography and furthermore the presence of meteorological and climatological phenomena at different scales, Mexico presents a great diversity of climates and high spatial and temporal variability in water resources availability.

2. Climatological and meteorological aspects

Some phenomena that have influence on intensity on intensity, as well as spatial and temporal distribution of precipitation, are severe convective storms, tropical cyclones, cold fronts, easterly waves, seasonal Intertropical Convergence Zone (ITCZ) migration, seasonal variability of jet streams and warming in the tropical regions, also called dynamics of the East Pacific Warm Pool.

Although tropical cyclones contribute to surface and ground water recharge, due to the atmospheric moisture transport from the oceans to the continental regions that increase the rainfall during the tropical cyclone season, **Figure 1**, they also lead to severe damage to exposed and vulnerable population centres affected by floods.

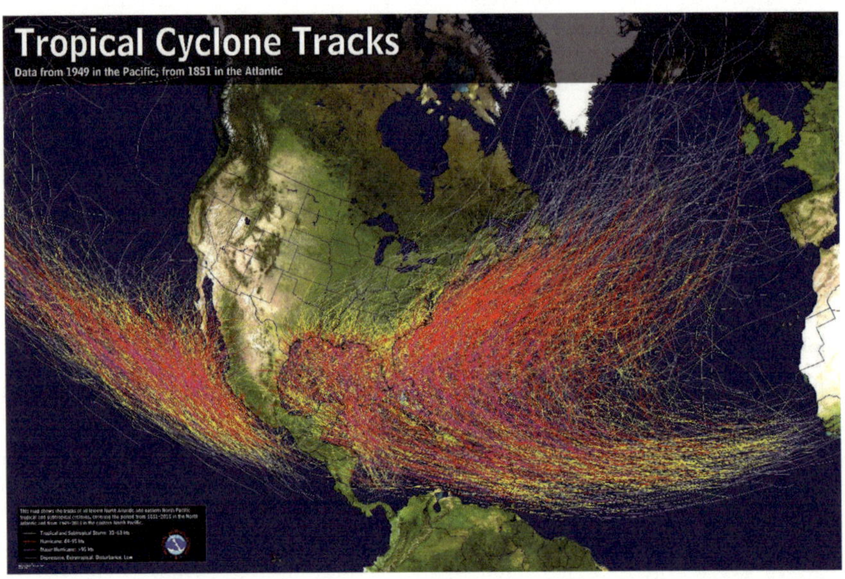

Figure 1. Historical tracks of tropical cyclones, from 1949 to 2015 in the Pacific and 1851 to 2015 in the Atlantic. Source: NOAA [19].

http://dx.doi.org/10.5772/intechopen.69834

In Mexico, according to the recorded paths, during the period 2000–2016 in the Pacific Ocean, an average of 17.6 tropical cyclones were generated, an average of 15.4 acquired name and 3.8 made landfall. While, in the Atlantic Ocean during the same period, an average of 15.4 tropical cyclones were generated, an average of 14.8 acquired name and 2.6 made landfall.

Cold fronts also have great influence; they are present between September and May. Their behaviour includes the dragging of cold and humid air masses that can reach as far as the southeast of the country.

3. Climate change impact

Considering global climate change impact on the selection, design and implementation of flood control measures, it represents a major challenge since the level of certainty regarding its influence on the variables involved remains insufficient.

The vulnerability can be defined as the degree to which a system (in this case to water resources) is susceptible to adverse effects, and according to the Intergovernmental Panel on Climate Change (IPCC) on his Fourth Assessment Report, vulnerability depends on degree of exposure, sensitivity and adaptive capacity [16].

Based on this definition, an estimate of the level of the social vulnerability to climate change in Mexico municipalities has been made. The following contains a description of the publication.

3.1. Atlas of Water vulnerability to climate change in Mexico (Mexican Institute of Water Technology, IMTA, 2016)

IMTA coordinates the analysis, updating and publication of climatic scenarios in Mexico, based on the General Circulation Models (GCM) of the Coupled Model Intercomparison Project phase 5 (CMIP5) experiment for the historical period 1961– 2000 and greenhouse gas emission projections denominated Representative Concentration Pathways 6.0 and 8.5 (RCP6.0 and RCP8.5), for two periods of the twenty-first century, from 2015 to 2039 and from 2075 to 2099, **Figure 2**. The Atlas of Water vulnerability to climate change in Mexico includes scenarios of the effect on maximum temperature, minimum temperature, average temperature and precipitation. Furthermore, it includes an estimate of the municipal risk for rainy and tropical cyclone seasons, **Figure 3**.

Maps are one of the most useful tools in flood risk management, since they concentrate and synthesize large amounts of information analysed and processed using a geographic information system. They play a central role in the delineation of floodplains for different return periods, visualization of change in flow depth and velocity, strategic planning, early warning systems, impact analysis on infrastructure and population and damage assessments.

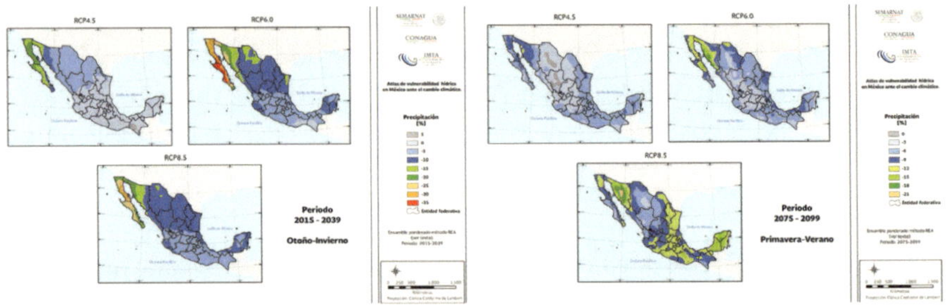

Figure 2. Projected precipitation change for the periods 2015–2039 and 2075–2099, due to climate change. Source: IMTA [13].

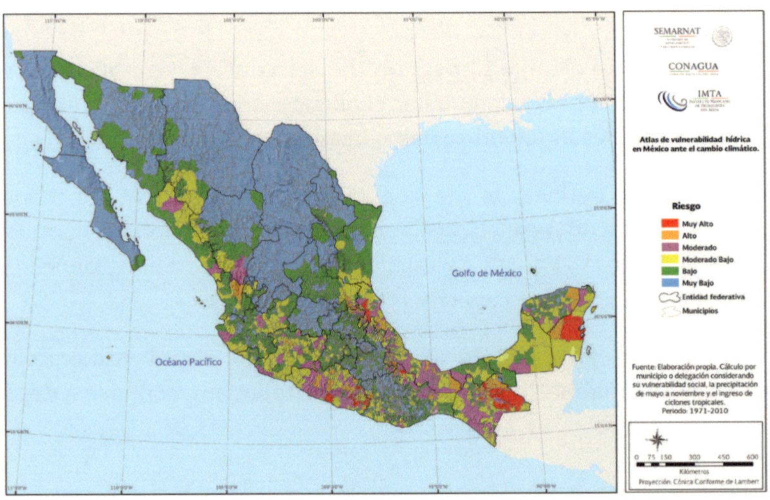

Figure 3. Municipal Risk for rainy and tropical cyclone season in Mexico. Source: IMTA [13].

4. Area susceptible to flooding

The IPCC defines flood as 'the overflowing of the normal confines of a stream or other body of water, or the accumulation of water over areas that are not normally submerged'. Floods include river (fluvial) floods, flash floods, urban floods, pluvial floods, sewer floods, coastal floods and glacial lake outburst floods [16].

Some causes of flooding are precipitation associated with hydrometeorological and climatic phenomena, snowmelt, drainage conditions in watersheds, deforestation, urbanization, poor drainage systems without regular maintenance, inadequate management operation of dam systems, failure of constructed or landslide dams, high tide or some combination of these.

An example of landslide damming of river at risk of failure occurred in 2007 in the Grijalva River, the second largest river in the country by the volume of water it discharges. A wedge

http://dx.doi.org/10.5772/intechopen.69834

of rock and clay material, at 55 Mm^3, slid and dammed the stream. The dam was estimated to have a height of 80 m, a length of 800 m and a width of 300 m, located downstream of the Malpaso Dam and upstream the Peñitas Dam. The town of Juan de Grijalva was located on the right bank of the river and was affected by the landslide and the flood wave that followed, both caused the death of 25 people. An estimated volume of 15 mm^3 was filling the Grijalva River valley and creating a landslide dam which impounded the river. The potential risk of failure of this natural dam posed a grave threat to the Peñitas Dam and to more than 3 million downstream inhabitants of cities, such as Villahermosa, Cárdenas, Comalcalco and Huimanguillo (state of Tabasco). The problem was solved by constructing a channel on the material slid, so the river was returned to its natural course.

From the analysis of several floods in Mexico, it has been concluded that long-term heavy rainfall in large basins is associated with river flooding, whereas short-term heavy rainfall in small basins is related to pluvial flooding. The flood management strategies must also consider the presence of debris flow.

The flood problem becomes more complex for the wide range of factors involved and their variability, for example, precipitation intensity is already being influenced by climate change. In addition, inappropriate development policies encourage unplanned urbanization, piping of channels and/or engineering projects that reduce the flow capacity of the rivers.

It has been estimated that 162,000 km^2 of the Mexican territory are susceptible to flooding, see **Figure 4**. Although this accounted for only 8% of its territory, the socio-economic impact may be equivalent up to billions of dollars, according to the affected urban zones, their population density, economic activities, existing infrastructure and, above all, their vulnerability and resilience.

Figure 4. Location and delineation of the main wetlands in Mexico. Source: INEGI [14].

5. Severe floods in Mexico

There is information about floods in the city of Tenochtitlan until 1521 and in New Spain in later years. Since then, structural and non-structural measures for flood control have been adopted. In **Table 1**, some of the worst floods recorded in central Mexico are listed.

Some other major floods are presented in **Table 2**.

Year	Event
1446	First great flood in Tenochtitlán, some sectors of the city were flooded due to heavy rainfall [5]
1449	Tenochtitlán suffered another flood and a 16 kilometer long embankment was constructed [20]
1499	Another flood occurred in Tenochtitlán that covered streets and squares and later reached the lake through the canals. As a result, the water level of the lake raised and it was necessary to build new dykes [17]
1555	On 10th October, intense precipitation began in the Valley of Mexico; after 4 days, people were forced to transport in canoes [17]
1604	The city of Mexico suffered floods that persisted during months, the only way of water removal was by evaporation. It was then decided to build an artificial outlet to drain excess water into the Tula river basin [6]
1607	Heavy rainfall occurred in June in the capital city, which resulted in the most severe flood in the country's capital since the Spanish occupation [17]
1629	Called the Great Flood, it was the most severe in the history of the city. Heavy rains were combined and little progress was made in the construction of the General Drainage of the city, it is estimated that 30,000 people died [6]

Table 1. Floods in central Mexico.

Year	Event
1760	Silao River overflowing caused floods in the city of Guanajuato, with the presence of silt under the bridges that reduced the river capacity [18]
1909	The worst flood recorded in the city of Monterrey as a result of the overflowing of Santa Catarina River that caused the death of 5,000 people [12]
1982	Due to the effect of Hurricane Paul, numerous floods occurred in several cities in the state of Sinaloa [2]
1985	There were 64 floods in 16 states of the Mexican Republic [2]
1988–1992	Flooding in the state of Veracruz left 75,000 victims in 1988; 52,546 in 1989, 129,565 in 1990, 67,470 in 1991 and 60,000 in 1992 [2]
1995	Hurricanes Ismael, Roxanne and Opal affected the states of Tabasco, Campeche, Quintana Roo, Yucatan and Veracruz [2]. Between 1 June and 31 October, in the state of Tabasco, it rained 1792 mm [1]
1997	Hurricane Paulina caused flooding in the states of Chiapas, Oaxaca and Guerrero [2]
1998	The floods affected 29,000 people in Chiapas and there were 4840 people affected in Oaxaca, leaving 700,000 people without electricity [2]

http://dx.doi.org/10.5772/intechopen.69834

Year	Event
1999	The combination of several meteorological phenomena caused one of the worst floods in the city of Villahermosa, which resulted in the construction of levees around the city and along urban rivers, as well as the implementation of the Comprehensive Flood Control Program (PICI) [1].
2005	In July, Hurricane Emily hit the Atlantic coast of Mexico, Quintana Roo, Yucatán, Tamaulipas and Nuevo León were the most seriously affected states. In October, the states of Quintana Roo and Yucatan were affected by Hurricane Wilma [8]
2007	On 4 October, the city of Tapachula in Chiapas, suffered the worst disaster in its history, as a result of prolonged and heavy rainfall associated with Hurricane Stan [1]
2008	The combined effect of Hurricane Noel and two cold fronts, caused two-thirds of the city of Villahermosa, was flooded during 40 days. Together with the phenomenon of landslide damming of Grijalva River, represent two of the greatest natural disasters in recorded history of Mexico [1]
2010	There were floods in the states of Zacatecas, Chihuahua, Veracruz and Tabasco. In relation to the latter, the PICI Flood Management Plan was redefined, for what would be the future Comprehensive Water Program of Tabasco (PHIT) [1]
2011	This was the second rainiest year in the country, with local flooding in Mexico City and many others due to the effects of hurricanes Alex, Karl and Matthew, the cost of damages exceeded the 4000 mdd [7]
2013	Hurricane Jova affected the states of Colima and Jalisco, whereas Arlene affected the state of Veracruz and the central part of the country and the state of Hidalgo. For the fifth year in a row, the state of Tabasco suffered severe floods [9]
2014	As a result of the incidence of two simultaneous tropical cyclones, Ingrid in the Gulf of Mexico and Manuel in the Pacific Ocean, the state of Guerrero was seriously affected. Besides, hurricane Barbara hit the pacific coast in state of Chiapas [11]
2015	Heavy damage occurred in the state of Veracruz during October. In addition, the state of Tamaulipas and Chihuahua were flooded [10]

Table 2. Major floods in Mexico.

Urban developments in flood plains are historical in Mexico, the Tabasqueña Plain and the lower basin of Bravo River (Tamaulipas), Pánuco River (Tamaulipas and Veracruz), Coatzacoalcos River (Veracruz), Papaloapan River (Veracruz); the Coast of Chiapas, the Atoyac, Jamapa, Tecolutla, Nautla and Antigua Rivers (Veracruz) and Tulancingo River (Hidalgo) [1] are clear examples of areas where frequent flooding can be expected.

6. Additional contributing factors to flood events

In addition to the factors mentioned above, in Mexico, floods occur on low-lying areas of minor capital gain which are likely to be inhabited by the poorest sectors of the population. Also there exist a number of misperceptions, even of the authorities. For example, the idea that not using a territory during the dry season that is only flooded during the rainy season is a waste.

Furthermore, there is a land-use regulation by different laws and institutions from the three levels of government, which makes its implementation difficult, and a society with a low level of insurance culture, which leads the State becoming the insurer.

7. Cost of flood damage in Mexico

The cost of flood damage includes those associated with the impact on infrastructure at urban and rural areas; infrastructure for agricultural, industrial or commercial activities, as well as transport, communication and public services. This cost also includes indirect damages, which are those related to damaged or lost resources-dependent activities, for example, temporary or permanent job losses and stopping production chains.

Direct and indirect costs include tangible and intangible impacts, the latter are extremely difficult to assess and a common example is human health.

In Mexico, the National Center for Disaster Prevention (CENAPRED) collects information from the public and private sector and estimates the cost of damages due to natural and human-induced hazards, including those associated with hydrometeorological events.

In the period 2000–2015, damages from hydrometeorological phenomena have represented between 60 and 99% of total damages and losses at national level due to natural and socio-organizational events. Particularly noteworthy is the scale of the damages due to hydrometeorological phenomena in 2010, which amounts to approximately $4100 million, **Figure 5**.

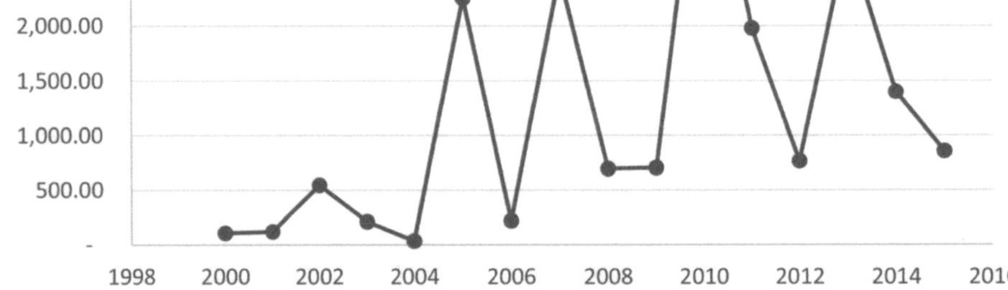

Figure 5. Cost of flood damage due to hydrometerological phenomena (mdd). Data Source: García et al., [7].

8. Flood control infrastructure

In order to mitigate the risk of flooding in more than 639,000 km of rivers, a large number of flood control works have been built. In Mexico, there are more than 5000 km of compacted clay (in some cases sand) levees to protect population centres, industrial and agricultural areas against overflowing. When the flood plain conditions allows it, permanent diversion works are constructed, including relief channels for diverting excess flow into the sea, a lake or another stream. In places where there are large capacity channels for irrigation, they have been used for diverting [21].

In some cases when the neighbourhood of watercourses is lower, it is feasible to use them for flood retention even though this land is intended for livestock farming and agriculture; this option has been chosen as long as the damage is less than what would occur in protected regions. Another measure for flood control is to restore river conditions to recover the capacity of transport and suppressing of meanders [21].

River canalisation and piping of streams have caused problems in urban areas; however, in many instances, this kind of measures is necessary. Nevertheless, due to their size, dams offer greater protection; furthermore, they may have multiple purposes such as water diversion, water supply of urban areas, irrigation and power generation. It is worth mentioning that some are built for the sole purpose of flood protection. In Mexico, dams for flood and siltation control have been built to complement flood control actions. There are 810 large dams (classification of the International Commission of Large Dams); a total of 5700 have been inventoried and it has been estimated to be of the order of 8000 small dams and borders that remain unrecorded [4].

In Mexico, 54% of the dams are earth or rock fill dams, the second type in terms of the quantity is gravity section with 21%, followed by the buttresses representing only 5%, **Figure 6**. The highest dam is Chicoasén Dam, which belongs to the Grijalva Hydroelectric System with

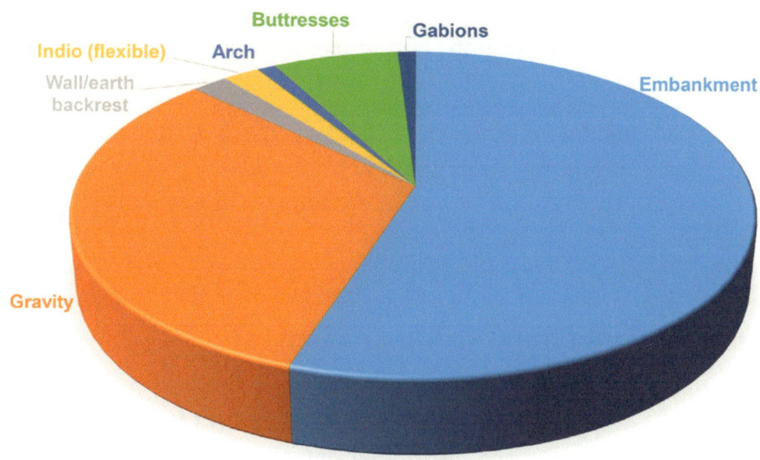

Figure 6. Types of dams in Mexico.

261 meters high, whereas the longest one is Falcon Dam, located between the U.S. state of Texas and the Mexican state of Tamaulipas, with 8014 meters length, although there are longer levees for flood protection in the city of Villahermosa, Tabasco.

9. Flood control actions

As previously outlined, since pre-Hispanic times Mexico has suffered from floods and has built infrastructure to deal with them; however, these solutions have generally been reactive and scattered actions. It is from the 1990s when comprehensive flood control programs have been defined such as those implemented in the state of Tabasco.

Later in 2014, the National Program against Hydraulic Contingencies was implemented, characterized by a preventive nature rather than reactive and promoting a coordinated participation of the governmental institutions responsible for flood management as well as the involvement of civil society.

The program is based on Integrated Flood Management and considers the hydrological basin and risk management, besides the adoption of the best possible combination of political, administrative, financial and physical strategies and a participatory approach.

A typical weak point in this type of programs is the uncertainty associated to the hydrometeorological information, and especially to climatic projections in Mexico, it is therefore required to:

- Prioritize the maintenance of hydrometeorological monitoring network.
- Have a national database of validated meteorological and hydrometric data.
- Generate scenarios and disseminate them in support of public awareness about flood hazard and risk.
- Develop the capacity of the institutions and implement technology focused on solving the national flood problem.
- Build resilience and sustainability in cities, both focused on delivering short-, medium- and long-term results, designing strategies and implementing actions to maintain essential activities and services during and immediately after a disaster without jeopardizing the availability of resources for future generations.

10. Use of technology in flood risk management

There are several institutions that have developed flood control technology, including IMTA, a decentralized public agency focused on solving national and regional problems associated with water resource management through research and technological developments.

Since 1986, IMTA has carried out projects in all the hydrological regions of Mexico, many of which are requested by one or more of the three levels of government and in other cases by non-governmental organizations, educational institutions or private companies.

Some of the projects developed by IMTA focused on flood risk management are as follows:

- Methodology for flood maps generation: a two-dimensional hydrological model, which allowed to visualize flow depth and velocity, was used to define levels of hazard, vulnerability and risk; it helped to optimize structural and non-structural measures as proposed.
- Flow characterization in urban areas: flood risk analysis in urban areas considered existing storm drainage infrastructure and runoff in streets by hydrologic and hydraulic modelling.
- Hydrological modelling: it was applied in hydrological forecasting systems for 24, 48 and 72 h that allowed to identify potential flood events and implementing timely measures to minimize damages and losses.
- Two-dimensional modelling using MDE (from LIDAR or Mexican Elevation Continuous Database CEM): IMTA has a collection of satellite and cartographic information from all over the country. Among its applications, there is a Flood Event Damage Estimation Module that assesses the damage for each MDE pixel.

11. Conclusions

Floods in Mexico lead to high losses due to the vulnerability and exposure of the population, rather than the extension of areas which are susceptible to flooding. Despite severe floods records in several regions of the country, their impact is communally underestimated.

Hydrometeorological monitoring and hydraulic modelling allow to predict the occurrence of floods and estimate the risk for a population centre and then design appropriate precautionary and control measures before the disaster occurs. However, particular attention should be given to the improvement of monitoring networks through maintenance, calibration and replacement, as well as the establishment of training and updating programs for stakeholders in flood management.

A global approach to water resources management should enable the identification of the causes, return period, extension of affected region and expected flood damage; the actual and potential effect of urbanization, as well as the planning, installation and implementation of structural and non-structural measures required for its control.

Among the non-structural measures is the implementation of a flood early warning system, the relocation of vulnerable and exposed buildings, the monitoring of changes in the watershed related to flood risk and the environmental effects of flood mitigation measures. Furthermore, inter-agency coordination is imperative to achieve adequate legislation.

The involvement of science, innovation and research in the design of flood control measures is essential to propose an effective combination of structural and non-structural measures; still the particular conditions of each case and weaknesses should be carefully considered, there is not yet an overall strategy.

The dissemination of flood risks and their impact, as well as the necessary actions for their management and positive Results, between civil society and authorities, is recommended.

One of the main reasons is the tendency to underestimate the magnitude of flood risk and not to consider it as a priority when it comes to allocation of resources.

Learning-based flood control and management measures designed to withstand flooding are part of building urban resilience, which is a necessity in the face of the unplanned urban growth proliferation, since this leads to an increase in social vulnerability, exposure and greater losses in case of disaster.

Author details

Felipe I. Arreguín-Cortés and Claudia Elizabeth Cervantes-Jaimes*

Address all correspondence to: elizabeth_cervantes@tlaloc.imta.mx

Instituto Mexicano de Tecnología del Agua, Mexico

References

[1] Arreguín-Cortés FI, Rubio-Gutiérrez H, Domínguez-Mora R, Luna-Cruz FD. Análisis de las inundaciones en la planicie tabasqueña en el periodo 1995-2010. Tecnología y ciencias del agua. 2014;**5**(3):5-32

[2] Bitrán D. Características e impacto socioeconómico de los principales desastres ocurridos en México en el período 1980 – 99. CENAPRED; 2000

[3] Comisión Nacional del Agua, editor. Estadísticas del agua en México, 2016. México: 2017a. Secretaría de Medio Ambiente y Recursos Naturales editor. Manual para el control de inundaciones. 1st ed. México D: F.: Comision Nacional del Agua; 2011. p. 326

[4] Comisión Nacional del Agua. Sistema de Seguridad de Presas [Internet]. 2017b. Available from: http://201.116.60.136/inventario/hinicio.aspx [Accessed: Inventario de presas]

[5] Cruz RMS. Las inundaciones un problema persistente en Cd. de México durante el último siglo colonial. Fuentes. 1991;**2**:83-89

[6] Domínguez R. Las Inundaciones en la Ciudad de México, Problemática y Alternativas de Solución. Revista Digital Universitaria, UNAM. 2000;**1**(2). http://www.revista.unam.mx/vol.1/num2/proyec1/ Access: April 24 2017

[7] García AN, Marín CR, Méndez EK, Nava SS, Vazquez BF. Impacto Socioeconómico de los Desastres en México durante 2015. 1st ed. México: CENAPRED; 2016a. p. 18

[8] García AN, Marín CR, Méndez EK, Bitrán D, Matías RLG, Jiménez EM, et al. Características e impacto socioeconómico de los principales desastres ocurridos durante 2005. México: CENAPRED; 2006. p. 487

[9] García AN, Marín CR, Méndez EK, Troncoso AN, Salas SM., Robles MT, et al. Características e impacto socioeconómico de los principales desastres ocurridos en la república mexicana en el año 2011. México: CENAPRED; 2012. p. 400

[10] García AN, Méndez EK, Nava SS, Vazquez BF. Impacto socioeconómico de los principales desastres ocurridos en la república mexicana en 2014. Mexico: CENAPRED; 2016b. p. 64

[11] García AN, Méndez EK, Reyes RR, Marín CR. Impacto socioeconómico de los principales desastres ocurridos en la república mexicana en 2013. México: CENAPRED; 2015. p. 80

[12] HernándezLGE, Lozano CGV. Historia de las crisis del agua en el área metropolitana de Monterrey (AMM), previa a la llegada de las grandes represas (1597-1955). Ciencia UANL. 2014;**67**(May-Jun)

[13] Instituto Mexicano de Tecnología del Agua, editor. Atlas de vulnerabilidad hídrica en México ante el cambio climático. Efectos del cambio climático en el recurso hídrico de México. Morelos México: IMTA; 2015. p. 150

[14] INEGI. INEGI Recursos Naturales [Internet]. Available from: http://www.inegi.org.mx/geo/contenidos/recnat/humedales/ Access: April 10 2017

[15] INEGI. Encuesta Intercensal Tabulados básicos [Internet]. 2015. Available from: www.inegi.org.mx April 24 2017

[16] IPCC. Glossary of terms. In: Field CB, Barros V, Stocker TF, Qin D, Dokken DJ, Ebi KL, Mastrandrea MD, Mach KJ, Plattner, Allen SK, editors. Managing the Risks of Extreme Events and Disasters to Advance Climate Change Adaptation. New York, USA: Cambridge University Press; 2012. pp. 555-564

[17] Mathes WM. To save a city: The Desague of Mexico-Huehuetoca, 1607. The Americas. 1970;**26**(04):419-438

[18] Nieto MDCC. Desastre en Guanajuato: La inundación del 5 de julio de 1760. Tzintzun. Revista de Estudios Históricos. 2007;**45**:11-32

[19] NOAA. Tropical Cyclone History Map for Atlantic and Eastern Pacific [Internet]. 2016. Available from: https://www. weather.gov/images/lot/tropicalsystems/TStracks.png [Accessed: 2017]

[20] Vázquez MR. Diseño de cimentación de la bocatoma unidad desde la obra de toma del proyecto hidroeléctrico La Yesca [thesis]. Mexico. 2012. Available from: http://132.248.52.100:8080/xmlui/handle/132.248.52.100/2448

[21] CONAGUA, editor. Manual para el control de inundaciones. 2011

Chapter 14

Estimating Flood Quantiles on the Basis of Multi-Event Rainfall Simulation

Elżbieta Jarosińska and Katarzyna Pierzga

Additional information is available at the end of the chapter

http://dx.doi.org/10.5772/intechopen.68648

Abstract

This chapter provides an insight into a new approach to estimating the flood quantiles based on rainfall-runoff modelling using multiple rainfall events. The approach is based on the prior knowledge about the probability distribution of annual maximum daily totals of rainfall in catchments, random disaggregation of the totals into hourly values and rainfall-runoff modelling. The new presented method called MESEF (*Multi-Event Simulation of Extreme Flood*) combines design event method based on single-rainfall event modelling and continuous simulation method used for estimating the maximum discharges of a given exceedance probability using rainfall-runoff models. The MESEF method considers varied moisture conditions in model catchment before the occurrence of rainfalls. To verify the efficiency of the proposed method, a comparison was carried out between the values of flood quantiles estimated by the MESEF method and the flood quantiles estimated by direct method. The proposed approach was tested in two catchments in the Upper Vistula River basin. The results of the MESEF method in both catchments were satisfactory; however, in order to verify its effectiveness, more research is needed within catchments of diverse features and landscape. Special attention should be paid to the proportion of moisture conditions that is a crucial factor in future use of the MESEF method in uncontrolled catchments.

Keywords: rainfall event, precipitation generating, rainfall-runoff modelling, probability distribution of annual maximum discharges, antecedent runoff conditions (ARC), flood quantiles

1. Introduction

Flood quantiles represent important hydrological features. Their estimation determines decisions on the size of planned hydrotechnical facilities, levees, dams or bridges. It is common practice to determine the quantiles based on many years of hydrological observations and statistical analysis of maximum discharges. At present, the problem of flood quantiles estimation is facing new challenges. According to recent forecasts, unfavourable changes in hydrographic conditions in Poland are expected to occur by 2030. The main factors behind them are said to be climate change [1] and the accelerated process of catchment sealing [2]. The United Nations Framework Convention [3] defines climate change as identifiable (for instance, by statistical tests) changes of the climate condition as well as changes in the significance of climatic elements that persist for a longer period (10 years or more). This relates to every change that occurs in climate, regardless of whether it is caused by nature or comes about as a result of human activity. It has been noted that rainfall characteristics change and rainfall becomes an increasingly random event; it is often shorter and more intense than ever before. The number of days with daily precipitation ≥50 mm has increased, and precipitation takes the form of heavy and storm rains. This change is expected to become more present especially in southern Poland with heavy rainfalls of over 20 mm/day. What has also been observed is longer dry periods. According to forecasts, heavy rainfall-induced flash floods are likely to occur more often and damage areas of poor land management. These variations are expected to intensify especially in years 2011–2030 [1] and will likely be the cause of increased frequency and violence of floods in Poland. What is more, the number of floods is expected to be on the increase in the years to follow.

Flood hazard is present in the whole of Poland and is also related to anthropogenic factors. Cities currently experience a rapid process of changes in land use which, in effect, results in most of the land becoming sealed. This factor poses additional flood hazard yet has so far been disregarded as the main cause of flood by authorities deciding on flood protection measures [4]. In addition, catchment areas tend to be developed in ways that prevent them from anticipating the precise effects of the extent of that development in the future [5].

At the moment, there are a number of methods available in Poland to estimate flood quantiles Q_p. Their application is dependent on the availability of hydrometric data. In controlled catchments, it is customary to apply direct statistical methods (SM) based on long sequences of observation of values of annual maximum discharges ($N \geq 30$) [6]. In uncontrolled catchments, on the other hand, the choice is made from among indirect methods and it depends on catchment surface area and its location in Poland. For catchments located in the region of Upper Vistula River, it is customary to use the Punzet formulas [7]. In case of uncontrolled catchments located in: (1) the Upper and Middle Odra river region, it is common to apply the Wołoszyn formulas [8], (2) the Middle and Upper Vistula River region quantiles Q_p are calculated with the use of area regression equation [9, 10]. Meanwhile, in controlled catchments of surface area over 50 km^2 located in the middle and northern Poland, quantiles Q_p are calculated based on snowmelt equation [9]. A common equation to estimate quantiles Q_p in catchments up to 50 km^2 in Poland is rainfall formula [11]. Unfortunately, these methods are now outdated and often generate unreliable results burdened with significant errors. Considering the above as well as the ever-growing demand, it is crucial to search for new methods of estimating flood quantiles Q_p. A recent example of such investigation in the Upper Vistula River region is a method based on regional flood frequency analysis [12]. The abovementioned demand also includes new methods of estimating flood quantiles that would use rainfall-runoff modelling.

http://dx.doi.org/10.5772/intechopen.68648

2. Rationale for the study

The authors present a new approach to estimating flood quantiles—the MESEF (*Multi-Event Simulation of Extreme Flood*) method. This approach makes it possible to consider climate changes, future changes in catchment urban planning as well as consider the changeable character of precipitation. The MESEF method is currently being developed in small and middle-sized catchments located in the Upper Vistula River catchment.

The Upper Vistula River basin is located in South Poland and forms part of the Vistula drainage basin (**Figure 1**). Length-wise and catchment area–wise, the Vistula river is one of the largest river in Europe [13]. The Upper Vistula River at Zawichost gauge takes up a total of 50,731.8 km^2. The flood hazard in this region of Poland is the highest due to several factors: its topographic features (mountains, highlands and basins), geological conditions as well as the land use developments located in river and stream valleys. The most common cause

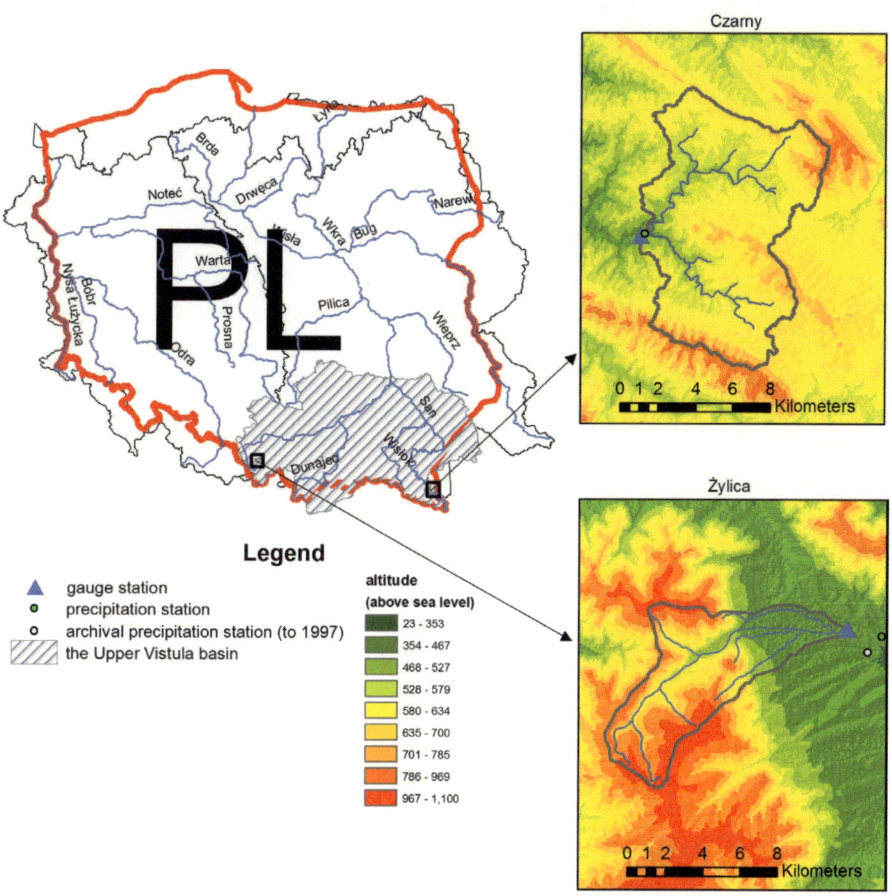

Figure 1. The natural catchment of Czarny River and the semi-urbanised catchment of Żylica with the networks of precipitation stations.

of flood in this part of Poland is heavy rainfall events. In consequence, because the MESEF method is dedicated to the southern part of the country, it was based on rainfall. Another factor in favour of this methodology is wider availability of data. Thanks to the dense network of rainfall gauging stations in this region, it is easier to obtain reliable rainfall values information.

The MESEF method uses the possibilities of the rainfall-runoff modelling. This type of modelling allows for simulation of catchment response to the impulse in form of rainfall (more specifically, rainfall distribution in time). That response, in case of the rainfall-runoff models, is the runoff information in form of runoff hydrograph, that is, temporal runoff distribution in section at the modelled catchment. The modelling allows for obtaining catchment's response to a specific precipitation event, under specific antecedent moisture conditions in catchment (reflected in the values of specific parameters) and under specified parameters of transformation of rainfall into runoff.

The rainfall-runoff modelling allowed for developing a common practical approach the so-called design event. This approach is premised on assumption that a design discharge of a given exceedance probability is created as a result of modelling a rainfall of the same exceedance probability while usually assuming normal moisture conditions in catchment before the occurrence of design event. The rainfall duration is usually assumed to be equal to or greater than the catchment concentration time and the rules of rainfall total disaggregation into smaller time steps are also specified. Undoubtedly, the simplicity of the method and straightforward results interpretation are one of the most significant advantages of this approach. What is more, not only does it allow for obtaining the value of the peak discharge but also the volume of the flood wave. This, in turn, is significant in some of the applications of the method, for instance, when designing reservoirs or planning activities in floodplain areas. The vulnerability of the design event method, on the other hand, lies in the simplified assumptions required when one hyetograph is to ensure obtaining a hydrograph of specific frequency of occurrence. In real-life scenario, a flood of peak discharge of, for example, exceedance probability $p = 1\%$ may be caused by an infinite number of combinations of catchment conditions, including rainfall and moisture conditions before the occurrence of rainfall. In the design event method, however, in order to obtain the desired result, it is necessary to define only one combination of varied input conditions. So far, no comprehensive evaluation of how this task should be carried out based on the conditions present in Poland has been published. In practice, it is most common to assume modelling of a rainfall event of duration of 24 h, represented by a daily total value of specified exceedance probability divided into hourly time steps according to the DVWK scenario [14] and normal moisture conditions in catchment before the occurrence of rainfall.

A somewhat different approach to estimate project discharges using modelling is continuous simulation approach [15, 16]. This method requires the use of stochastic rainfall generator and a model whose structure allows for modelling not only rainfall events but also the periods in between them. Peak discharges taken from long sequence of results obtained from this type of modelling are later used to formulate the maximum discharges probability curve and define the values of the desired quantiles. There are also other approaches that function, to an extent, as hybrids of the two described above, that is, the Schadex method [17], an approach presented by Francés et al. [18], or various examples of the Monte Carlo method applications. The proposed MESEF method also belongs to this group of methods.

3. Study areas and hydrow-meteorological data

The MESEF method was applied in two catchments situated in south-eastern Poland: (1) the natural catchment of Czarny River, with an area of 95.20 km^2 up to the water gauge section Polana [19] and (2) the semi-urbanised Żylica, with an area of 52.56 km^2 up to the water gauge section of Łodygowice [20]. Both of them are located in the Upper Vistula River basin (**Figure 1**).

Considering that the results of the MESEF method need to be verified against the results obtained using a direct (statistical) method, both research catchments are closed by a gauging cross-section.

The features of the Czarny catchment are considered natural as the land use changes did not affect the discharge values. This is why it was chosen as a representative. Forestland makes up over 80% of this catchment, the elevation differences reach 600 m, and the catchment planning did not change in a significant way throughout the period from which the data were sourced. Therefore, the selected catchment can be seen as a solid starting point for analysing the influence of catchment planning changes on changes in discharge maximum values.

The Żylica catchment, by contrast, is influenced by tourism which results in additional land development. The elevation differences reach 708 m. Regarding the land use, forestland still covers most of the mountain slopes; however, its overall area is gradually decreasing. The lower parts of the catchment are mostly covered by agricultural land and wastelands. The Żylica catchment was chosen as a semi-urbanised catchment because of the observable process of its diminishing proportion of forestland in favour of housing developments.

The previous evaluations of the Czarny River used meteorological data from Polana precipitation station and hydrological data from Polana gauge station on the Czarny River. The data included sequences of annual maximum daily rainfall totals P_0 and annual maximum discharges Q_0 from a multiannual period (1977–2012). For calibration of the rainfall-runoff model, flood information from the following years was used: 1997, 2007, and 2008.

For the Żylica River, the data used included a sequence of annual maximum daily rainfall totals P_0 from the period of 1972 to 1996 from Żylica precipitation station and a sequence of annual maximum discharges Q_0 from the period of 1972 to 2011. The model was calibrated based on the data from years 2006, 2007 and 2008.

It was assumed that the observed annual maximum daily rainfall totals P_0 show the three-parameter Weibull distribution W(λ, κ, γ), where $\lambda, \kappa > 0$ and the density function is expressed by Eq. (1):

$$f_W(p_o; \lambda, k, \gamma) = \begin{cases} \frac{k}{\lambda}\left(\frac{p_o-\gamma}{\lambda}\right)^{k-1} exp\left(-\left(\frac{p_o-\gamma}{\lambda}\right)^{k}\right) & \text{for } p_o \geq \gamma \\ 0 \text{ for } p_o < \gamma. \end{cases} \quad (1)$$

The parameters of this distribution estimated using the maximum likelihood method are shown in **Table 1**.

River/station	n	λ	κ	γ
Czarny/Polana	36	1.7107	36.137	22.827
Żylica/Łodygowice	25	1.4294	28.481	18.915

Table 1. Parameters of Weibull distribution 3p for the observed annual maximum daily rainfall totals P_o.

For the observed annual maximum discharges Q_o, it was assumed that they have a log-normal distribution. The density function of the log-normal distribution LN(μ, σ), where $\mu, \sigma > 0$, was expressed by Eq. (2):

$$f(q_o, \mu, \sigma) = \begin{cases} \frac{1}{q_o \sigma \sqrt{2\pi}} \exp\left[-\frac{(ln(q_o) - \mu)^2}{2\sigma^2}\right] \text{ for } q_o > 0 \\ 0 \text{ for } q_o \leq 0. \end{cases} \tag{2}$$

The parameters of this distribution were estimated using the maximum likelihood method (**Table 2**).

River/station	n	μ	σ
Czarny/Polana	36	3.4149	0.71172
Żylica/Łodygowice	40	1.4294	28.481

Table 2. Log-normal distribution parameters for the observed annual maximum discharges Q_o.

4. MESEF method

The MESEF method is proposed for estimating flood quantiles Q_p in small and medium catchments located in the Upper Vistula River region. It is based on the annual maximum daily rainfall totals P_o. The fundamental assumption of the MESEF method is as follows: the rainfall-runoff modelling repeated for many rainfall events from annual maximum daily rainfall totals P_o distribution allows for obtaining exceedance probability distribution of annual maximum discharges Q_p matching the probability distribution of observed discharges Q_o [19]. Each rainfall event that becomes an entry in the rainfall-runoff model in the MESEF method comes from a rainfall generator [21]. For each of the generated rainfall event, a rainfall-runoff modelling is carried out considering all kinds of moisture conditions ARC (*Antecedent Runoff Conditions*) in catchment [22]. The final probability distribution Q_p is obtained for optimal proportion of the ARC moisture conditions from many hydrograph peak values obtained from rainfall-runoff modelling. The diagram of the MESEF method (**Figure 2**) presents the following stages: (1) rainfall generation, (2) rainfall-runoff modelling, (3) estimating optimal proportion of the ARC moisture conditions, (4) establishing maximum discharges exceedance probability distribution.

http://dx.doi.org/10.5772/intechopen.68648

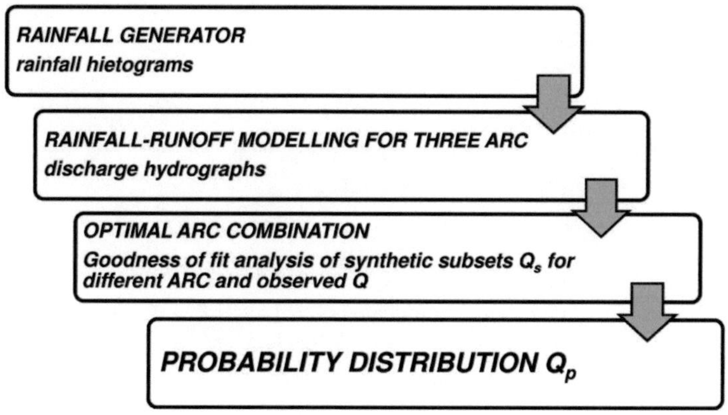

Figure 2. MESEF method action—scheme (source: own).

4.1. Rainfall generation

A generator is intended to generate maximum daily rainfall totals P_{oi} and, subsequently, disaggregate them into hourly values using beta distribution to the P_{oi} parameters α, β. For rainfall generation, it was assumed that the probability distribution P_o is known as well as the two-dimensional frequency distribution of parameters α, β for the beta distribution.

In order to disaggregate the P_{oi} into hourly values, a density function of the beta distribution f_B $(x; \alpha, \beta)$ was applied, represented by Eq. (3):

$$f_B(x;\alpha,\beta) = \begin{cases} \frac{1}{B(\alpha,\beta)} x^{\alpha-1}(1-x)^{\beta-1} \text{ dla } x \in (0,1) \\ 0 \text{ dla } x \notin (0,1), \end{cases} \tag{3}$$

where parameters α, $\beta > 0$, and $B(\alpha, \beta)$ is the Euler beta function [23]. The properties of this distribution, that is, random asymmetry (depending on the values of parameters α, β) and two-sided limitation, make its application useful in disaggregation of daily rainfall into values of smaller time steps [24]. Disaggregation of P_{oi} into hourly values requires assuming that a day, that is, 24 h, constitutes the x-coordinate value and the area under the curve of density function $f_B(x; \alpha, \beta)$ equals P_{oi}.

Considering that the beta distribution parameters α, β have significant influence on the mode of disaggregation P_o into hourly values (**Figure 3**), as verified by Wałęga et al. [25], it was necessary to estimate possible numerical values of those parameters.

To this end, two-dimensional frequency of the beta distribution parameters α, β occurrence estimated by matching the beta distribution density function with the observed data of an hourly time step rainfall in Kraków from the period of 1961 to 1985 was used. The results showed that the values of parameters α and β are in the range of 0–60. Both α and β usually show values of ranges (0.1), (1.2), (2.5) and (20.30). The values of parameter β are significantly more frequent within the range of (10.20) and the values of parameter α within (20.30) and (30.60) (**Table 3, Figure 4**).

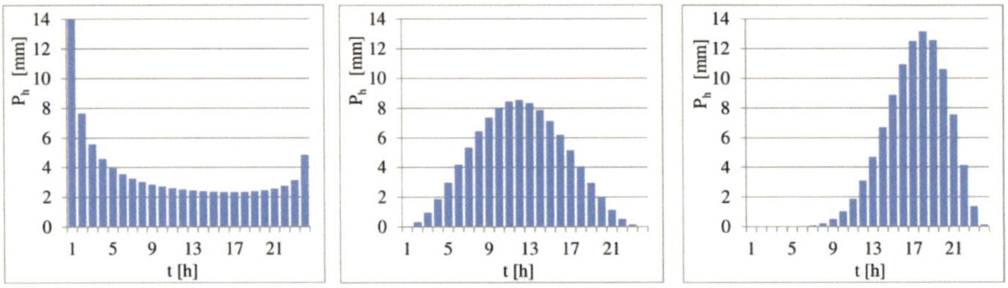

Figure 3. Exemplary hyetographs of annual maximum daily rainfall totals P_o = 100 mm, broken down into 24 hourly values, for which the values of the beta parameters (α, β) are: (a) $\alpha = 0.377$, $\beta = 0.702$; (b) $\alpha = 3.403$, $\beta = 3.689$; (c) $\alpha = 9.102$, $\beta = 3.969$ (source: own).

Generator operation consists of several stages: (1) generating synthetic value P_{oi} from the probability distribution of maximum daily rainfall totals Po, (2) generating pairs parameters α, β from their two-dimensional frequency distribution, (3) creating a hyetograph of an hourly time step by disaggregating synthetic value P_{oi}, (4) *n*-fold repetition of steps 1–3. Applying the density function of the beta distribution f_B $(x;\alpha,\beta)$ to the synthetic values P_{oi} allowed for obtaining *n* rainfall hyetographs (**Figure 5.**)

4.2. Rainfall-runoff modelling

For runoff modelling in the MESEF method, the authors used the HEC-HMS (*Hydrologic Modelling System*) model, version 3.5 [26]. The Soil Conservation Service (now the Natural Resources Conservation Service) Curve Number loss method (SCS CN), based on the knowledge of total precipitation, soil type, land cover type and soil moisture at the beginning of the rainfall, was used

α	**B**							
	(0, 1) r.f. [%]	**[1, 2) r.f. [%]**	**[2, 5) r.f. [%]**	**[5, 10) r.f. [%]**	**[10, 20) r.f. [%]**	**[20, 30)**	**[30, 60) r.f. [%]**	**Total r.f. [%]**
(0, 1)	6.6	2.6	2.2	0	0	0	0	11.3
[1, 2)	0.7	5.1	4.7	1.8	0	0.4	0	12.8
[2, 5)	1.1	2.6	6.2	4.4	2.2	0.7	0.7	17.9
[5, 10)	0.4	4.0	3.3	2.6	1.8	0.4	1.5	13.9
[10, 20)	0	0.7	2.6	2.2	3.3	2.6	1.1	12.4
[20, 30)	0	0	0.7	2.6	6.9	6.9	1.1	18.2
[30, 60)	0	0	0.4	1.1	9.9	2.2	0	13.5
Total	8.8	15	20.1	14.6	24.1	13.1	4.4	100

*The calculations were performed in 2013 by Stanisław Węglarczyk, Cracow University of Technology, Institute of Water Engineering and Water Management.

r.f.: relative frequency.

Table 3. Frequency of the beta distribution parameters (α, β) occurrence in the (0.60) × (0.60) domain*.

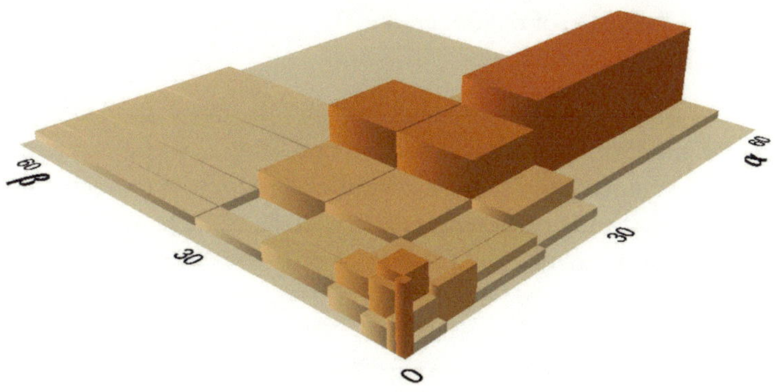

Figure 4. 2D histogram of (α, β) values in the (0.60) × (0.60) domain.

to determine the value of effective rainfall. All of these factors are accounted in the CN parameter [27]. The Unit Hydrograph method (SCS UH) used to determine the value of the peak discharge, total runoff volume, hydrograph shape and time history was chosen for the rainfall-runoff transformation. To determine the baseflow, the Recession Method was used. It allows the approximation of typical streamflow behaviour also after the rainfall event. This situation—descending part of the hydrograph—is depicted in the form of the exponential recession curve [26].

The choice of these simple methods was driven by their widespread use, small number of parameters, as well as their applicability in ungauged catchments due to the possibility of parameter estimation on the basis of the catchment characteristics. Moreover, the loss and transformation models allow to diversify the parameter values of the HEC-HMS model according to the antecedent conditions of runoff in catchment which have influence on the values of peak discharges. This is significant for the MESEF method proposed in this chapter.

The calibration procedure was conducted using HEC-HMS software. Five parameters from the model underwent calibration: the initial abstraction, CN parameter, T_{Lag} parameter, baseflow threshold coefficient and recession constant. Different ARCs, which had impact on the value of

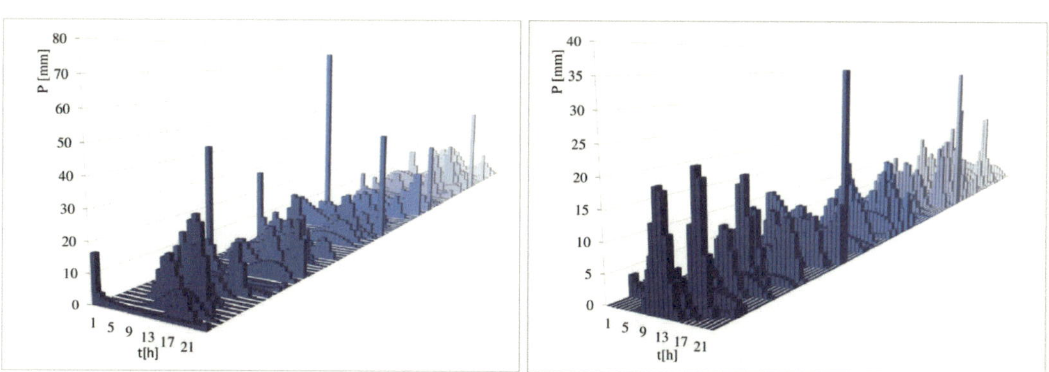

Figure 5. Examples of generated hyetographs P_{oi}, disaggregated into hourly values in the Czarny catchment (on the left) and Żylica catchment (on the right) (source: own).

the CN parameter and the value of the dependent T_{Lag} parameter, were allowed in the calibration procedure.

A model of specified parameters can be used for rainfall-runoff modelling using generated rainfall hyetographs. Rainfall-runoff modelling is conducted for a set of *n* rainfall events, each event for three kinds of antecedent runoff conditions (ARC) in the catchment, that is, soil moisture levels at the beginning of the rainfall: dry (ARC I), normal (ARC II), and wet (ARC III). Thus, $n \times 3$ hydrographs are obtained and from those hydrographs subsequent $n \times 3$ peak discharge values (**Figure 6**).

4.3. Exceedance probability distribution of maximum discharges Q_s

It was assumed that probability distribution of observed annual maximum discharges Q_o is realistic distribution, and the simulated data Q_s were aimed at showing maximum equality with this distribution. In order to do this, it became essential to search for optimal proportion of moisture conditions ARC that would make it possible.

4.3.1. Searching for optimal proportion of ARC

From each of the *n*-element data set of discharges obtained for each of the three conditions ARC, a specific number of values was selected randomly, creating, in effect, *n*-element sequences of random discharge values mixed together. The random selection was performed for 38 possible combinations, thus creating 38 *n*-element sequences of discharge values. For example, combination 1-2-0 consisting of 100 elements means that from the data set of discharge values for ARC I, there were 33 values selected randomly which constitute one-third of elements in the whole sequence, and from the dataset for ARC II, there were 67 values that constitute the remaining two-third elements of the whole sequence. In this case, there was no random selection performed from the data set for ARC III.

The next stage is to verify the correlation of all created data sets with theoretical probability distribution for observed data Q_o. In order to assess the match of both distributions, three equality tests are necessary to be performed: Kolmogorov-Smirnov (K-S), Anderson-Darling

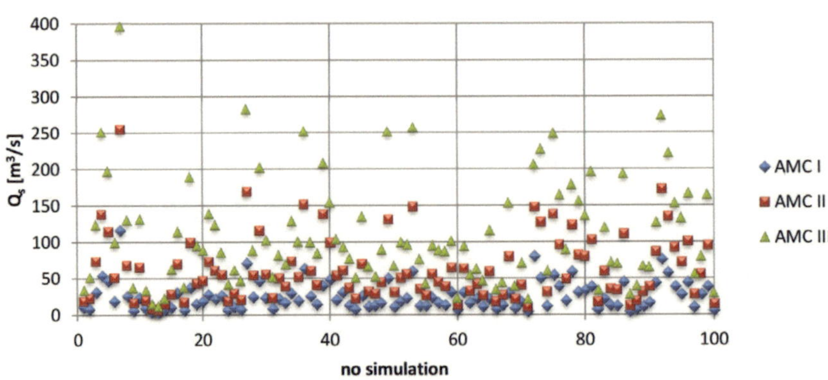

Figure 6. Synthetic peak discharge values for three types of ARC.

(A-D), and χ^2 Pearson's (χ^2 P) test. The best match to the theoretical probability distribution of discharges Q_o has the lowest statistical values.

Based on the test of statistic values, it can be concluded that the synthetic discharge values in proportions **2-3-0** (in the Czarny River) and **3-2-0** (in the Żylica River) of the ARC conditions in catchment demonstrate the best compatibility with the observed data (**Figures 7** and **8**). In these optimal combinations, the synthetic discharge values come from dry (from 33 to 40%) and normal (from 60 to 67%) in the Czarny River and from dry (60%) and normal (40%)—in

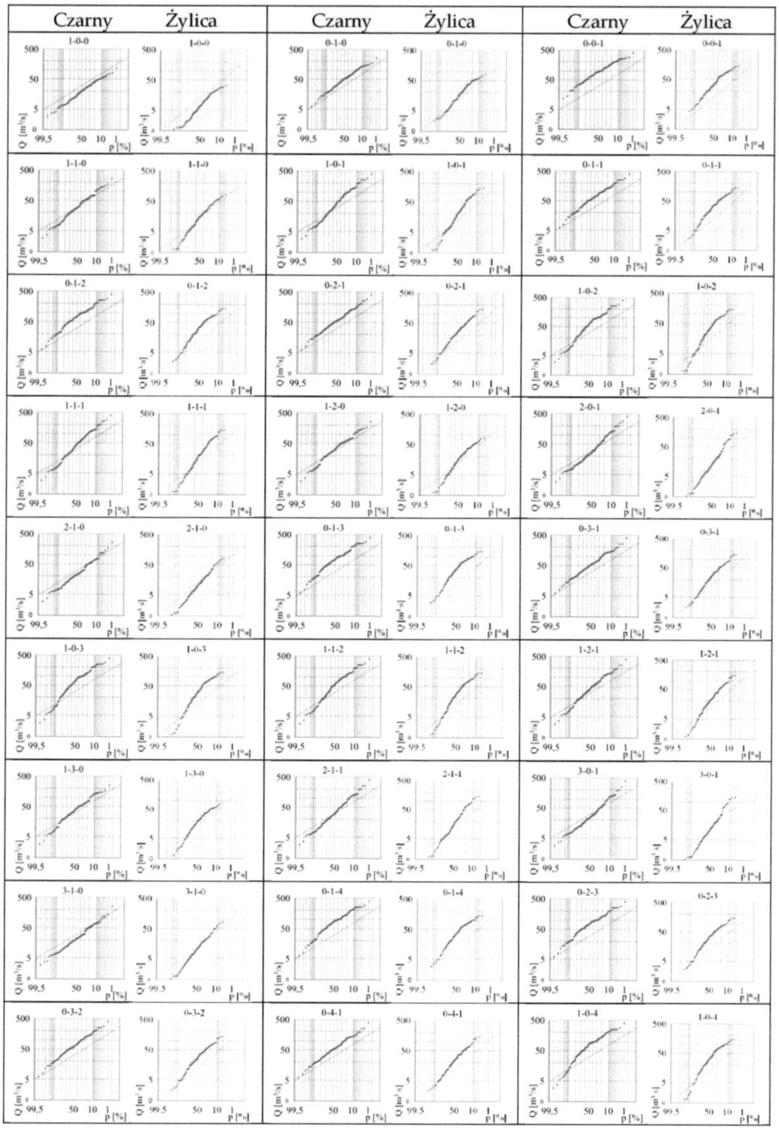

Figure 7. Exceedance probability curves of discharges Q_s (circles) and Q_o (solid lines) in the Czarny catchment (on the left) and in the Żylica catchment (on the right) for different proportion of conditions ARC. Combinations for which both catchments revealed best distributional equality are marked in frame (source: own).

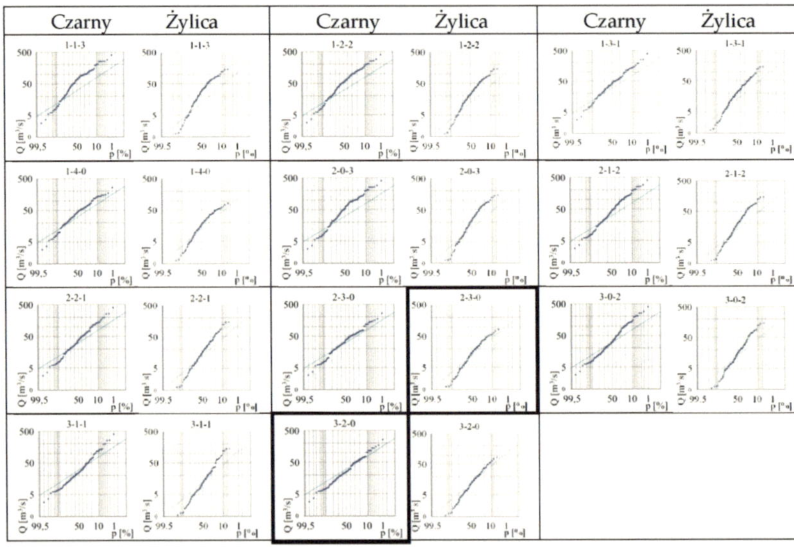

Figure 8. Exceedance probability curves of discharges Q_s (circles) and Q_o (solid lines) in the Czarny catchment (on the left) and in the Żylica catchment (on the right) for different proportion of conditions ARC. Combinations for which both catchments revealed best distributional equality are marked in frame (source: own) - continuation.

the Żylica River—antecedent moisture conditions in catchment. Wet antecedent runoff conditions (ARC III) do not affect the synthetic discharge values in both cases.

4.4. Exceedance probability distribution of maximum discharges Q_s for optimal conditions ARC

For a data set of discharges at optimal proportion of moisture conditions ARC, a Q_s exceedance probability curve is created. In order to consider high-frequency quantiles reliable, it is necessary to create the curve using data from several thousands of simulations, more specifically an empirical probability curve. In consequence, a comparison was made between the values of simulated quantiles Q_s and the observed quantiles Q_o for the specified values of probability p (**Figure 9**).

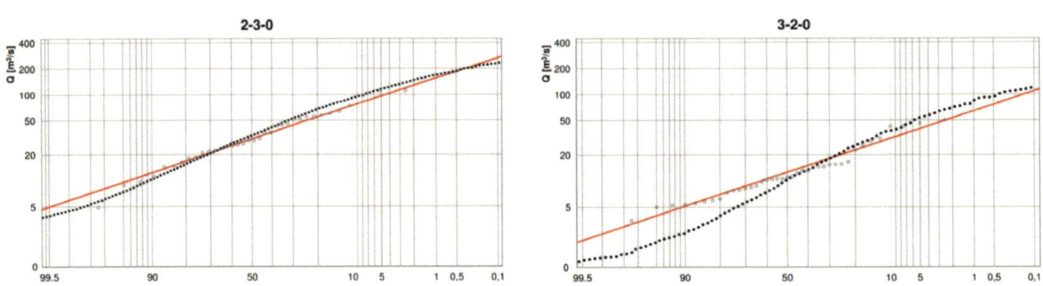

Figure 9. Exceedance probability of maximum discharges Q_s curves from the MESEF method (dotted line), empirical (circles) in the Czarny catchment (on the left) and the Żylica catchment (on the right). The solid line marks the probability curves for the remaining observed Q_o.

http://dx.doi.org/10.5772/intechopen.68648

5. Comparing flood quantiles

In order to verify the usefulness (effectiveness) of the MESEF method, the obtained quantiles Q_p (for optimal conditions ARC) need to be compared with the quantiles Q_p obtained using a statistical direct method (SM).

For the Czarny catchment, a comparison was carried out for an optimal combination 2-3-0 (**Table 4**), and for Żylica catchment, a comparison was carried out for an optimal combination 3-2-0 (**Table 5**).

As it can be observed, in the Czarny catchment, the flood quantiles estimated using the MESEF method reveal slightly higher values (**Table 4**) than those estimated using the statistical method (for $p \geq 0.5\%$). Only for $p = 0.1\%$, the flood quantile estimated using the MESEF method is slightly lower than the quantile estimated using the statistical method SM. It can be observed here that for the **2-3-0** proportion, the relative error reveals values from 0.8 to 24.6%.

In the Żylica catchment, it was observed that the quantiles estimated using the MESEF method reveal slightly higher values of the quantiles estimated using the SM method (for $p \leq 20\%$), and only for $p = 30$ and 50%, they are slightly lower (**Table 5**). What is more, relative error (for the specified values p) is within the range of 0.9–33.6%.

To sum up, in both catchments, the natural (Czarny) and the semi-urbanised (Żylica), the flood quantiles estimated using the MESEF method are comparable with the flood quantiles estimated using the SM method (for $0.5\% \leq p \leq 20\%$) obtained from observed data. Based on the above, it can be concluded that the flood quantiles estimated using the MESEF method are, in both analysed cases, similar to the observed quantiles. For example, the flood quantile of a $p = 1\%$ exceedance probability was obtained with the relative error of 15.7% (Czarny) and 10.3% (Żylica) using the MESEF method. The comparison showed the usefulness of the

p [%]	Q_p (SM) [m^3/s]	$Q_{p(2\text{-}3\text{-}0)}$ (MESEF) [m^3/s]	Difference Δ_1 (2-3-0) [m^3/s][a]	Relative error δ_1 [%][b]
0.1	242.0	236.0	6.0	2.5
0.5	190.2	191.8	−1.5	0.8
1	147.2	170.4	−23.2	15.7
2	123.2	148.8	−25.7	20.9
5	94.1	117.0	−22.9	24.3
10	74.0	92.2	−18.2	24.6
20	55.2	67.9	−12.7	23.0
30	44.5	52.7	−8.2	18.4
50	31.1	33.3	−2.2	7.1

[a] $\Delta_1 = Q_p(\text{SM}) - Q_{p(2\text{-}3\text{-}0)}(\text{MESEF})$.
[b] $\delta_1 = \Delta_1/Q_p(\text{SM}) \times 100$.

Table 4. Comparison of flood quantiles estimated using the SM method and the MESEF method for a combination 2-3-0 (Czarny catchment).

p [%]	Q_p(SM) [m³/s]	$Q_{p(3\text{-}2\text{-}0)}$ (MESEF) [m³/s]	Difference Δ_2 (3-2-0) [m³/s][a]	Relative error δ_2 [%][b]
0.1	109.7	120.2	−10.5	9.6
0.2	94.5	110.4	−15.9	16.8
0.5	76.5	92.9	−16.4	1.7
1	64.2	84.4	−20.2	10.3
2	53.0	69.7	−16.7	8.6
3	47.0	62.7	−15.7	18.2
5	39.8	53.2	−13.4	33.6
10	30.9	37.7	−6.8	22.0
20	22.7	25.3	−2.6	11.7
30	18.2	18.0	0.2	0.9
50	12.6	10.3	2.3	17.8

[a] $\Delta_1 = Q_p(\text{SM}) - Q_{p(2\text{-}3\text{-}0)}(\text{MESEF})$.
[b] $\delta_1 = \Delta_1/Q_p(\text{SM}) \times 100$.

Table 5. Comparison of flood quantiles estimated using the SM method and the MESEF method for a combination 3-2-0 (Żylica catchment).

MESEF method; therefore, it could prove to be a good alternative for estimating the maximum discharges of a given exceedance probability using rainfall-runoff models.

6. Strengths and weaknesses of the MESEF method

Good results obtained using the MESEF method in two different catchments encourage to continue further research. The method, however, next to its strengths, has also some weaknesses that could be further verified.

This section summarises the strengths and weaknesses of the MESEF method. Undoubtedly, the greatest strength of the MESEF method is that it is based on the values of precipitation whose network of measuring stations is far more dense than the network of water gauges. This allows for a wide range of applications. The MESEF method based on rainfall also has the potential to take into account changes in climate. It requires, of course, earlier identification of quantitative and qualitative influence of the changes on the rainfall distribution; especially, when it comes to the future values of annual maximum daily rainfall totals. On the other hand, the qualitative influence refers to the future temporal distribution of daily rainfall changed due to climate change. The application of rainfall-runoff modelling in the MESEF method opens up possibilities for the method to implement changes in developments in the catchment area.

A particular strength of the method is the possibility of performing simulation of the influence of existing flood protection facilities in catchment (e.g., a reservoir) or non-technical flood protection measures connected with volume reduction (afforestation) on flood quantiles.

The weaknesses of the MESEF method, according to the authors, are the uncertainty associated with the nature of point measurements of precipitation in conjunction with the aerial

http://dx.doi.org/10.5772/intechopen.68648

character of real phenomenon that requires consideration in further studies and a small number of tests with the MESEF method in different catchments. It would be interesting to see whether the same proportion of ARC would be confirmed in other catchments. Confirming the same proportion of ARC in more catchments (e.g. by region) could provide a possibility of using the MESEF method for estimating flood quantiles in uncontrolled catchments.

7. Summary

The application of the MESEF method to estimate flood quantiles gave good results. The obtained values Q_p were similar to the observed values. A comparison showed usefulness of the MESEF method. Based on the results obtained so far, it could be concluded that the proposed MESEF method is an effective approach and could provide a good alternative to the currently used method of estimating flood quantiles Q_p in small catchments in the Upper Vistula River basin. This could be of significant importance in those applications where evaluations are performed of activities carried out in catchments related to changes in volume or the shape of flood wave as a result of, for instance, location of reservoir or the afforestation of a part of catchment. In order to ultimately confirm the effectiveness of the method, it would be necessary to apply it in different controlled catchments which are planned in the future work. The proposed method involves several weaknesses that would need to be resolved in the course of future research.

Author details

Elżbieta Jarosińska[1]* and Katarzyna Pierzga[2]

*Address all correspondence to: ejarosin3@gmail.com

1 Cracow University of Technology, Cracow, Poland

2 Hydromodel Center of Hydrology and Environmental Engineering, Brzezna, Poland

References

[1] SPA 2020. Polish National Strategy for Adaptation to Climate Change (SPA 2020). Warszawa, Poland: Ministry of Environment; 2013. Available from: http://klimada.mos.gov.pl/wp-content/uploads/2013/10/SPA2020.pdf (in Polish)

[2] KPZK 2030. National Spatial Development Concept 2030. Polish Monitor. Warszawa; 2012. 27.04.2012. Item. 252. Available from: http://isap.sejm.gov.pl/DetailsServlet?id=WMP20120000252 (in Polish)

[3] ACT. The Regulation of the Council of Ministers of 18 October 2016 on adopting a flood risk management Plan for the region of the Vistula River basin. Journal of Laws. 2016. Item 1841. Available from: http://www.dziennikustaw.gov.pl/du/2016/1841/1 (in Polish)

[4] Kowalczak P. Climate variability and the causes of the floods in Poland. Poznań, Poland: Institute for Agricultural and Forest Environment, Polish Academy of Sciences. 2011. Available from: http://www.umweltaktion.de/pics/medien/1_1324305988/Vortrag_Kowalczak.pdf (in Polish)

[5] Jarosińska E. Local flooding in the USA, Europe and Poland—An overview of strategies and actions in face of climate change and urbanisation. Infrastructure and Ecology of Rural Areas. 2016;III:801-821. Available from: http://dx.medra.org/10.14597/infraeco.2016.3.1.059

[6] Więzik B, Madzia M. Comparison of maximum discharges of a given exceedance probability calculated using empirical methods in the Skawica catchment. In: Więzik B, editor. The Problems of Estimating Extreme Discharges in Controlled and Uncontrolled Catchments. Warszawa: Monographs of Committee of Environmental Engineering PAS; 2013 (in Polish)

[7] Association of Polish Hydrologists. Methodology of estimating discharges and maximum rainfalls of specified exceedance probability for controlled and uncontrolled catchments and identifying models of transforming rainfall into runoff. Warszawa. 2009. Available from: https://pl.scribd.com/document/97351208/03-hydrologia-Metodyka-Qmax-Raport-koncowy (in Polish)

[8] Rogala R, Dołęga J. Supporting materials for hydrology calculations. Wrocław: Wroclaw University of Technology; 1973 (in Polish)

[9] Stachý J, Fal B, Czarnecka H. Estimating maximum discharges of a given exceedance probability. Warszawa: IMGW; 1998 (in Polish)

[10] Regulation No. 4/2014 of Director of Regional Water Management Board in Kraków (the RZGW in Kraków) of 16 January 2014 on conditions to use water of the Upper Vistula river region. Kraków, Poland.: Regional Water Management Board. Available from: http://bip.malopolska.pl/rzgwkrakow,a,848610,rozporzadzenie-nr-42014-dyrektora-regionalnego-zarzadu-gospodarki-wodnej-w-krakowie-z-dnia-16-styczn.html (in Polish)

[11] Ciepielowski A, Dąbkowski SL. Methods of estimating maximum discharges in small river catchments. Bydgoszcz: Oficyna Wydawnicza Projprzem-EKO; 2006 (in Polish)

[12] Rutkowska A, Żelazny M, Kohnová S, Łyp M, Banasik K. Regional L-moment-based flood frequency analysis in the Upper Vistula River basin, Poland. Pure and Applied Geophysics. 2016

[13] Kundzewicz ZW et al. Flood risk in the Upper Vistula Basin. GeoPlanet: Earth and Planetary Sciences. 2016. DOI: 10.1007/978-3-319-41923-7_1

[14] DVWK. Arbeitsanleitung zur Anwedung von Niederschlag-Abfluss Modellen in kleinen Einzugsgebieten. Teil II: Synthese, Deutscher Verband für Wasserwirtschaft und Kulturbau. Vol. 113. Hamburg: Paul Parey Verlag; 1984 (in German)

[15] Boughton W, Droop O. Continuous simulation for design flood estimation: A review. Environmental Modelling & Software. 2003;**18**(4):309-318. DOI: 10.1016/S1364-8152(03)00004-5

[16] Calver A, Lamb R. Flood frequency estimation using continuous rainfall-runoff modelling. Physics and Chemistry of the Earth. 1995;**20**(5-6):479-483. DOI: 10.1016/S0079-1946(96)00010-9

[17] Paquet E, Garavaglia F, Garçon R, Gailhard J. The SCHADEX method: A semi-continuous rainfall-runoff simulation for extreme flood estimation. Journal of Hydrology. 2013;**495**:23-37. DOI: 10.1016/j.jhydrol.2013.04.045

[18] Francés F, García-Bartual R, Ortiz E, Salazar S, Miralles J, Blöschl G, et al. CRUE Research Report No. I-6: Efficiency of Non-structural Flood Mitigation Measures: "Room for the River" and "Retaining Water in the Landscape". London: CRUE Funding Initiative on Flood Risk Management Research; 2008. pp. 172-213

[19] Jarosińska E, Pierzga K. Estimating flood quantiles on the basis of multi-event rainfall simulation: Case study. Acta Geophysica. 2015;**63**(6):1639-1663. DOI: 10.1515/acgeo-2015-0046, ISSN 1895-6572

[20] Pierzga K, Jarosińska E. Estimating projected discharges on the basis of modelling many hypothetical rainfall events in semi-urbanised catchment. In: Hydrology of Urban Areas. Vol. 39. Warszawa: Monographs of Committee of Environmental Engineering PAS; 2016. pp. 201-212. ISSN 0867-7816

[21] Pierzga K, Jarosińska E. Annual maximum daily rainfall generator. In: Magnuszewski A, editor. Hydrology in the Protection and Landscape Environment. II National Hydrological Congress. Vol. 2. Warszawa: Monographs of Committee of Environmental Engineering PAS; 2014. Association of Polish Hydrologists and the Warsaw University of Life Sciences – SGGW

[22] Hawkins RH, Ward TJ, Woodward DE, van Mullem JA, editors. Curve Number Hydrology: State of the Practice. ASCE/EWRI Curve Number Hydrology Task Committee. Reston: American Society of Civil Engineers; 2009. DOI: 10.1061/9780784410042

[23] Węglarczyk S. Statistics in Environmental Engineering. Cracow: Cracow University of Technology Publisher; 2010 (in Polish)

[24] Więzik B. Annual maximum discharges of given exceedance probability in small ungauged catchments. In: Więzik B, editor. Hydrology in Engineering and the Water Management. Vol. 68. Warszawa. Monographs of Committee of Environmental Engineering PAS; 2010. pp. 153-165 (in Polish)

[25] Wałęga A, Cupak A, Miernik W. Influence of entrance parameters on maximum flow quantity receive from NRCS-UH model. Infrastructure and Ecology of Rural Areas. 2011;**7**:85-95

[26] Scharffenberg W, Fleming M. Hydrologic Modeling System HEC-HMS v.3.5 User's Manual. U.S. Army Corps of Engineers. USA: Davis; 2010

[27] Mishra SK, Singh VP. Soil Conservation Service Curve Number (SCS-CN) Methodology. Water Science and Technology Library. Vol. 42. Dordrecht: Kluwer Academic Publishers; 2003. p. 516. DOI: 10.1007/978-94-017-0147-1

Chapter 15

Towards the Development of a Capability Assessment System for Flood Risk Management

Mohammad Hijji, Saad Amin, Wayne Harrop and Rahat Iqbal

Additional information is available at the end of the chapter

http://dx.doi.org/10.5772/intechopen.68784

Abstract

Having in place adequate levels of emergency management capabilities (EMCs) underpins a managed civil emergency response, especially during a flooding event(s). Good EMC is either built on having the right internal capabilities or by exploiting existing emergency capabilities from other responders. In some countries, such as Saudi Arabia, there is a noted lack of decision-making in the Civil Defence (CD) Authority about generating effective mutual-aid requests. Three core areas of EMC include having the right types and levels of response equipment to hand, ensuring sufficient Human Resources, can be maintained throughout a sustained event, and developing adequate Training capabilities. Other factors impacting on Saudi Arabia include both stress and a lack of work experience. In this chapter, we examine the effectiveness of a prototype IT System in the case of Saudi CD Authority as a tool for addressing the availability and adequacy of mutual-aid for EMC, Human Resources (HR), and training capabilities against scalable levels of flood risk event(s). The proposed IT System is built using the 'fuzzy expert system' approach.

Keywords: decision support systems, expert system, emergency management technology, information technology systems, capability preparedness, capability mutual-aid

1. Introduction

In the context of 'decision-making' in 'Flood Risk Management', a lack of decisions related to the required level of preparedness across known EMC within disaster and emergency management organisations could critically increase the likelihood and/or impact of a flood response failure, thus increasing the rate of losses to properties and lives. From a decision-makers' point of view, there are several factors involved; however, factors such as (a) pressure to respond quickly when a flooding alert is given, (b) the lack of experience attributed to decision-making, and (c) a lack of information from other parties [1] could directly affect the timing and quality of decisions [2]. Many studies have agreed that there is a high level of vagueness and uncertainty involved in the decision-making process especially in assessment and determination of needs against current and future flood risk events. In this context, Information Technology (IT) Systems, such as Fuzzy Expert Systems (FES), have created a potential contribution in decision-making for the field of 'Flood Risk Management'.

There are several studies which address the issue of vagueness and uncertainty involved in the decision-making process. For example, Virtual Immersive Reality Training System (VIRTS) is an IT system that trains and assesses emergency responders through visualizing '3D scenarios' [3]. VIRTS enables an end-user to design and modify various risk scenarios, however it does not provide risk scenarios related to natural disasters such as flood and earthquake scenarios. Another study proposed an IT system for detecting and evaluating problems in the equipment capability during operational use [4]. A proposed IT system for mass evacuation [6], which aims to aid decision-makers in understanding complexity within the pre-alert situation and evaluates the adequacy of evacuation plans for areas, however the system has not been tested and validated.

There remains no 'IT system able to evaluate precisely the current preparedness of EMC against scalable levels of flood risk events, and address needed EMC to measure adequate levels of preparedness, as well as adequate 'mutual-aid' requirements. In the case of flooding, areas at risk can be divided into zones (or a ser ies of geographical areas). A EMC zonal assessment can then be undertaken both before and during a flood event, and the assessment can help determined if any active zones need urgent support from other zones (even the wider regional resources) to better manage serve flooding events. **Figure 1** represents a scenario requiring mutual-aid provision.

Figure 1 shows four zones (A, B, C and D), each zone has certain amounts and types of training capabilities (for example skill A), and each zone may be subject to variable levels of flood risk. One of the major issues facing most of the dtecision-makers before a flood event is the ability to assess in context and between zones individual capabilities. Decision-makers often have no rationale model for determining the (re)distribution of limited resources, training, and mutual-aid requirements between zones.

In this chapter, a new prototype and purpose built IT system is presented and examined to determine whether flood response can be optimised using a virtual assessment of 'Capabilities Preparedness' in the context of Saudi Arabian 'Flood Risk Management' scenarios. The IT system performs intelligently by digitalising the process of decision-making in the problem domain. The proposed IT system is called 'The Intelligent Capability Preparedness (ICP) System'. The ICP system focuses on three types of EMC categories: (a) Training; (b) Human Resources (HR), and (c) Equipment. The Effectiveness of using the ICP system has been examined in the case study of Saudi CD Authority, Saudi Arabia. This case study is selected due to

http://dx.doi.org/10.5772/intechopen.68784

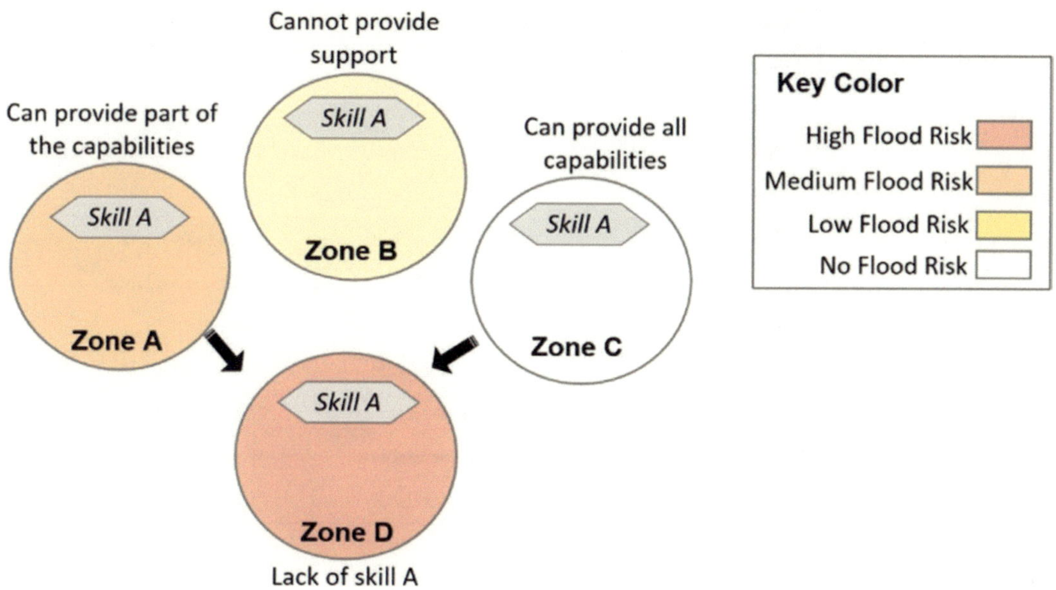

Figure 1. Mutual-aid and flood risk (Source: Authors).

the major problem of flood risk facing the Saudi CD Authority in the city of Jeddah. The chapter is structured as follows. First, a theoretical background on 'what is fuzzy expert systems (FESs) will be presented to understand the principles behind the IT System. Second, an overview of comparable applications like FES is highlighted. Third, the typical methodology of building a FES is highlighted. Fourth, the design of the ICP system is outlined and explained. This is followed by the results of evaluating the effectiveness of the ICP system after consultation and input from the Saudi CD Authority.

2. Saudi Arabian and flash-flooding: background

Saudi Arabia with a population of around '30,770,375' million and an area of approximately '2,149,690 sq. km' is the biggest country in the Middle East [5]. Over the last decade, Saudi Arabia has been marked as one of the foremost countries in the Middle East's facing various kinds of serious disasters, particularly flash-flooding. Flash-flooding is a type of flood which occurs suddenly, as a result of intense rainfall within a short period of time [6].

Generally, flood hazard is considered as the most frequently occurring type of disaster in Saudi Arabia, and over the years, it has been increasing in impact [7], this is mainly because of the nature of the countries' topographical and geographical characteristics as well as the changing global weather [6]. **Table 1** shows a timeline of some of the major flash flooding disasters that have taken place in the country.

City	Year	Killed people	Affected people
Several parts of the country	1964	20	1000
Northwestern	1985	At least 32	5000
Jizan	2004	5	430
Mecca	2005	29	17
Jeddah	2009	163	More than 10,000
Jeddah	2010	More than 122	*Data unavailable
Jeddah	2011	More than 10	144
Tabuk	2013	*Data unavailable	*Data unavailable
Mecca	2014	*Unveiled data	*Data unavailable
Jeddah	2015	*Data unavailable	*Data unavailable
Aseer	2017	*Data unavailable	*Data unavailable

Table 1. Flood-event occurrences in Saudi Arabia [7].

3. Study context and hypothesis

The primary investigation conducted with the Saudi CD Authority in 2013 demonstrated variable flash-flooding capabilities for each of the country's regions. The investigation proposed human resources, training, and equipment capabilities could better incorporate local hazards and vulnerabilities based on a decision-makers experience and perception. Up until this point, it was noted how flash-flooding events had no joined up analysis process or a clear evaluative framework [8].

Furthermore, there were noted concerns by decision-makers about how effective flash-flood response capabilities remained in the face of unequitable access and distribution of flood resources throughout the flooding episode(s). The issues were especially acute about human resources, equipment and appropriate levels of training by first responders. In some cases, the resources were overly committed, or not appropriately allocated across the impacted flooding areas.

Saudi CD Authority is determined to improving the effectiveness of its flood risk management [9]. However, one of the factors hindering it from achieving this objective is the lack of analysis, readiness, and optimized capabilities spanning all flood zones [10]. Therefore, this study aims to develop a new ICP system aimed at better supporting the Saudi CD Authority and its affiliated government sectors manage flash-flooding events.

The proposed ICP system is meant to be used as a reference and a standard for evaluating the level of three types of flood management capabilities (i.e. training, HR and equipment) and to help prepare them to an adequate level. The ICER system can be used as an assistance and assurance tool of the Saudi CD Authority to better assess and prepare its own EMC. It is expected that this will give the Saudi CD Authority confidence and involve other stakeholders against various levels of flood risk (high, medium, low categories of flooding in each geographical/

http://dx.doi.org/10.5772/intechopen.68784

zonal area). This approach should offer mitigation and preparedness actions against current and future flash-flood risk events. The ICER system will assist the Saudi CD Authority to computerise and share the processes involved in the developed framework. It is envisaged that clear objectives can be established in the framework to unify local hazard and vulnerability data, with optimal resource allocation/capabilities and flood risk forecasting.

4. What is fuzzy expert systems (FESs)?

Fuzzy expert systems (FESs) are a term used to refer to any IT system that works using the principles of 'Fuzzy logic' and 'Expert systems'. Expert systems are an important subset of the Artificial Intelligence (AI) field. They were first introduced by the AI community in the 1960s [11]. Over the years, several definitions of expert systems have been proposed. Some of the most popular definitions include:

a. *'An intelligent computer program that uses knowledge and inference procedures to solve problems that were difficult enough to require significant human expertise for their solutions'* [12].

b. *'A computer program designed to model the problem-solving ability of a human expert.'* [13].

c. *'A system that uses human knowledge captured in a computer to solve problems that ordinarily requires human expertise.'* [11].

On the other hand, Fuzzy Logic is a component of fuzzy expert systems. It is a mathematical approach primarily developed in the 1960's by Lotfi Zadeh to analyse problems that involve a degree of truth or partial truth [14]. The concept of fuzzy logic is based on the assumption that the truth of anything can be expressed as a matter of degrees. For example, beauty, health, or distance – in other words they can be assigned to a sliding scale.

Over the years, several definitions of Fuzzy Logic have been proposed. Among them include:

a. *'…as a set of mathematical principles for knowledge representation based on degrees of membership rather than on crisp membership of classical binary logic'* [15].

b. *'Mathematical technique for dealing with imprecise or incomplete information in a specified scenario'* [16].

The underlying principle of the fuzzy logic approach is based on the fuzzy sets. A fuzzy set can be explained by using the definition of a classical set. A classical set is that which contains every element within a domain or excludes every element within the domain. Simply put, an element X of a universal set U (domain) can either be in two sets, a set A or a set not A. For instance, consider a set of 'old people' as an example. The health organisations in UK define 'old people' as those people that have reached the age of '65'. Therefore, this can be represented by the classical set as shown in **Figure 2**.

Accordingly, any person that is of '65 years' of age can be regarded as 'old'; and any person less than that (64 and less) is 'not old'. In a classical set, an element of any of the sets can

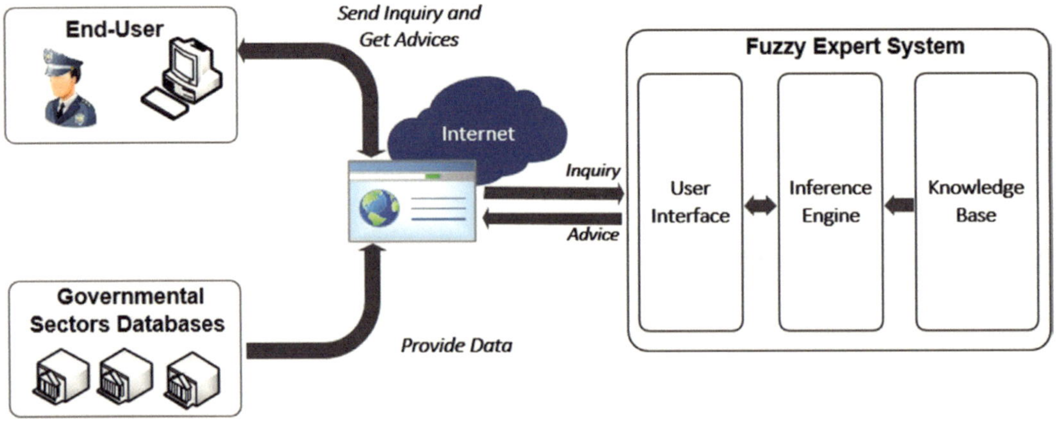

Figure 2. An example of the classical set boundary and fuzzy set boundary for the case of old people.

be categorised as 'yes' or 'no' depending on the boundary of demarcation (which is crisp). However, such boundaries are not clearly defined in reality across all societies. In some societies a person of '55, 65 or 75' can be considered to be old [17–18]. Fuzzy sets make it possible to describe the range of all the possible ages that are considered to make up the 'old peoples' category.

To clearly explain the functioning of the FES, it is best to first understand the architecture of the expert systems, including it various components. The two main components of the FES (as explained previously) are the 'Expert System' and the 'Fuzzy Logic'. Each of the two components also has their own constituent's components. For the typical 'FES, it is made up of six main constituents, these are:

iv. Rule-base

v. Inference

vi. Fuzzification

vii. Defuzzification

viii. User-interface

ix. Users

Figure 3 shows the individual constituents and their relationships. As both components (Expert System and Fuzzy Logic) involve the Rule based and inference, these are only represented once in the fuzzy expert system architecture. The next subsections will give a brief description of the individual components.

4.1. Rule-base

The FES Rule-Base component is that part which holds the pool of rules and facts relating to that specific domain. These are generally presented as a set of fuzzy 'IF-THEN' rules. The information used in developing these rules is derived from experts in that domain of the problem, for

http://dx.doi.org/10.5772/intechopen.68784

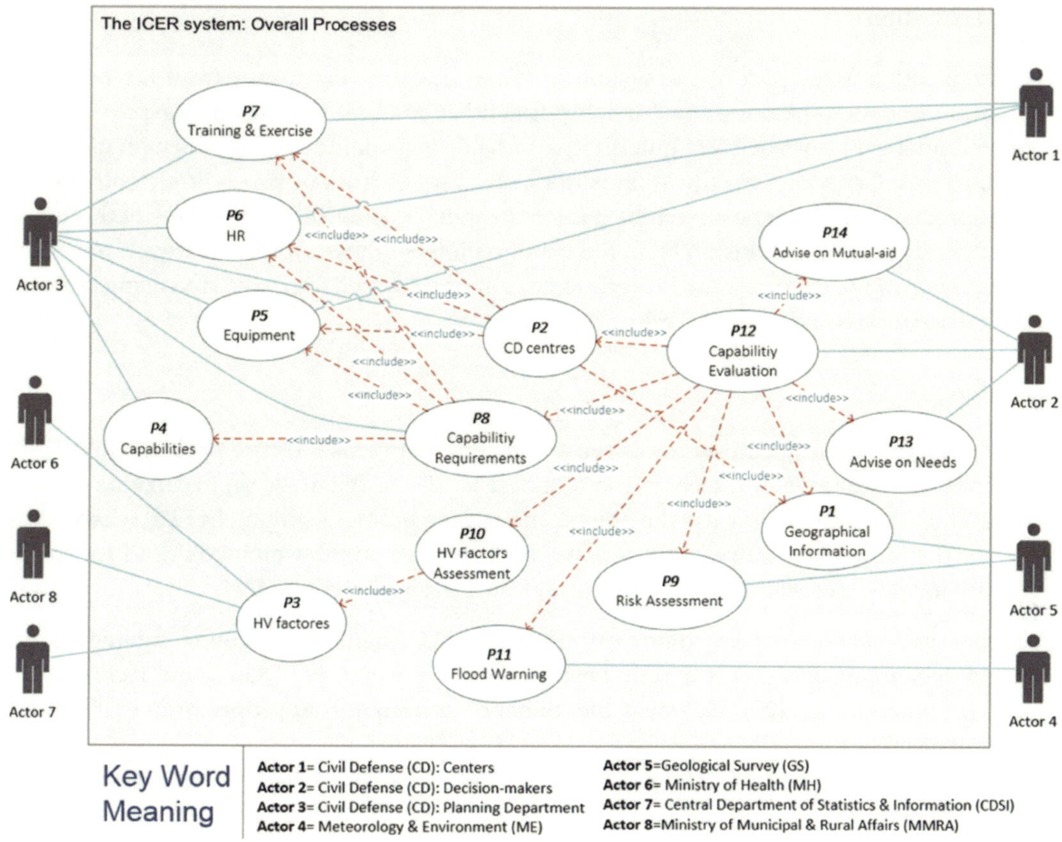

Figure 3. FES complements [12].

example, for the development of an FES relating to 'public awareness campaign'. The contents of such associated rules should typically include issues such as the community location, community size, or campaign schedule. Again, as such rules are continuously evolving and undergoing change, this means that the system should also make a provision to enable these rules to be updated and modified as required. Furthermore, when such input variables, output variables and membership functions have been established, the rule-base needs to be designed in such a way that it will convert the input variables to output variables as represented below [19]:

'IF <Conditions> THEN <Conclusions>'

These rules can be defined subject to the number (and the possible values) of qualitative inputs and outputs. Furthermore, having more rules will also mean a greater degree of reliability from the inferences (even for the same number of variables). Again, it should be noted that, even though rules in general are derived from the expert(s), the whole knowledge does not have to be translated into rules, as in some cases some rules might be irrelevant. One possible approach for designing the rule-base and combinations of inputs is by using the decision matrix. This way the rule-base is interpreted based on the degree of the output's membership by using fuzzy reasoning [11].

4.2. Fuzzification

The fuzzification is the part of the system involved in extracting inputs from the no-fuzzy inputs/outputs, with the use of membership functions. In other words, it is the part used to convert numerical values of every quantitative variable to a qualitative variable with membership functions. Generally, fuzzification is not a strict set of rules or procedure; contrarily, it is an approach that works partly on the basis of insight, experience and analyses of the rules, whereupon it infers a conclusion from the combinations of these inputs. However, for effective operation of the fuzzification, it needs to be calibrated, tested and validated using realistic and accurate inputs and outputs [19].

4.3. Inference

Within the FES, the part that serves as the engine and performs the core problem-solving is the inference. The inference uses both the rule-base as well as the fuzzy set in carrying out the problem-solving. The inference is the engine and core of problem-solving in FES, when mapping out the inputs into outputs. Generally, the two most popular methods used for fuzzy inference include the 'Mamdami technique and the Sugeno technique [21].

In reality, most IT developers commonly used the 'Mamdami' technique compared to the 'Sugeno' technique, and this is mainly because of the fact that the 'Mamdami' technique is closer to human inputs [20]. However the 'Sugeno' technique is appropriate to mathematically analysis [21].

4.4. Defuzzification

The Defuzzification is a similar component to the 'Fuzzification'. This component also involves the conversion of fuzzy inputs to output based on membership functions. These can be achieved by means of various techniques and graphical examples to obtain the final value of the output. The defuzzification method offers a degree of flexibility with which experts can combine their knowledge (with a higher degree of sensibility) on how the results conform to reality. However, the choice of using the defuzzification method will depend on the context of the decision problem [19].

4.5. User-interface

The user inference is the main component for interaction between the end-user and the system. Furthermore, apart from being the main communication link between the system and the user, the success of the whole system (like any other IT system) will depend on the design of the user interface. The simpler and more engaging the user interface, the greater the chances that it will be used by the users [22].

From a user's perspective, the quality of the interface is determined by what the user senses or sees, what the user needs to know to understand the system, and the actions the user needs to take to obtain required results from the system.

The process of providing an interface of the suitable quality is a complex task that requires both technological and psychological factors as well as other associated physical and influencing factors. Some of the important factors that determine the quality of a user interface include [22]:

- The screen design and layout
- The human-machine interaction sequence
- Use of colours
- Information density
- Use icons and symbols
- Information display format

4.6. User-interface

Generally, in the development of an expert system, three main classes of user are considered who directly and indirectly interact with the system. These are the End-Users, Experts and Knowledge Engineers [22]. End-Users are the individuals (non-experts) that make use of the system after it is fully developed. They are the main beneficiary of the system, and whose decision-making processes are aided by using the system. Knowledge Engineers are the professionals that are technically involved in building the advanced underlying logic, and with which the expert system uses to mimic the human decision-making process and high-level cognitive tasks. The knowledge engineers provide the system with all or part of the 'knowledge' that enables its operation. Experts are the individuals that provide the knowledge with which the system use to solve problems within the specific domain. However, other means of obtaining this knowledge are also available in developing the fuzzy expert system [11].

5. Fuzzy expert systems (FES) in other application

Many researchers have worked to identify the most suitable areas or applications for FES implementation. There are several wide-ranging applications. The following are some applications of FES in other areas:

a. Estimating

In this application, FES will ask the user to provide it with the required data to any domain and compare it against the expert knowledge and historical data, for example, ecological characteristics of marine fishes. The FES needs such data to estimate their intrinsic vulnerability to fishing. The rules of such a FES are extracted heuristic rules (expressed in 'IF-THEN' clauses). The rules are describing the relationships between biological characteristics and vulnerability. Input and output variables are defined by fuzzy sets which deal explicitly with the uncertainty associated with qualitative knowledge. Conclusions from different lines of evidence are combined through fuzzy inference and defuzzification processes. This type of FES can be used as a decision support tool in management and marine conservation planning [23].

b. Risk assessment

The FES in this area will ask the user to provide it with the required data to any targeted standard, for example, ISO/IEC Information Security Management System (ISMS) standard, and measure the value of risk based on the risk assessment method and establish a set of FESs. In the meantime, the FES may then provide a recommended acceptable value 'at' or 'of' risk for facilitating and assisting decision-makers through practical aspects [24].

c. Site planning

The FES in this situation would be able to determine the location of equipment for carrying out a certain job, and also location of materials and support facilities at a given site. For example, such a FES could forecast the wind speed at a wind energy conversion system site and the electrical power that will be generated. The FES asks the user to define the forecast horizon, which can range from some minutes up to several hours ahead. The system can make reliable wind speed forecasts in real time [25].

d. Project scheduling

Some of the tasks that would be expected from a FES in this area will include providing the user with information on time or duration of activities. For example, management problems related to the estimated duration of an activity can be solved by using the FES. To implement the FES, various membership functions need to be estimated using good judgement and assisted by experts. One of the downsides of such a system is that it is not sensitive to small variations in the membership values and it can be easily implemented in existing computer programs for project scheduling. This is a very desirable property. However, the method is sensitive to the choice of the fuzzy relations [26].

e. Human resource management

This area is considered as a very important area of consideration when using FES. This type of system aids the user, especially modern and global managers, to meet pressing business challenges. For example, managers are required to possess a set of competencies or multiple intelligences in order to meet pressing business challenges. Hence, expanding global competencies is becoming an important issue. Many scholars and specialists have proposed various competency models containing a list of required competencies. But it is hard for someone to master a broad set of competencies at the same time. Here arises an imperative issue on how to enrich global competencies by way of segmenting a set of competencies into some portions in order to facilitate competency development with a stepwise mode. To solve this issue involving the vagueness of human judgements, good types of FES can be an effective solution [27].

f. Operational problems in constructed facilities

Operational problems can accrue in constructed facilities. FES can solve facilities related problems by giving causes and remedial actions for functional failures such as leaking, poor ventilation, and temperature control. Also it can provide causes and remedial actions for structural failures such as foundation settlement and cracking [28].

g. Training and development

FES can be an excellent tool for inexperienced project staff to improve their project management skills and techniques. It can also form an important part of the training function and can be used as an aid to training programs. The advantage of this would be that the trainee can have access to the expert's knowledge virtually at any time as expert systems can be mounted on desk top PCs. This will give the opportunity 'to' or 'for' trainees to improve on their skills by going through the training expert system within the work environment. Some other areas include constructability evaluation, material management and legal issues [29].

6. How to build a fuzzy expert system (FES)?

In order to build or develop a FES, five key steps need to occur, and **Figure 4** shows each step [22, 19].

Step 1. Specification of the problem and defining the linguistic variables

This is a key step required for determining the input and output variables as well as their ranges. This will involve a critical assessment of the problem as will be described by a 'knowledge engineer'.

Step 2. Determining the fuzzy sets

Fuzzy sets can be represented in various shapes, but most commonly, a triangular or trapezoid shape is enough to represent the expert knowledge, thereby considerably simplifying the computation process.

Step 3. Elicit and construct fuzzy rules

The next step is the development of the fuzzy rules. This can be achieved by getting the expert to make use of the defined fuzzy linguistic variables in describing the problem-solving process. On the other hand, the knowledge engineer will need to do that if the necessary knowledge is to be obtained from other sources like observed human behaviour, computer databases, books or flow charts.

Step 4. Encoding

After the fuzzy sets and fuzzy rules are established, the next step is the 'encoding', which is the actual building of the fuzzy expert system. These can be achieved in two ways:

- Building the system by the use of programming language (e.g. C#, Java or C).
- Applying a fuzzy logic development tool like Fuzzy Knowledge Builder™ or MATLAB Fuzzy Logic Toolbox.

In most cases, experienced developers prefer to use C/C++ programming language as it has more flexibility. On the other hand, when rapid prototyping and development of the fuzzy expert system are required, the fuzzy logic development tool is the best choice because it offers a complete building and testing environment.

Figure 4. Typical processes of building a FES [19].

Step 5. Evaluating and fine-tuning the system

Finally, the last and most difficult step is the evaluation and fine-tuning of the system. This step determines if the system meets the requirement defined at the beginning of the project. This evaluation or test is generally dependant on the type of variables considered in the development of the system.

After the test, if the developers are not fully satisfied with the performance of the system, it is possible to improve the performance by adding more sets or by extending the rule base according to the FAM (Fuzzy Associative Memory), which is usually referred to as tuning. Generally, to tune an expert system requires much more effort and time than it takes to determine the initial fuzzy sets, and to construct the fuzzy rules. It is always better to successfully develop the solution of the problem on the first trial, with the initial series of fuzzy sets and fuzzy rules.

Another method of implementing FES is by using shells. Shells are defined as 'a software package that facilitates the building of knowledge-based expert systems by providing a knowledge representation scheme and an inference engine' [30]. Therefore, it refers to the software module that contains the (i) interface, (ii) inference engine (iii) and a structured skeleton of a knowledge base (in its empty state) as well as those appropriate facilities for representing the knowledge. Some examples of shell include: CLIPS, EMYCIN, AION-DS and JESS [31].

7. The proposed intelligent capability preparedness (ICP) system

In this section, the Intelligent Capability Preparedness System (ICP) tool is presented, which is constructed based upon the literature review and on data collected via interviews with the domain experts in the Saudi CD Authority. The ICER system is implemented as a 'Prototype' in order to test and evaluate its effectiveness became a fully develop system. The implementation of the ICER system is conducted in 'Visual Studio.net' Framework. The rationale of selecting the '.Net framework' is because its compatibility with the current IT infrastructure used by the Saudi CD Authority; moreover, its flexibility and features are required for implementing such as a system. C Sharp (C#) is the programming language used for coding the system, and also 'Microsoft SQL Server' is used as the database management platform. Furthermore, there are additional tools that are used in the implementation, in order to enhance the functionality and presentation of the results, as well as enhancing the end-user's experience. The following are a brief description and usage of each tool:

- High-Charts: it is an interactive JavaScript charts for web page application. It is free open source java script library for rendering a lot of different types of charts used in the system's reports or outputs.
- Google Maps: it is used to address geographical data such as regions, cities and districts, also it aids in computing the distances between each zone (which is needed in the case of mutual-aid).

7.1. Architecture of ICP system

The architecture of the ICP system is made up of three key components (see **Figure 5**).

Starting on the right-hand side of the diagram, there is the 'fuzzy expert system' component, 'end-user' component and finally 'Governmental sectors' component. The ICP system requires many types of inputs from various location and parties. Therefore, it is necessary to have a worldwide network of computer systems, and for this reason, the Internet is an essential element within the ICP system. Using the Internet will support the ICP system in terms of [32, 33]:

- Providing and obtaining data from the various governmental sectors that are involved in the emergency management.
- Providing the ability to access to the ICP system remotely from any locations, in order to review and discuss the evaluation and recommendations provided by the ICP system.
- Increasing the access to real-time information and up-to-date data relating to flood risk events.

The governmental sectors are the component which will intensively feed the 'Knowledge Base' within the 'Fuzzy Expert System' component, by providing the required input data. The

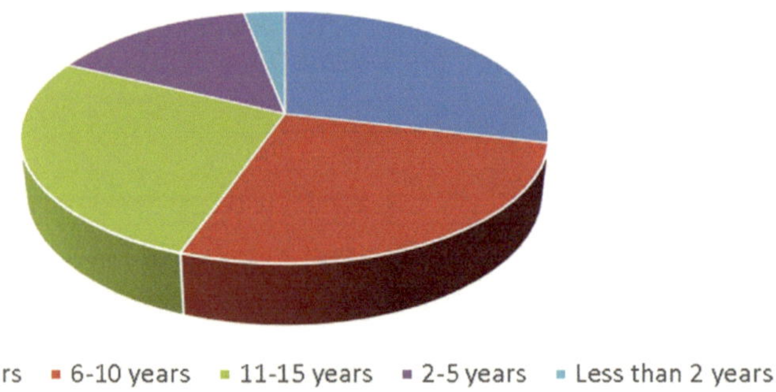

Figure 5. Architecture of the ICP system.

'Knowledge Base' is considered as the warehouse for all the data, information, rules, cases and relationships. The ICP system stores all these types of inputs here, in order to use them to solve the problems in the domain. The 'Knowledge Base' is designed based on the interview conducted with the domain experts in the Saudi CD Authority. The 'Knowledge Base' includes two types of data: (a) facts, and (b) rules. The facts are the part that contains regular inputs such as types of available training, HR and equipment provided, as well as the types and quantity of capabilities within each centre. It also houses the regional information, ID and statue of each emergency responders, and information about Hazard and Vulnerability (HV) factors within each zone. These rules are a more advanced type of input that is provided by experts in the domain to capture their experience. The Fuzzy Expert System includes two types of rules. The first type uses the 'fuzzy set' variables and is called the 'fuzzy rules', for examples, the fuzzy rules, which are used for making decisions relating to capabilities evaluation. The second type of rules does not use 'fuzzy set' variables, for example, which are the capability requirements for flood response missions, and types of exercises required to refresh each training?

A database subsystem is designed within the ICP system to store these inputs or data (see **Figure 6**). The purpose of the designed database subsystem is to organise and manage, systematically, the inputs within the database. As shown in **Figure 6,** the database subsystem comprises of six groups in the model, namely: Geographical Information (T1); Rule and Policy (T2); Flood Risk and HV factors (T3); Training Capability (T4); HR Capability (T5); and Equipment Capability (T6). Although, storing the data in six groups takes more time and effort, it is also very useful to avoid complex data structure. Doing this makes the designed database more professional and allows easy data management.

The end-user will be able to communicate and interact with the ICP system through the 'User-Interface', by sending inquiries and receiving feedback from the ICP system. The design of the 'User-Interface' and how the ICP system should interact with the end-users is a very significant issue, as it influences how efficiently the end-user understands and uses the outcomes as they deal with the problem domain. Therefore, during the process of designing the user-interface of the ICP system, many meetings were conducted with the domain experts in the Saudi CD Authority, as a result, the final design of the main 'User-Interface' for decision-making was developed, which is referred to as the 'Dashboard'.

http://dx.doi.org/10.5772/intechopen.68784

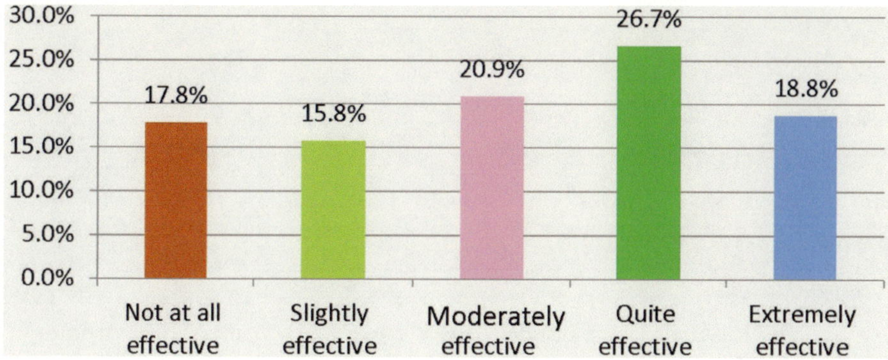

Figure 6. The ICP system's database.

The design of the ICP system involves '14' key processes as shown in the user-case diagram in **Figure 7**. Some of these processes are used for providing inputs data, such as 'P1', 'P2', 'P3', 'P4', 'P5', 'P6', 'P7', 'P8', 'P9' and 'P11'. Some are used for computing and providing outputs, such as 'P12', 'P13' and 'P14'. The use-case diagram shows the interaction between and among the processes. For example, P12 is considered the most complex process in the system; also it interacts with six other major processes. The use-case diagram is also used to clearly show the role of each actor [34, 35].

7.2. ICP system's dashboard for decision-makers

Through the 'Dashboard' interface, decision-makers will be able to communicate and interact with the ICER system, where it displays the needed outputs for decision-making processes. As shown in **Figure 8**, the dashboard of the ICP system comprises three main sections.

- Section 1: This section of the dashboard presents the overall outputs relating to how prepared the current capability is across all zones. The section also provides other significant information such as the up to date flood status, which are presented next to each zone. The details of the capability evaluation for each capability's type are also shown in this section. In addition, it provides outputs relating to the local HV factors for each zone.

- Section 2: This section of the dashboard presents the outputs relating to how prepared the capability need to be. This section of the system provides recommendations on the needed capability for each zone, which should be considered by decision-makers to reach an adequate level of readiness capability across all types of defined capabilities within this zone.

- Section 3: This section of the dashboard presents the outputs relating to the available mutual-aid (or support). This is an important output particularly, when there is major weakness of capability readiness within a zone, and it is under a warning and exposed to flooding. As shown in **Figure 8**, the section provides a list of zones that could provide all/or parts of the missing capabilities; furthermore, it provides other vital information for decision-makers, such as the required time to deliver the support.

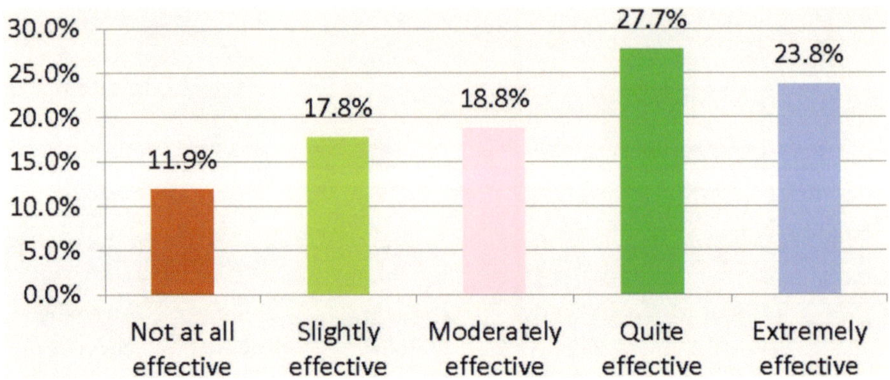

Figure 7. The ICP system's overall processes.

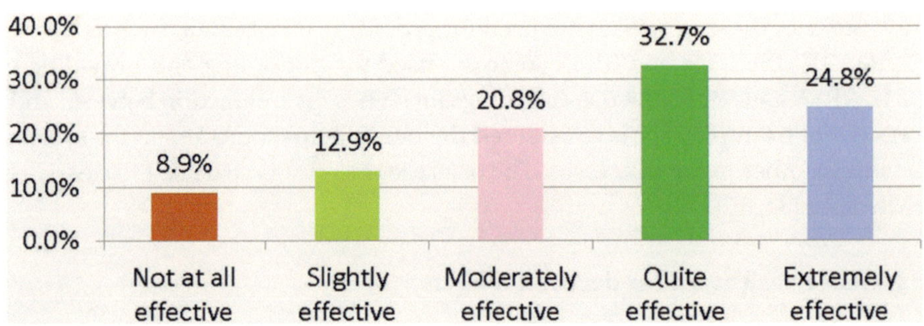

Figure 8. The dashboard of the ICP system.

8. The ICP system V.S EMC preparedness

This section presents the outcome of examining the ICP system in the case study of the CD Authority, Saudi Arabia. The aim of conducting the examination is to evaluate the effectiveness of using such an IT system on the general performance of the Saudi CD Authority, particularly in EMC preparedness for Flood Risk. The examination of ICP system is conducted by using questionnaire and structured-interview. The examination included more than 30 Key Performance Indicators (KPIs); however, only some of them are covered in this section.

It should be mentioned that all the participants who were involved in the questionnaire and structured-interview tested the ICP system before they started the questionnaire and structured-interview. The targeted audience in the structured-interview was focused only on high-level decision-makers in the Saudi CD Authority; however, the questionnaire was targeted at officers and experts. The sample in the questionnaire had '100' participants selected randomly from different locations and '5' participants in the structured-interview. The following subsections describe the analysis and results found via the examination in the field study. **Figure 9** shows the frequency distribution of the participants who were involved in the questionnaire according to their experience. The following subsections describe the analysis and results found via the examination undertaken in the field study.

8.1. Evaluation of the ICP system via questionnaire

The following points are some of the questions that were raised in the questionnaire:

1. How effective is the ICP system at evaluating the readiness of the training capabilities?

Figure 10 shows that 17.8% of the respondents think that the system is extremely effective to evaluate the readiness of the training capabilities, 26.7% of the respondents think that the system is quite effective, 20.9% think that the system is moderately effective, while 15.8% of the respondents think that the system is slightly effective. On the other hand, 18.8% of the respondents think that the system is not at all effective to evaluate the readiness of the training capabilities.

2. How effective is the ICP system at evaluating the readiness of the HR capabilities?

As shown in **Figure 11**, the results reveal that 23.8% of the respondents think that it is extremely effective, 27.7% think that it is quite effective, 18.8% think that it is moderately effective, and 17.8% think that it is slightly effective, while 11.9% of the respondents think that the system is not at all effective.

3. How effective is the ICP system at evaluating the readiness of the equipment capabilities? As shown in **Figure 12**, the results reveal that 24.8% think that it is extremely effective, 32.7% think that the system is quite effective, 20.8% think that it is moderately effective, 12.9% think that it is slightly effective, and only 8.9% of the respondents think that the system is not at all effective.

4. How effective is the ICP system at aiding experts in flood preparedness?

As shown in **Figure 13**, the results reveal that 39.6% of the respondents think that the system is extremely helpful to aid experts in flood preparedness, 35.6% think that it is very helpful, 19.8% think that it is moderately helpful, 3.0% think that it is slightly helpful, and only 2.0% of the respondents think that the system is not at all helpful to aid experts in flood preparedness.

5. How helpful is the ICP system at aiding experts in flood response?

As it shown in **Figure 14**, the results reveal that 36.6% of the respondents think that the system is extremely helpful, 34.7% think that it is very helpful, 18.8% think that it is moderately helpful, 8.0% think that it is slightly helpful, and only 1.0% of the respondents think that the system is not at all helpful to aid experts in flood response.

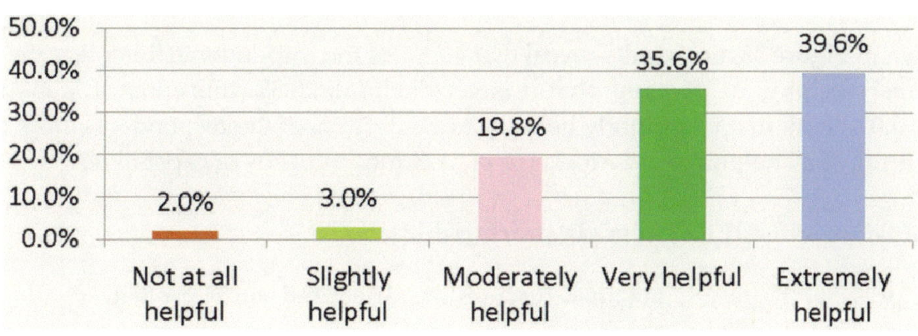

Figure 9. Participants' experience.

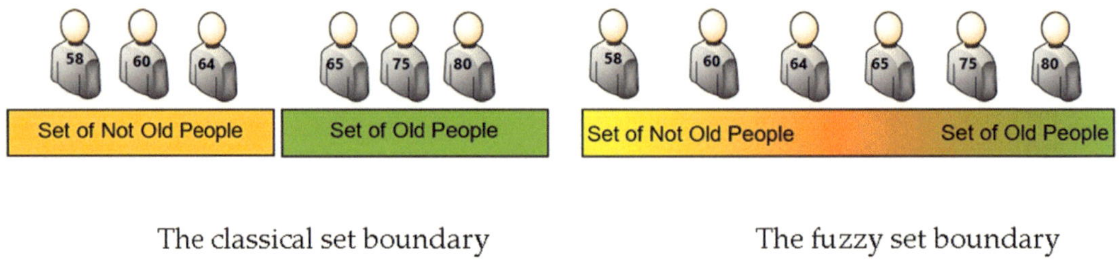

Figure 10. Result for Question 1.

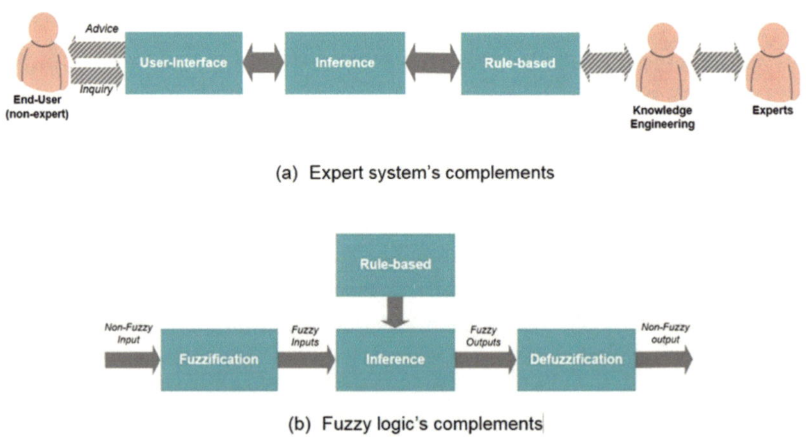

Figure 11. Result for Question 2.

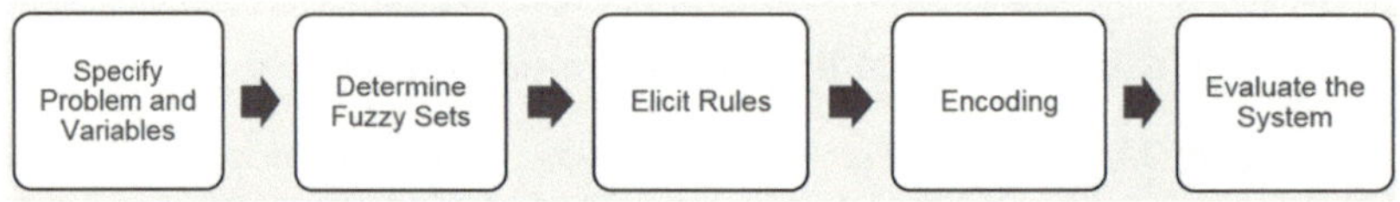

Figure 12. Result for Question 3.

6. How helpful is the ICP system at aiding experts to predict future capabilities needed?

As shown in **Figure 15**, the results reveal that 40.6% of the respondents think that the system is extremely helpful, 34.7% think that it is very helpful, 20.8% think that it is moderately helpful, 3.0% think that it is slightly helpful, and only 1.0% of the respondents think that the system is not at all helpful to aid experts to predict futures needs of capabilities.

8.2. Evaluation of the ICP system via structured-interview

Table 1 presents the results obtained through the structured-interview with five high-level decision-makers. As can be seen from **Table 1**, the ICP system made a marked improvement in all the aspects of the KPIs except for 'viable risk reduction options as identified, evaluated,

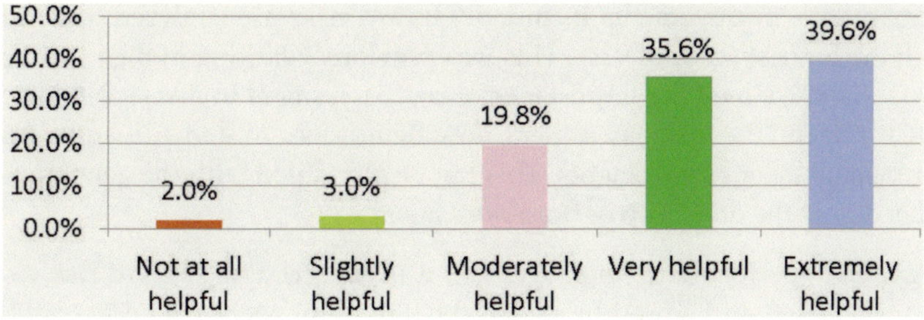

Figure 13. Result for Question 4.

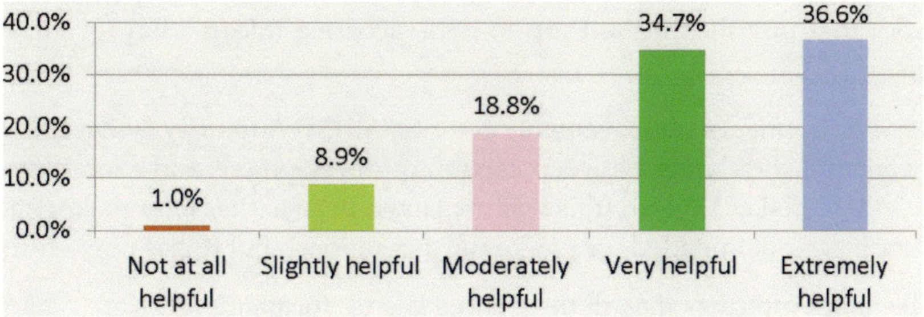

Figure 14. Result for Question 5.

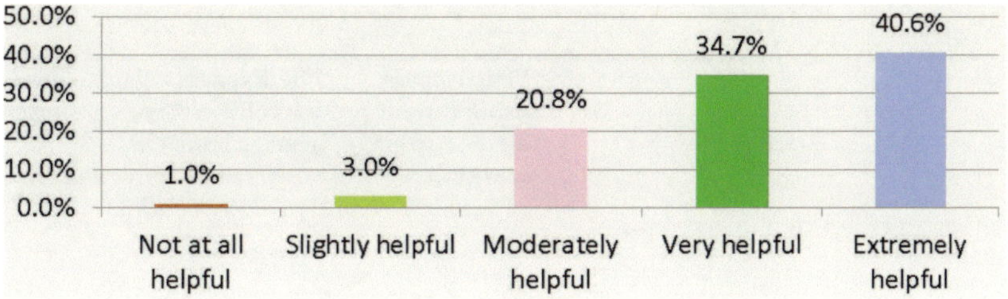

Figure 15. Result for Question 6.

and used to inform planning' and 'warning systems are in place and are maintained and effective', which showed only a slight improvement in performance.

The weak improvement in the first KPI is not surprising, given that, at the moment, the system is designed for the experts to suggest the risk reduction and mitigation options, this will take time to fine tune and improve, and it will also be recommended as a future study to develop a comprehensive data base of response strategies for each identified risk.

Regarding the second weakest area in the KPI improvement (warning systems are in place and are maintained and effective), it is currently out of the scope of this study but it will also be recommended for future studies and development.

The most noticeable improvements included 'Hazard risks are analysed to determine local impact' with the highest score of 92%. This was previously lacking in the system, and it was initially scored at 8%. Given the importance of risk assessment in disaster risk reduction, by using the ICP system, the risk assessment now determines all the subsequent activities of the disaster management requirements. This can be accredited with the general performance improvement across the other key performance areas.

Another significant performance improvement was observed in 'Hazard risk research, and information is collated and stored in a central database or register and is readily available to all stakeholders' this was initially scored as 4% as there was no central system for all the stakeholders to have up-to-date instant information on hazards, risks and vulnerabilities, which is the basis of the planning requirements. The current system now offers a central repository that provides instant, up-to-date, accurate information for all the relevant stakeholders.

Another major performance improvement on the Saudi CD Authority is that the system now allows 'a deliberate correlation between capability development and exercising objectives' this was initially scored at 12% too (this was explained in an earlier interview where previous exercises were just recommended for day to day operations), but this is now scored at 92%.

The above feedback indicates that all these three key performance areas are now in a mature stage of capability readiness. It also highlights visible performance improvements and impacts on all the identified key performance areas. This shows that the system is perceived by end-users to have a significant impact on the flash flood disaster management planning and response in Saudi Arabia (**Table 2**).

KPIs	Effectiveness using current method, average score (%)	Effectiveness using the ICER system, average score (%)	Increased improvement (%)
Emergency Management research is undertaken, assessed, and analysed	16	80	64
Hazard risk research and information is collated and stored in a central database or register and is readily available to all stakeholders	4	88	84
Hazard risks are analysed to determine local impact	8	100	92
Hazards, vulnerabilities, and risks are monitored on an on-going basis	16	68	52
The organisation uses hazard risk information to identify gaps within existing organisational plans, and prioritises planned expenditure	12	72	60
There is a process for monitoring gaps in individual/ organisational capability with regard to managing emergency operational functions	8	76	68
Viable risk reduction options are identified, evaluated, and used to inform planning	32	52	20

http://dx.doi.org/10.5772/intechopen.68784

KPIs	Effectiveness using current method, average score (%)	Effectiveness using the ICER system, average score (%)	Increased improvement (%)
Implementation of risk reduction programmes is inclusive and coordinated	16	48	32
Capability development strategy and programs are developed according to organisational needs	16	72	56
The Saudi CD's centres and member work together cooperatively and collaboratively	12	80	68

Table 2. Improvement in KPIs before and after using the ICP system (Source: Authors).

9. Chapter summary

In this chapter, we introduced the FES and discussed the philosophical ideas behind it. Fuzzy logic is a logic that describes fuzziness. As fuzzy logic attempts to model humans' sense of words, decision-making and common sense, it is leading to more human intelligent machines. Fuzzy logic is a set of mathematical principles for knowledge representation based on degrees of membership rather than on the crisp membership of classical binary logic.

This chapter has also discussed and explained the benefits of the ICP system, and how it contributes to the EMC readiness of the Saudi CD Authority in flood risk management (relating to appropriate levels of flash flood equipment, HR and training provisions). The ICER system is implemented in the 'Visual Studio.net' framework because it requires flexible features for implementing the system. The 'dashboard' interface for decision-makers is designed and implemented to display the three types of outputs required for decision-making processes.

Most of the participants in the study considered the ICP system to be effective in aiding experts to improve EMC preparedness and deploy better flood risk responses. Regarding the usefulness of the ICP system to aid in predicting future needs of EMC, the results showed that almost all the respondents found the system easy to use, and only a small minority (less than 5%) considered the system not to be effective. Most importantly, all of the respondents believed that there is a demand for the ICP system in the Saudi CD Authority, and that they would recommend the ICP system to be adopted and used across the CD centres. The most significant recommendation for improvement from the survey was that the system should include other types of EMC and then update the ICP system to aid in management of other disasters. Subsequently, the evaluation of the performance improvement from the use of the system showed that there is significant improvement in almost all the aspects of the identified KPIs compared to the evaluations before deployment of the IT solution.

Author details

Mohammad Hijji*, Saad Amin, Wayne Harrop and Rahat Iqbal

*Address all correspondence to: e-tool@hotmail.com

Faculty of Engineering, Environment and Computing, Coventry University, Coventry, United Kingdom

References

[1] R. H. Flin, P. O'Connor, and M. Crichton, Safety at the sharp end: a guide to non-technical skills: Ashgate Publishing, Ltd., Aldershot, 2008.

[2] J. W. Dean and M. P. Sharfman, Does decision process matter? A study of strategic decision-making effectiveness, Academy of management journal, vol. 39, pp. 368-392, 1996.

[3] X V R. (2013, 29 Sep). Immersive Visualization Systems. Available: http://www.vrmedia.it/en/xvr.html

[4] M. Simões-Marques, R. A. Ribeiro, and A. Gameiro-Marques, A fuzzy decision support system for equipment repair under battle conditions, Fuzzy sets and Systems, vol. 115, pp. 141-157, 2000.

[5] General Authority for Statistics. (2014, 08 May). The total number of the population in 2014 (people). Available: http://www.cdsi.gov.sa/ar/indicators/1

[6] W. Solecki, R. Leichenko, and K. O'Brien, Climate change adaptation strategies and disaster risk reduction in cities: connections, contentions, and synergies, Current Opinion in Environmental Sustainability, vol. 3, pp. 135-141, 2011.

[7] Y. Alamri, Emergency Management in Saudi Arabia: Past, Present and Future, Un. Of Christchurch report, New Zealand, p. 21, 2010

[8] S. Mashael Al, Assessment of flood hazard of Jeddah area 2009, Saudi Arabia, Journal of Water Resource and Protection, vol. 2010, 2010.

[9] S. A. Alshehri, Y. Rezgui, and H. Li, Public perception of the risk of disasters in a developing economy: the case of Saudi Arabia, Natural hazards, vol. 65, pp. 1813-1830, 2013.

[10] N. M. Momani and A. Salmi, Preparedness of schools in the Province of Jeddah to deal with earthquakes risks, Disaster Prevention and Management: An International Journal, vol. 21, pp. 463-473, 2012.

[11] T. Efraim, E. A. Jay, T.-P. Liang, and R. McCarthy, Decision support systems and intelligent systems, Upper Saddle River, NJ, Prentice Hall, 2001.

[12] G. I. Doukidis and R. J. Paul, Research into expert systems to aid simulation model formulation, Journal of the Operational Research Society, pp. 319-325, 1985.

[13] J. Durkin, Expert systems: catalog of applications: Intelligent Computer Systems, 1993.

[14] R. Gelbard and A. Meged, Handling Fuzzy Similarity for Data Classification, ed, 2009.

[15] L. A. Zadeh, Fuzzy sets, Information and control, vol. 8, pp. 338-353, 1965.

[16] K. Kolomvatsos and S. Hadjiefthymiades, Time Constraints for Sellers in Electronic Markets, Encyclopedia of E-Business Development and Management in the Global Economy, p. 265, 2010.

[17] J. Roebuck, When Does Old Age Begin?: The Evolution of the English Definition, Journal of Social History, vol. 12, pp. 416-428, 1979.

[18] M. Gorman, Development and the rights of older people: London: Earthscan Publications Ltd., 1999.

[19] H.-J. Zimmermann, Fuzzy set theory—and its applications: Springer Science & Business Media, New York, 2001.

[20] J. C. Bezdek, A review of probabilistic, fuzzy, and neural models for pattern recognition, Journal of Intelligent & Fuzzy Systems, vol. 1, pp. 1-25, 1993.

[21] H. Ying, The Takagi-Sugeno fuzzy controllers using the simplified linear control rules are nonlinear variable gain controllers, Automatica, vol. 34, pp. 157-167, 1998.

[22] M. G. Helander, Handbook of human-computer interaction: Elsevier. Amsterdam, 2014.

[23] B. M. Ayyub and A. Haldar, Project scheduling using fuzzy set concepts, Journal of Construction Engineering and Management, vol. 110, pp. 189-204, 1984.

[24] T.-P. Hong and C.-Y. Lee, Induction of fuzzy rules and membership functions from training examples, Fuzzy sets and Systems, vol. 84, pp. 33-47, 1996.

[25] J. Liebowitz, The handbook of applied expert systems: cRc Press, 1997.

[26] I. G. Damousis and P. Dokopoulos, A fuzzy expert system for the forecasting of wind speed and power generation in wind farms, in Power Industry Computer Applications, 2001. PICA 2001. Innovative Computing for Power-Electric Energy Meets the Market. 22nd IEEE Power Engineering Society International Conference on, 2001, pp. 63-69.

[27] A. R. Fayek and Z. Sun, A fuzzy expert system for design performance prediction and evaluation, Canadian Journal of Civil Engineering, vol. 28, pp. 1-25, 2001.

[28] W. W. Cheung, T. J. Pitcher, and D. Pauly, A fuzzy logic expert system to estimate intrinsic extinction vulnerabilities of marine fishes to fishing, Biological conservation, vol. 124, pp. 97-111, 2005.

[29] M. Hadjimichael, A fuzzy expert system for aviation risk assessment, Expert Systems with Applications, vol. 36, pp. 6512-6519, 2009.

[30] Artificial Intelligence. (2010 18 Feb). WHAT IS EXPERT SYSTEM? Available: http://intelligence.worldofcomputing.net/ai-branches/expert-systems.html#.VsY7_fKLTIU

[31] E. C. Ogu and Y. Adekunle, Basic Concepts of Expert System Shells and an Efficient Model for Knowledge Acquisition, International Journal of Science and Research Volume 2 Issue, vol. 4, pp. 554-559, 2013.

[32] R. Want, B. N. Schilit, and S. Jenson, Enabling the internet of things, Computer, vol. 48, pp. 28-35, 2015.

[33] S. K. Lippert and C. Govindarajulu, Technological, organizational, and environmental antecedents to web services adoption, Communications of the IIMA, vol. 6, p. 14, 2015.

[34] Hijji, M., et al. The Significance of Using Expert System to Assess the Preparedness of Training Capabilities against Different Flash Flood Scenarios. Lecture Notes on Software Engineering 3.3 (2015): 214.

[35] Hijji, Mohammad, et al. A Critical Evaluation of the Rational Need for an IT Management System for Flash Flood Events in Jeddah, Saudi Arabia. Developments in eSystems Engineering (DeSE), 2013 Sixth International Conference on. IEEE, 2013.